Energy:
TECHNOLOGY AND
DIRECTIONS
FOR THE FUTURE

Energy:
TECHNOLOGY AND
DIRECTIONS
FOR THE FUTURE

By
John R. Fanchi

ELSEVIER
ACADEMIC
PRESS

Amsterdam   Boston   Heidelberg   London   New York   Oxford
Paris   San Diego   San Francisco   Singapore   Sydney   Tokyo

Academic Press is an imprint of Elsevier

Elsevier Academic Press
200 Wheeler Road, Sixth Floor, Burlington, MA 01803, USA
84 Theobald's Road, London WC1X 8RR, UK

This book is printed on acid-free paper. ∞

Library of Congress Cataloging-in-Publication Data
Application submitted

British Library Cataloguing in Publication Data
A catalogue record for this book is available from the British Library

ISBN: 0-12-248291-3

For all information on all Academic Press publications
visit our Web site at www.academicpress.com

Printed in the United States of America
03 04 05 06 07 08   9 8 7 6 5 4 3 2 1

To the pioneers in the emerging energy industry—for the benefit of future generations.

Contents

CHAPTER SIX

CHAPTER THIRTEEN

Endnotes 456

Exercises 456

Preface

Energy: Technology and Directions for the Future presents the fundamentals of energy for scientists and engineers. It recognizes that society's dependence on fossil energy in the early twenty-first century is in a state of transition to a broader energy mix. Forecasts of the twenty-first-century energy mix show that a range of scenarios is possible. The more likely scenarios anticipate the use of several different energy sources. Energy professionals of the future will need to understand the origin and interactions of these energy sources to thrive in an energy industry that is evolving from an industry dominated by fossil fuels to an industry working with many energy sources.

Energy: Technology and Directions for the Future is a survey of energy sources that will be available this century. It is designed to help the reader understand and appreciate the role of alternative energy components in the energy mix. To achieve this objective, the reader will learn about the history and science of energy sources as well as energy technology. A familiarity with the material presented in the text will help the reader better assess the viability of emerging energy technologies and the role they will play in the future.

SCOPE OF TEXT

Energy: Technology and Directions for the Future begins by introducing the historical context of twenty-first-century energy, and then presents the concept of energy transformations. The modern system of distribution of energy in the form of electricity is then discussed, followed by a review of heat and thermodynamic concepts. This background sets the stage for the study of specific energy types.

The first energy type to be considered is geothermal energy. It allows us to introduce basic concepts of planetary formation and geology in anticipation of our study of fossil fuels. The discussion of planetary formation and geology explains the source of geothermal energy and illustrates our approach to presenting an energy source in the broad context of the leading theories of several relevant sciences. Among these theories are Big Bang cosmology, the Kant-Laplace hypothesis, plate tectonics, the Oparin-Haldane hypothesis for the origin of life, and the synthetic theory of evolution. These theories provide a context that should help the reader understand existing mainstream ideas and be prepared to assess competing theories that may become significant over the course of a career.

Once geothermal energy has been studied, we consider fossil fuels, which provide the majority of energy consumed today. This is followed by a discussion of solar energy, nuclear energy, alternative energy (wind, water, biomass, synfuels), and then hydrogen. In several cases, such as the Battle of Currents, the development of the modern oil industry, and the discovery and development of nuclear energy, the historical context of the technology is used to put the technical material in a social context. The reader is exposed to the role of energy in society, including economic, ethical, and environmental considerations. The final topic is a discussion of energy forecasts and the trend toward a hydrogen economy.

THEMES AND STRANDS

Energy: Technology and Directions for the Future is presented as a set of thematic modules. Themes include electricity distribution, geothermal energy, fossil fuels, solar energy, nuclear energy, alternative energy, the hydrogen economy, energy and society, and energy forecasting. The technical level of presentation presumes that readers have completed college-level physics with calculus and mathematics through calculus of several variables. Given this level of preparation, it is possible to present a more sophisticated discussion of energy topics.

Technical strands run through the thematic modules to help prepare the reader for increasingly sophisticated concepts. For example, the Lagrangian is introduced in the review of energy transformations and then used to discuss the development and interpretation of quantum mechanics and relativistic quantum mechanics using path integrals. Quantum mechanics is used to help the reader understand tunneling (as in nuclear decay), and to explain the free electron theory of metals. The latter theory is useful for discussing the photoelectric effect and photovoltaics, which have applications in solar energy. Relativistic quantum mechanics is presented to help the reader understand the concept of mass–energy transformation and develop a more sophisticated understanding of nucleosynthesis and nuclear energy (both fission and fusion). The use of technical strands lets the reader become familiar with mainstream scientific concepts and the technical themes demonstrate the significance of each energy option in a broad social context.

MATHEMATICS

The focus of *Energy: Technology and Directions for the Future* is on concepts, facts, and exercises that can be solved without advanced mathematics. Exercises are included in the book to complement the text and enhance its value as a textbook. The exercises range in difficulty from practice at converting units or verifying material in the text to analyses of multifaceted problems. Many of the exercises guide the reader through a detailed analysis of important aspects of contemporary energy issues. As a rule, energy statistics are included primarily in exercises and in the text for historical insight.

Energy: Technology and Directions for the Future contains sections that presume college-level calculus. When new mathematics is presented, such as the definition of conditional probability or the Einstein summation convention, it is presented at a level that is suitable for readers with a calculus background. For example, readers are exposed to differential equations, but are not expected to solve them. If a solution to a differential equation is needed to solve a problem, the solution will be provided. In some exercises, readers are asked to substitute the solution into the equation to verify its validity. As another example, integrals are used to show how to obtain the fundamental equations of quantum mechanics. Solutions to the integrals are provided.

The level of mathematics used in *Energy: Technology and Directions for the Future* facilitates a sophisticated exposure to many topics without exceeding the level of preparation of readers majoring in technical subjects. Indeed, the mathematics can be used to refresh a reader's mathematical knowledge. If readers would like more mathematical preparation, they can refer to standard mathematics textbooks or the book *Math Refresher for Scientists and Engineers*, 2nd Edition, written by the author and published by Wiley-Interscience in 2000.

PROFESSIONAL AUDIENCE

Every energy professional should be exposed to the material presented in *Energy: Technology and Directions for the Future*. The selection of topics is designed to provide energy professionals with an introduction to the language, concepts, and techniques used in all major energy components that are expected to contribute to the twenty-first-century energy mix. A familiarity with material in the text should facilitate discussions between energy professionals with different specialties.

A WORD TO FACULTY

The text is suitable as the primary or supplemental reference for energy survey courses. I have used the text in a 3–semester-hour course (approximately 48 contact hours). The course is designed to help undergraduate and graduate students expand their knowledge of the relative merits of different energy sources, and select an energy source for specialized study. Energy survey courses at this level can be taught by a faculty team or a faculty member who may be supported by guest speakers. The selection of topics provides a suitable background for relatively sophisticated classroom discussions. Some of the material in the text can be skipped on first reading if you have limited time or would prefer to add topics of your own. For example, Chapter 10 is an enrichment chapter that covers nucleosynthesis and could be skipped or left to more advanced students.

Acknowledgments

I would like to thank my students and guest speakers for providing useful comments about the text. The Department of Petroleum Engineering at the Colorado School of Mines helped fund the development of the course concept. Colleagues from other departments have provided encouragement for the development of an energy survey course. I want to thank Kathy Fanchi and Tony Fanchi for their support and efforts in the production of this book. I am, of course, responsible for any errors or omissions and would appreciate any comments the reader may have.

John R. Fanchi
October 2003

About the Author

John R. Fanchi is a Professor of Petroleum Engineering at the Colorado School of Mines. He has worked in the technology centers of three major oil and gas companies, and served as an international consultant. Dr. Fanchi's publications include software systems for the U.S. Department of Energy, numerous articles, and several books, including *Shared Earth Modeling* (Butterworth–Heinemann, 2002), *Principles of Applied Reservoir Simulation*, 2nd Edition (Butterworth–Heinemann, 2001), *Integrated Flow Modeling* (Elsevier, 2000), *Math Refresher for Scientists and Engineers*, 2nd Edition (Wiley, 2000), and *Parametrized Relativistic Quantum Theory* (Kluwer, 1993). He has a Ph.D. in physics from the University of Houston, studied geothermal storage of solar energy as a post-doctoral fellow, and is president of the International Association for Relativistic Dynamics.

OTHER BOOKS BY JOHN R. FANCHI

Shared Earth Modeling, Butterworth–Heinemann, Woburn, MA and Oxford, UK (2002).

Principles of Applied Reservoir Simulation, Second Edition, Butterworth–Heinemann, Woburn, MA and Oxford, UK (2001). Chinese edition published in 2001.

Math Refresher for Scientists and Engineers, Second Edition, J. Wiley — Interscience, New York (2000).

Integrated Flow Modeling, Elsevier, Amsterdam (2000).

Parametrized Relativistic Quantum Theory, Kluwer Academic Publishers, Dordrecht, The Netherlands (1993).

CHAPTER ONE

Introduction

Energy demand is expected to grow in the twenty-first century as more countries seek a better quality of life for their citizens. The energy demand will be met by a global energy mix that is undergoing a transition from an energy portfolio dominated by fossil fuels to an energy portfolio that includes a range of fuel types. Fossil fuels such as coal, oil, and gas were the fuel of choice during the last half of the twentieth century. Forecasts of the twenty-first century energy mix show a gradual transition from the current dominance of fossil fuels to a more balanced distribution of energy sources. Forecasts are discussed in more detail in Chapter 15.

This chapter introduces the history of energy consumption, explains the importance of energy to the quality of human life, presents the concepts of energy and energy transformation in classical mechanics, and discusses the need for energy professionals to function as stewards of the earth's natural resources.

1.1   UNITS AND DIMENSIONAL ANALYSIS

Energy may be expressed in many different units, ranging from the British Thermal Unit (BTU) in the English system to the erg in the cgs system and the joule in SI (System Internationale) units. We have selected the SI unit as the primary system of units because of its global use in science and to make it easier to compare different energy sources.

SI BASE UNITS

The SI base units and their symbols are presented in Table 1-1. All other units can be derived from the SI base units.

An example of a derived unit is the joule. The joule is the SI unit of energy. It is a derived quantity with $1 \text{ J} = 1 \text{ kg·m}^2/\text{s}^2$. The watt is the SI unit of power. One watt is equal to one joule per second. Another useful

Table 1-1
SI base units

Physical quantity	Base unit	Symbol
Length	meter	m
Mass	kilogram	kg
Time	second	s
Electric current	ampere	A
Temperature	Kelvin	K
Amount of substance	mole	mol
Luminous intensity	candela	cd

derived unit is the unit of solar intensity (watt/m^2). Solar intensity measures the amount of solar energy passing through a cross-sectional area in a given time duration. Solar energy is the energy of light from the sun. For example, one watt/m^2 is equal to one joule of energy passing through one square meter in one second. The unit watt/m^2 can be used to compare different energy types, as is illustrated in several exercises in this book.

The physical quantities of interest here have values that range from very small to very large. It is therefore necessary to use scientific notation to express the values of many physical quantities encountered in the study of energy. In some cases, the value may be expressed in a more traditional form by using SI prefixes. The most common prefixes used in SI are presented in Table 1-2.

In some parts of the book, we use units from other unit systems if they are typical of the units you will encounter in the literature. For example, the

Table 1-2
Common SI prefixes

Factor	Prefix	Symbol	Factor	Prefix	Symbol
10^1	deka	da	10^{-1}	deci	d
10^2	hecto	h	10^{-2}	centi	c
10^3	kilo	k	10^{-3}	milli	m
10^6	mega	M	10^{-6}	micro	μ
10^9	giga	G	10^{-9}	nano	n
10^{12}	tera	T	10^{-12}	pico	p
10^{15}	peta	P	10^{-15}	femto	f
10^{18}	exa	E	10^{-18}	atto	a

energy content of food is often expressed in Calories. One food Calorie is equivalent to 1000 calories or 4.184×10^3 J. Another important unit is the quad. A quad is a unit of energy that is often used in discussions of global energy because it is comparable in value to global energy values. One quad equals one quadrillion BTU (10^{15} BTU or approximately 10^{18} J).

It would be easier to learn about energy and compare different energy types if everyone used a single system of units, but the use of a variety of energy units is common. Many exercises are designed to help you improve your skills in converting from one system of units to another. Appendix C contains a selection of unit conversion factors that should help you complete the exercises. A more complete set of conversion factors can be found in the literature.

DIMENSIONAL ANALYSIS

The dimension of a quantity may be expressed in terms of the SI base units shown in Table 1-1. In many cases, the conversion of units between different unit systems or the solution of a problem can be most readily achieved by analyzing the dimensions of the physical quantities. The most effective way to avoid careless errors with units is to carefully write both the value and the unit of each physical quantity involved in a calculation.

1.2   A BRIEF HISTORY OF ENERGY CONSUMPTION

Energy consumption increased as society evolved. Table 1-3 shows an estimate of daily human energy consumption at six periods of societal development [Cook, 1971]. Cook assumed the only source of energy consumed by a human during the period labeled "Primitive" was food. Cook's energy estimate was for an East African who lived about one million years ago. Energy is essential for life, and food was the first source of energy. Humans require approximately 2000 Calories ($\approx$ 8 MJ) of food per day.

The ability to control fire during the hunting period let people use wood to heat and cook. Fire provided light at night and could illuminate caves. Firewood was the first source of energy for consumption in a residential setting. Cook's energy estimate was for Europe about 100,000 years ago.

The primitive agricultural period was characterized by the domestication of animals. Humans were able to use animals to help them grow crops and cultivate their fields. The ability to grow more food than they needed became the impetus for creating an agricultural industry.

Table 1-3
Historical energy consumption

Period	Daily per capita consumption (1000 kcal)				
	Food	H & C*	I & A**	Transportation	Total
Primitive	2				2
Hunting	3	2			5
Primitive Agricultural	4	4	4		12
Advanced Agricultural	6	12	7	1	26
Industrial	7	32	24	14	77
Technological	10	66	91	63	230

* H & C = Home and Commerce.
** I & A = Industry and Agriculture.
Source: Cook, 1971.

Cook's energy estimate was for the Fertile Crescent circa 5000 B.C.E. (Before Common Era).

More energy was consumed during the advanced agricultural period, when people learned to use coal and built machines to harvest the wind and water. By the early Renaissance, people were using wind to push sailing ships, water to drive mills, and wood and coal for generating heat. Transportation became a significant component of energy consumption by humans. Cook's energy estimate was for northwestern Europe circa 1400 C.E. (Common Era).

The steam engine ushered in the industrial period. It provided a means of transforming heat energy to mechanical energy. Wood was the first source of energy for generating steam in steam engines. Coal, a fossil fuel, eventually replaced wood and hay as the primary energy source in industrialized nations. Coal is easier to store and transport than wood and hay, which are bulky and awkward. Coal was useful as a fuel source for large vehicles, such as trains and ships, but of limited use for personal transportation. Oil, another fossil fuel, was a liquid that contained about the same amount of energy per unit mass as coal but could flow through pipelines and into tanks. People just needed a machine to convert the energy in oil to a more useful form. Cook's energy estimate was for England circa 1875 C.E.

The modern technological period is associated with the development of internal combustion engines and applications of electricity. Internal combustion engines can vary widely in size and they use oil. The internal combustion engine could be scaled to fit on a wagon to create "horseless carriages." The transportation system in use today evolved as a result of the development of internal combustion engines. Electricity, by contrast, is generated from primary energy sources such as fossil fuels. Electricity generation and distribution systems made the widespread use of electric motors and electric lights possible. One advantage of electricity as an energy source is that it can be transported easily, but electricity is difficult to store. Cook's energy estimate was for the United States circa 1970 C.E.

1.3 ENERGY CONSUMPTION AND THE QUALITY OF LIFE

A relationship between energy consumption and quality of life has been reported in the literature. Quality of life is a subjective concept that can be quantified in several ways. The United Nations calculates a quantity called the Human Development Index (HDI) to provide a quantitative measure of the quality of life. The HDI measures human development in a country using three categories: life expectancy, education, and gross domestic product (GDP). Gross domestic product accounts for the total output of goods and services from a nation and is a measure of the economic growth of the nation. The HDI is a fraction that varies from zero to one. A value of HDI that approaches zero is considered a relatively low quality of life, and a value of HDI that approaches one is considered a high quality of life.

A plot of HDI versus per capita electricity consumption for all nations with a population of at least one million people is shown in Figure 1-1.

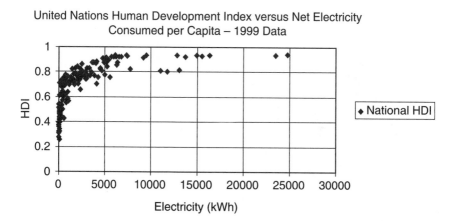

Figure 1-1. Human development and annual electricity consumption.

Per capita electricity consumption is the total amount of electricity consumed by the nation divided by the population of the nation. It represents an average amount of electricity consumed by each individual in the nation. The calculation of per capita electricity consumption establishes a common basis for comparing the consumption of electricity between nations with large populations and nations with small populations. The HDI data are 1999 data from the 2001 United Nations Human Development Report [UNDP, 2001], and annual per capita electricity consumption data are 1999 data reported by the Energy Information Administration of the United States Department of Energy [EIA Table 6.2, 2002].

Figure 1-1 shows that quality of life, as measured by HDI, increases as per capita electricity consumption increases. It also shows that the increase is not linear; the improvement in quality of life begins to level off when per capita electricity consumption rises to about 4000 kWh. A similar plot can be prepared for per capita energy consumption (Figure 1-2).

Figure 1-2 is a plot of HDI versus per capita energy consumption in all nations with a population of at least one million people. The HDI data are from the 2001 United Nations Human Development Report [UNDP, 2001], and annual per capita energy consumption data are 1999 data reported by the Energy Information Administration of the United States Department of Energy [EIA Table E.1, 2002]. The figure shows that quality of life increases as per capita energy consumption increases. As in Figure 1-1, the increase is not linear; the improvement in quality of life begins to level off when per capita energy consumption rises to about 200,000 MJ.

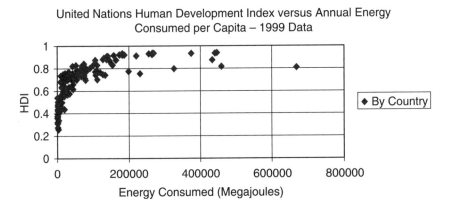

Figure 1-2. Human development and annual energy consumption.

It is interesting to note that some countries have relatively low HDI values, on the order of 80%, despite a relatively large per capita consumption of electricity and energy. These countries include Kuwait (13082 kWh, HDI $= 0.818$), Qatar (11851 kWh, HDI $= 0.801$), and the United Arab Emirates (11039 kWh, HDI $= 0.809$). All of these countries have relatively small populations (less than 3 million people each in 1999). In addition to their citizenry, the populations in these countries include relatively large, poor, service classes. The countries with the largest HDI values, in excess of 90%, are nations with relatively mature economies, such as western European nations, Canada, Australia, the United Kingdom, Japan, and the United States. These countries tend to have relatively large middle classes. The three countries with the largest per capita consumption of electricity in Figure 1-1 are Norway (24773 kWh, HDI $= 0.939$), Iceland (23486 kWh, HDI $= 0.932$), and Canada (16315 kWh, HDI $= 0.936$). The per capita consumption of electricity in the United States (HDI $= 0.934$) was 12838 kWh in 1999.

The data used to prepare Figures 1-1 and 1-2 can also be used to make a quick forecast of energy demand. Suppose we assume that the world population will stabilize at approximately 8 billion people in the twenty-first century and that all people will want the quality of life represented by an HDI value of 0.9 (which is approximately the HDI value achieved by Italy, Spain, and Israel). In this scenario, the per capita energy demand from Figure 1-2 is approximately 200,000 MJ per person, or 1.6×10^{15} MJ $\approx$ 1500 quads. The world population of approximately 6 billion people consumed approximately 387 quads of energy in 1999. According to this scenario, worldwide energy demand will almost quadruple the energy consumption level in 1999 by the end of the twenty-first century. Per capita energy consumption will have to increase from an average of 68,000 MJ per person in 1999 to the desired value of 200,000 MJ per person in 2100. This calculation illustrates the types of assumptions that must be made to prepare forecasts of energy demand. At the very least, a forecast of demand for energy at the end of the twenty-first century needs to provide an estimate of the size of the population and the per capita demand for energy at that time.

1.4   MECHANICAL ENERGY

The discussion in Section 1.3 demonstrates that energy consumption is an essential contributor to quality of life. The behavior of a physical or biological system depends on the energy of the system. The concept of energy is traditionally introduced in terms of the dynamic behavior of a simple physical particle. Newton's laws describe the dynamic behavior of a classical particle in terms of forces acting on the particle. Other descriptions of the dynamics of a classical particle rely more on energy than on force. Two such formulations underlie the science of energy and are used later to help us understand phenomena such as nuclear decay and solar photovoltaics. Before we discuss formulations of dynamics in terms of energy, we first provide a brief review of classical mechanics. Our review introduces notation that is used elsewhere in the text.

CLASSICAL MECHANICS

Classical physics has historically been the physical science of everyday life. The theory of classical physics is applicable to relatively large-scale natural phenomena. Some examples of physical systems within the purview of classical physics include planets moving in their orbits around stars, race cars navigating banked tracks, the operation of many common household appliances, and the flight of a tennis ball as it moves from one tennis racquet to another. These systems are often referred to as *macroscopic systems* because we can observe them with our unaided senses. Smaller-scale systems, such as the behavior of electrons in atoms, are called *microscopic systems* and do not always obey the "laws" of classical physics. Microscopic systems are described by quantum physics, which is introduced later. Classical energy is the concept of energy contained within classical physics.

Isaac Newton published the first authoritative expression of classical mechanics in 1687. Newton's treatise, *Mathematical Principles of Natural Philosophy,* contained a theory capable of describing the motion of macroscopic objects. His theory was based on three fundamental postulates that we know today as Newton's laws of motion. Before discussing these postulates, it is important to understand Newton's notions of space and time.

To Newton, time and space had several important qualities:

Q1-1. Absolute, true, and mathematical time, of itself, and from its own nature, flows equably without relation to anything external.... Absolute space, in its own nature, without relation to anything external, remains always similar and immovable. [Wolff, 1965, pg. 159]

Time and space are independent of each other. Like the Greek philosopher Aristotle, Newton believed that dimensions of time and space are permanent.

A belief in the permanence of space and time implies the existence of points in space and instants of time independent of objects or observers confined within the space and time domains. According to Newton, an observer moving at any speed would measure the passage of time at the same rate as a motionless observer. Time duration is the same regardless of the point of view of an individual measuring the time lapse. Furthermore, if the moving observer recorded two events occurring simultaneously in his frame of reference, the motionless observer would also record the simultaneous occurrence of the same two events.

For most practical purposes, Newton's conception of space and time is adequate. If you see the hood and trunk of your moving car pop up at the same time, someone standing nearby and watching would also see the hood and trunk open simultaneously. On the other hand, if we are moving at a rate approaching the speed of light, Newton's concepts become inadequate and we must use Albert Einstein's theory of relativity. Einstein's relativity is discussed later.

Building on his view of space and time, Newton formulated three postulates or "laws" to describe the motion of objects. According to his first law, the uniform, nonaccelerating motion of an object will continue indefinitely unless an outside force acts upon the object:

Q1-2. Law I: Every body continues in its state of rest, or of uniform motion in a right line (straight line), unless it is compelled to change that state by forces impressed upon it. [Wolff, 1965, pg. 166]

Newton's second law is the basis of his "force equals mass times acceleration" concept:

Q1-3. Law II: The change of motion is proportional to the motive force impressed; and is made in the direction of the right line in which that force is impressed. [Wolff, 1965, pg. 166]

Law I is a special case of Law II. The quantity of motion alluded to in Newton's statement of Laws I and II is a product of the mass of an object times the velocity of the object. This product—mass times velocity—is today called momentum. Law II says the change in momentum of an object during a given time interval depends on the forces acting on the object.

Newton's third law describes the interaction between objects:

Q1-4. Law III: To every action there is always opposed an equal reaction; or, the mutual actions of two bodies upon each other are always equal, and directed to contrary parts. [Wolff, 1965, pg. 167]

Newton's three laws and his notions of space and time are the essential elements of the theory known as nonrelativistic classical mechanics. Newton wrote the force equation of classical mechanics as

$$\vec{F} = \dot{\vec{p}} = \frac{d\vec{p}}{dt} \qquad (1.4.1)$$

where $\vec{F}$ is the force acting on the object, $\vec{p}$ is its momentum and t is absolute time. The dot over the momentum vector denotes differentiation with respect to time t. For an object with mass m and velocity $\vec{v}$ we have

$$\vec{p} = m\vec{v} \qquad (1.4.2)$$

If mass is constant with respect to time, Equation (1.4.1) becomes

$$\vec{F} = \frac{dm\vec{v}}{dt} = m\frac{d\vec{v}}{dt} = m\vec{a} \qquad (1.4.3)$$

where acceleration $\vec{a}$ is given by

$$\vec{a} = \frac{d\vec{v}}{dt} \qquad (1.4.4)$$

A position vector in a Cartesian coordinate system may be written as $\vec{r} = \{x, y, z\}$. Velocity and acceleration have the form

$$\vec{v} = \frac{d\vec{r}}{dt}, \quad \vec{a} = \frac{d\vec{v}}{dt} = \frac{d^2\vec{r}}{dt^2} \qquad (1.4.5)$$

and Equation (1.4.3) becomes

$$\vec{F} = m\frac{d^2\vec{r}}{dt^2} \tag{1.4.6}$$

Equation (1.4.6) is vector notation for the three differential equations

$$F_x = m\frac{d^2x}{dt^2}, \quad F_y = m\frac{d^2y}{dt^2}, \quad F_z = m\frac{d^2z}{dt^2} \tag{1.4.7}$$

Example 1.4.1: Free Particles

Suppose a particle with constant mass m is moving freely, that is, the particle is not subject to any external forces. In this case, $\vec{F} = 0$ and Equation (1.4.1) says that

$$\frac{d\vec{p}}{dt} = 0 \tag{1.4.8}$$

Equation (1.4.8) is satisfied when momentum is constant, thus

$$\vec{p} = m\vec{v} = \text{constant} = \vec{p}_f \tag{1.4.9}$$

where momentum $\vec{p}_f$ is the constant free particle momentum. Since mass is constant, we find the constant velocity

$$\vec{v}_f = \frac{\vec{p}_f}{m} \tag{1.4.10}$$

The position $\vec{r}_f$ of the free particle is found from the equation

$$\frac{d\vec{r}_f}{dt} = \vec{v}_f = \frac{\vec{p}_f}{m} \tag{1.4.11}$$

Equation (1.4.11) has the solution

$$\vec{r}_f = \vec{v}_f t + \vec{r}_0 \tag{1.4.12}$$

where $\vec{r}_0$ is the initial position of the particle, that is, it is the position at $t = 0$. We see from Equation (1.4.12) that a free particle will move in a

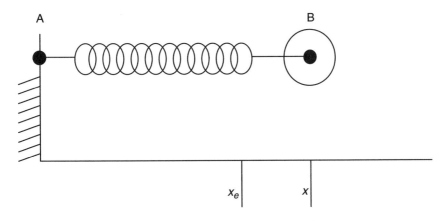

Figure 1-3. Harmonic oscillator.

straight line. An external force is needed to change the linear trajectory of the particle.

Example 1.4.2: Harmonic Oscillators

Figure 1-3 shows a spring with negligible mass that is connected to a ball with mass m at point B and to a stationary point A. The ball oscillates about an equilibrium position x_e along the x-axis.

We denote the displacement $q = x - x_e$. The force f on the ball along the x-axis obeys Hooke's law:

$$f = -kq \qquad (1.4.13)$$

The constant k is Hooke's constant. In terms of coordinates, the force equation becomes

$$f = m\ddot{x} = m\ddot{q} = -kq \qquad (1.4.14)$$

since

$$\ddot{q} = \frac{d^2 q}{dt^2} = \frac{d^2(x - x_e)}{dt^2} = \frac{d^2 x}{dt^2} = \ddot{x} \qquad (1.4.15)$$

Rearranging Equation (1.4.14) gives

$$\ddot{q} = -\frac{k}{m}q \tag{1.4.16}$$

If we define an angular frequency of oscillation ω as

$$\omega^2 = \frac{k}{m} \tag{1.4.17}$$

we can write Equation (1.4.16) as

$$\ddot{q} = -\omega^2 q \tag{1.4.18}$$

The solution of Equation (1.4.18) is

$$q = q_{max} \cos \omega t \tag{1.4.19}$$

where the constant q_{max} is the maximum displacement from x_e. Angular frequency ω in radians per second is related to frequency in Hertz (cycles per second) by $\omega = 2\pi f$.

Example 1.4.3: Newtonian Gravity

The magnitude of the gravitational force on a body A with mass m_A that is a distance r from a body B with mass m_B is

$$F_G = m_A \ddot{r} = -G\frac{m_A m_B}{r^2} \tag{1.4.20}$$

where G is the gravitational constant. The force is an attractive force along the line that connects the center of gravity of each body. The equation of motion is

$$m_A \ddot{r} = -\frac{G m_A m_B}{r^2} \tag{1.4.21}$$

The radial acceleration can be written as

$$\ddot{r} = \frac{d^2 r}{dt^2} = \frac{d\dot{r}}{dt} = \frac{dr}{dt}\frac{d\dot{r}}{dr} = \dot{r}\frac{d\dot{r}}{dr} \tag{1.4.22}$$

Substituting Equation (1.4.22) into Equation (1.4.21) gives

$$m_A \dot{r} \frac{d\dot{r}}{dr} = -\frac{Gm_A m_B}{r^2} \tag{1.4.23}$$

The integral of Equation (1.4.23) is

$$m_A \int \dot{r} d\dot{r} = -Gm_A m_B \int \frac{dr}{r^2} \tag{1.4.24}$$

or

$$\frac{1}{2} m_A \dot{r}^2 - \frac{Gm_A m_B}{r} = E \tag{1.4.25}$$

The constant of integration E is the sum of kinetic energy and potential energy. We show this by recognizing that $\dot{r}$ is radial velocity v_r. Equation (1.4.25) can be written as

$$E_K + V_G = E \tag{1.4.26}$$

where E_K is the kinetic energy of body A

$$E_K = \frac{1}{2} m_A v_r^2 \tag{1.4.27}$$

and V_G is the potential energy of the gravitational field acting on body A.

$$V_G = -\frac{Gm_A m_B}{r} \tag{1.4.28}$$

The force of gravity on body A at the surface of the earth is

$$-\frac{GMm_A}{R^2} = -m_A g \tag{1.4.29}$$

where M is the mass of the earth, R is the radius of the earth, and g is the acceleration of gravity on the surface of the earth. Simplifying Equation (1.4.29) gives

$$g = \frac{GM}{R^2} \tag{1.4.30}$$

FORCES OF NATURE

Classical physics epitomizes a deterministic view of nature. If we can measure the initial motion (or momentum) of an object and its initial location simultaneously, as Newton's theory allows, then we can, in principle, precisely compute the historic and future behavior of the object provided we know all of the forces that act upon the object during the time of interest to us. The mathematical equations for describing the behavior of the object are completely defined, though they may not be solvable, if we know the forces. What are the fundamental forces of nature?

Until the twentieth century only two fundamental forces were known: the gravitational force and the electromagnetic force. These two forces are the primary focus of classical physics. The gravitational force governs the motion of astronomical bodies. Every massive object experiences the pull of gravity.

Electricity and light are examples of electromagnetism. Electromagnetism is associated with objects that have an electrical charge, such as electrons. The electromagnetic force is the force that binds electrons to atomic nuclei. Electromagnetism is one of four fundamental forces. In addition to the gravitational force and the electromagnetic force, the two other forces are the weak force, and the strong force. The weak force is associated with the decay of elementary particles, and the strong force holds the nuclei of atoms together. It is possible to theoretically combine the electromagnetic force and the weak force to define a new force called the electroweak force. From this point of view, the four fundamental forces would be reduced to three: gravity, the electroweak force, and the strong force.

Today, physicists often talk about interactions rather than forces. The term *interaction* stems from the theoretical concept that what we perceive to be a force between two objects is actually an exchange of particles. For example, the earth and the moon are thought to exchange particles called gravitons. The physical consequence of the exchange of gravitons is an attraction between the earth and the moon.

1.5   DYNAMICS AND ENERGY TRANSFORMATIONS

One important property of energy is that it can change from one form to another. We begin our discussion of energy transformation by introducing two formulations of dynamics that depend on energy: Lagrange's equations and Hamilton's equations. The exchange of energy between the kinetic energy and potential energy of a harmonic oscillator is then studied as an illustration of energy transformation.

LAGRANGE'S EQUATIONS

French mathematician Joseph Louis Lagrange formulated the dynamic laws of motion in terms of a function of energy. Lagrange's formulation generalizes Newton's second law of motion, which can be written as

$$F_i = \frac{d}{dt} m\dot{x}_i, \quad i = 1, 2, 3 \tag{1.5.1}$$

The subscript i is used to denote the orthogonal components of a vector. For example, the components of the position vector in Cartesian coordinates are

$$x_1 = x, \quad x_2 = y, \quad x_3 = z \tag{1.5.2}$$

and the components of force are

$$F_1 = F_x, \quad F_2 = F_y, \quad F_3 = F_z \tag{1.5.3}$$

Lagrange introduced a function L that depends on coordinate position $\vec{q}$ and velocity $\dot{\vec{q}}$. The function L is called the Lagrangian function and has the form

$$L = L\left(\vec{q}, \dot{\vec{q}}\right) = T - V \tag{1.5.4}$$

where T is kinetic energy and V is potential energy. Kinetic energy T may depend on position and velocity, thus $T(\vec{q}, \dot{\vec{q}})$. Potential energy V may depend on position only for a conservative force. A conservative force is a force $\vec{F}$ that can be derived from the potential V by

$$F_i = -\frac{\partial V}{\partial q_i}, \quad i = 1, 2, 3 \tag{1.5.5}$$

or, in vector notation,

$$\vec{F} = -\nabla V \tag{1.5.6}$$

where ∇V is the gradient of the potential energy. Momentum is calculated from the Lagrangian as

$$p_i = \frac{\partial L}{\partial \dot{q}_i}, \quad i = 1, 2, 3 \tag{1.5.7}$$

and the force is

$$F_i = \frac{\partial L}{\partial q_i}, \quad i = 1, 2, 3 \tag{1.5.8}$$

The Lagrangian is a function that satisfies Lagrange's equations

$$\frac{d}{dt}\left(\frac{\partial L}{\partial \dot{q}_i}\right) - \frac{\partial L}{\partial q_i} = 0, \quad i = 1, 2, 3 \tag{1.5.9}$$

Lagrange's equations are equivalent to Newton's equations for a conservative potential, that is, a potential energy that depends on position only. Using Equations (1.5.7) and (1.5.8) in (1.5.9) gives Newton's second law

$$\dot{p}_i - F_i = 0, \quad i = 1, 2, 3 \tag{1.5.10}$$

Lagrange's equations retain the form shown in Equation (1.5.9) for an arbitrary transformation of coordinates.

Lagrange's equations can be derived from the variational principle, or principle of least action proposed by Scottish mathematician William R. Hamilton in 1834. Hamilton's variational principle states that a classical particle will follow a path that makes the action integral

$$S = \int_{t_1}^{t_2} L(q_i, \dot{q}_i, t) \, dt, \quad i = 1, 2, 3 \tag{1.5.11}$$

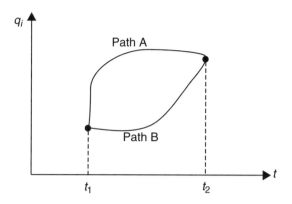

Figure 1-4. Paths of a classical system.

between times t_1 and t_2 an extremum. Figure 1-4 illustrates two possible paths of the system. The variational principle can be written mathematically as

$$\delta S = \delta \int_{t_1}^{t_2} L\, dt = 0 \qquad (1.5.12)$$

where δ denotes a variation. The action in Equation (1.5.11) is important in classical physics and also plays a significant role in quantum physics, as we discuss in Section 8.5.

Example 1.5.1: Lagrange's Equations for a Free Particle

The Lagrangian for a free particle moving in one spatial dimension is

$$L_f = \frac{1}{2} m \dot{q}_f^2 \qquad (1.5.13)$$

since

$$T_f = \frac{1}{2} m \dot{q}_f^2, \quad V_f = 0 \qquad (1.5.14)$$

The force on the particle is

$$F_f = \frac{\partial L_f}{\partial q_f} = 0 \tag{1.5.15}$$

and the momentum is

$$p_f = \frac{\partial L_f}{\partial \dot{q}_f} = m\dot{q}_f \tag{1.5.16}$$

since L_f does not depend explicitly on q_f. Lagrange's equation gives

$$\frac{d}{dt}\left(\frac{\partial L_f}{\partial \dot{q}_f}\right) - \frac{\partial L_f}{\partial q_f} = \frac{d}{dt}m\dot{q}_f = \frac{d}{dt}p_f = \dot{p}_f = 0 \tag{1.5.17}$$

Example 1.5.2: Lagrange's Equations for a Harmonic Oscillator

The kinetic energy T_{HO} and potential energy V_{HO} of a harmonic oscillator in one spatial dimension are

$$T_{HO} = \frac{1}{2}m\dot{q}_{HO}^2, \quad V = \frac{1}{2}kq_{HO}^2 \tag{1.5.18}$$

where m is the mass of the oscillating object and k is the spring constant. The corresponding Lagrangian for the harmonic oscillator is

$$L_{HO} = T_{HO} - V_{HO} = \frac{1}{2}m\dot{q}_{HO}^2 - \frac{1}{2}kq_{HO}^2 \tag{1.5.19}$$

The force on the oscillating object is

$$F_{HO} = \frac{\partial L_{HO}}{\partial q_{HO}} = -kq_{HO} \tag{1.5.20}$$

and the momentum is

$$p_{HO} = \frac{\partial L_{HO}}{\partial \dot{q}_{HO}} = -k\dot{q}_{HO} \tag{1.5.21}$$

Lagrange's equations give

$$\frac{d}{dt}\left(\frac{\partial L_{HO}}{\partial \dot{q}_{HO}}\right) - \frac{\partial L_{HO}}{\partial q_{HO}} = \frac{d}{dt}m\dot{q}_{HO} - (-kq_{HO}) = m\ddot{q}_{HO} + kq_{HO} = 0$$

$$(1.5.22)$$

which gives the expected force equation

$$m\ddot{q}_{HO} = -kq_{HO} \qquad (1.5.23)$$

HAMILTON'S EQUATIONS

Lagrange's equations and Newton's equations are second-order differential equations in time—that is, the equations depend on the differential operator d^2/dt^2. Hamilton's equations express the laws of classical dynamics as first-order differential equations in time—that is, Hamilton's equations depend on the differential operator d/dt. To obtain Hamilton's equations, we define the Hamiltonian function H that depends on momentum $\vec{p}$ and position $\vec{q}$, thus

$$H = H(p_i, q_i) = \sum_{i=1}^{3} p_i \dot{q}_i - L(q_i, \dot{q}_i) \qquad (1.5.24)$$

Hamilton's equations are

$$\frac{\partial H}{\partial p_i} = \dot{q}_i \qquad (1.5.25)$$

and

$$\frac{\partial H}{\partial q_i} = -\dot{p}_i \qquad (1.5.26)$$

For a conservative potential, the Hamiltonian equals the total energy, thus

$$H = T + V \qquad (1.5.27)$$

where $T = T(\vec{p}, \vec{q})$ and $V = V(\vec{q})$.

Example 1.5.3: Hamilton's Equations for a Free Particle

The kinetic energy and potential energy of a free particle with mass m moving in one spatial dimension are

$$T_f = \frac{p_f^2}{2m}, \quad V_f = 0 \tag{1.5.28}$$

The Hamiltonian is

$$H_f = T_f + V_f = \frac{p_f^2}{2m} \tag{1.5.29}$$

Hamilton's equations give the velocity

$$\frac{\partial H_f}{\partial p_f} = \frac{p_f}{m} = \dot{q}_f \tag{1.5.30}$$

and force

$$\frac{\partial H_f}{\partial q_f} = 0 = -\dot{p}_f \quad \text{or} \quad \dot{p}_f = 0 \tag{1.5.31}$$

The force equation, Equation (1.5.31), shows that the momentum of a free particle is constant.

Example 1.5.4: Hamilton's Equations for a Harmonic Oscillator

A body with mass m that is oscillating in one spatial dimension has the kinetic energy

$$T_{HO} = \frac{p_{HO}^2}{2m} \tag{1.5.32}$$

and potential energy

$$V_{HO} = \frac{1}{2}kq_{HO}^2 \tag{1.5.33}$$

with spring constant k. The Hamiltonian is

$$H_{HO} = T_{HO} + V_{HO} = \frac{p_{HO}^2}{2m} + \frac{1}{2}kq_{HO}^2 \qquad (1.5.34)$$

Hamilton's equations give the velocity

$$\frac{\partial H_{HO}}{\partial p_{HO}} = \frac{p_{HO}}{m} = \dot{q}_{HO} \qquad (1.5.35)$$

and force

$$\frac{\partial H_{HO}}{\partial q_{HO}} = kq_{HO} = -\dot{p}_{HO} \quad \text{or} \quad \dot{p}_{HO} = -kq_{HO} \qquad (1.5.36)$$

Combining Equations (1.5.35) and (1.5.36) gives

$$\ddot{q}_{HO} = -\frac{k}{m}q_{HO} = -\omega^2 q_{HO} \qquad (1.5.37)$$

with frequency of oscillation

$$\omega = \sqrt{\frac{k}{m}} \qquad (1.5.38)$$

ENERGY TRANSFORMATION IN A HARMONIC OSCILLATOR

The Hamiltonian in Equation (1.5.34) gives the total energy of a harmonic oscillator. The displacement of the harmonic oscillator from its initial position is given by Equations (1.4.17) and (1.4.19):

$$q_{HO} = q_{max} \cos(\omega t), \quad \omega = \sqrt{\frac{k}{m}} \qquad (1.5.39)$$

The rate of change of the displacement with respect to time is

$$\dot{q}_{HO} = -\omega q_{max} \sin(\omega t) \qquad (1.5.40)$$

The kinetic energy of the harmonic oscillator can be expressed as a function of time by substituting Equations (1.5.40) and (1.5.35) into (1.5.32).

The result is

$$T_{HO} = \frac{p_{HO}^2}{2m} = \frac{m}{2}\dot{q}_{HO}^2 = \frac{m}{2}\omega^2 q_{max}^2 \sin^2 \omega t = \frac{k}{2}q_{max}^2 \sin^2 \omega t \qquad (1.5.41)$$

A similar calculation for the potential energy of the harmonic oscillator using Equations (1.5.39) and (1.5.33) gives

$$V_{HO} = \frac{1}{2}kq_{HO}^2 = \frac{k}{2}q_{max}^2 \cos^2 \omega t \qquad (1.5.42)$$

The total energy of the harmonic oscillator is found by adding Equations (1.5.3) and (1.5.42) as in Equation (1.5.34). The resulting Hamiltonian is

$$H_{HO} = T_{HO} + V_{HO} = \frac{k}{2}q_{max}^2 \sin^2 \omega t + \frac{k}{2}q_{max}^2 \cos^2 \omega t = \frac{k}{2}q_{max}^2 \qquad (1.5.43)$$

We see from Equation (1.5.43) that the total energy of the harmonic oscillator is a constant with respect to time, even though the kinetic and potential energies depend on time. The kinetic, potential, and total energies of a harmonic oscillator with mass 2 kg, spring constant 40 N/m, and maximum displacement 0.1 m are shown in Figure 1-5. Kinetic energy and potential energy transform into one another in such a way that the total energy remains constant. Total energy is conserved.

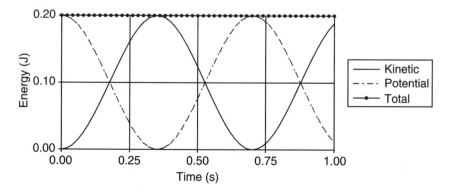

Figure 1-5. Harmonic oscillator energies.

1.6 ENERGY PROFESSIONALS

A student graduating today can expect a career to last to 2040 or beyond. The demand by society for petroleum fuels will continue at or above current levels for a number of years, but the trend seems clear. The global energy portfolio is undergoing a transition from an energy portfolio dominated by fossil fuels to an energy portfolio that includes a range of fuel types.

Increasing trends in population and consumption, price volatility, supply instability, and environmental concerns are motivating changes to the energy mix and energy strategies in the twenty-first century. In an attempt to respond to market realities, some oil and gas companies and electric power companies are beginning to transform themselves into energy companies, and the trend is expected to continue.[2] These companies are pioneers in an emerging energy industry. They will need a new type of professional to help them function at optimal levels.

Thumann and Mehta [1997] define energy engineering as a profession that "applies scientific knowledge for the improvement of the overall use of energy" (p. 1). Professionals in energy companies will need to understand and appreciate the role of alternative energy components in the energy mix. They will need to understand how energy can be transformed from one form of energy to another form, and the consequences of energy transformations. The creativity of future energy professionals and their ability to contribute to policy formation will be enhanced if they are able to identify and solve problems in the acquisition and environmentally acceptable use of several energy components.

Improvements in technology and an increasing reliance on information should require a high level of technical expertise to acquire resources on behalf of society. To meet the technical demands, Walesh [1995] predicted that future engineers would require periods of dedicated learning or retraining between periods of full-time employment throughout their careers. MacDonald [1999] wrote in the Society of Petroleum Engineers' career management series that people average a job change every five years, and that they occasionally change careers in the process. One way to help energy professionals prepare for job and career changes is to identify the requirements of an evolving profession and provide appropriate educational opportunities.

One of the objectives of an introductory energy course should be to help you understand and appreciate the role of alternative energy components in the energy mix. To achieve this objective, it is necessary to discuss the origin of energy sources and the technology of energy. By developing an understanding of the origin of energy sources, we can better assess the viability of emerging energy technologies and the role they will play in the future. This broad background will give you additional flexibility during your career and help you thrive in an energy industry that is evolving from an industry dominated by fossil fuels to an industry working with many energy sources.

ENDNOTES

1. A vast amount literature exists that discusses the way science is conducted. Some references that I have found to be most useful include Kuhn [1970], Gjertsen [1984], Traweek [1988], Root-Bernstein [1989], Wolpert [1992], Friedlander [1995], and Park [2000].
2. For more information, see the "Energy Mix" Scenario C in Schollnberger [1999], Rockefeller and Patrick's article [1999], and Hamto's discussion [2000] of fuel cell technology.

EXERCISES

1-1. A. Suppose a person can work at the rate of 100 W. How many hours must the person work to complete one kilowatt-hour (1 kWh) of work?

B. What is the person's horsepower?

1-2. A food Calorie is a kilocalorie of energy, that is, 1 Calorie = 1000 calories, where a calorie is a unit of heat energy. The thermodynamic calorie equals 4.184 J. Suppose a human being consumes 2000 Calories per day. Express this consumption in watts.

1-3. A. Calculate the energy used by a 1200-watt hair dryer that is operated for 15 minutes.

B. How long should a 100 W light bulb be left on to match the energy used in Part A? Express your answer in hours.

C. Suppose the light bulb is used only while the hair dryer in Part A is on. How much energy is used by the light bulb and hair dryer? Express your answer in joules and kilowatt-hours.

D. What percentage of total energy is used by the light bulb in Part C?

1-4. The work done by a heat engine is the amount of heat generated by the engine times its thermodynamic efficiency. Fill in the following table.

Period	Engine	Heat (J)	Thermodynamic efficiency (%)	Work (J)
Mid–1700s	Newcomen Steam	10,000	0.05	
Early–1800s	Walking-beam	10,000	4.0	
Mid–1800s	Corliss steam	10,000	20.0	

1-5. A. Consider a car that moves with a thermal power consumption of 100,000 W. Express the thermal power consumption in horsepower.
B. If it takes you 30 minutes to commute to work using the car in Part A, calculate the energy used (in kWh).

1-6. A. A nation with 100 million people consumes 30 quads of energy annually. Estimate the per capita energy consumption and express it in megajoules per person.
B. Estimate the per capita power consumption (energy consumption rate) and express it in watts per person.

1-7. A classical harmonic oscillator has mass 3 kg, spring constant 15 N/m, and maximum displacement 0.2 m. Plot kinetic energy, potential energy, and total energy of the harmonic oscillator. What is the angular frequency of oscillation (in s^{-1}) and total energy (in J)?

1-8. Estimate the universal constant of gravitation G if the acceleration of gravity at the surface of the Earth is $g = 9.8$ m/s^2. Hint: assume the Earth is a sphere and use the equatorial radius of the Earth as the radius of the sphere.

1-9. A sports car has a high-performance engine rated at 550 HP. During a road test, an average of 80% of the maximum power was used. The road test lasted 45 minutes. How much energy was output by the engine during the road test? Express your answer in megajoules.

1-10. A. Consider a country with a population of 20 million people. Suppose we need 200,000 MJ/person each year to support a satisfactory UN Human Development Index. How much energy (in MJ) is needed each year by the country?
B. How much power (in MW) is required?
C. How many 1000 MW power plants are needed?

D. How much energy (in MJ) is needed each day?

E. Suppose the energy is obtained by consuming crude oil with an energy density of 37,000 MJ/m^3. How many barrels of crude oil are required each day?

F. If the price of oil is US$20/bbl, what is the cost of oil per kWh of energy used each day? Hint: Calculate the energy used each day in kWh/day and the cost of oil per day in US$/day before calculating the cost of oil per kWh.

CHAPTER TWO

Electric Power Generation and Distribution

Power plants electrified the United States, and eventually the rest of the modern world, in only a century and a half. Although other types of energy are used around the world, electricity is the most versatile form for widespread distribution. The role of electric power plants is to generate electric current for distribution through a transmission grid.[1] The historical developments that led to the modern power generation and distribution system are outlined in the following section.[2] We then review the equations of electrodynamics, discuss the principles of electric power generation, and describe the system that has been developed to distribute electric power.

2.1 HISTORICAL DEVELOPMENT OF ELECTRIC POWER

People first used muscle energy to gather food and build shelters. Muscle energy was used to grind grain with stones, chop wood with hand axes, and propel oar-powered ships. In many instances in history, conquered people became slaves and provided muscle energy for their conquerors.

Stones, axes, and oars are examples of tools that were developed to make muscle energy more effective. Water wheels and windmills replaced muscle power for grinding grain as long ago as 100 B.C.E. Wind and sails replaced muscle energy and oar-powered ships. Early power stations were driven by wind and flowing water, and were built where wind and flowing water were available.

Furnaces use heat to smelt ore. Ore is rock that contains metals such as copper, tin, iron, and uranium. Heat from fire frees the metal atoms and allows them to be collected in a purified state. Copper and tin were the first metals smelted, and they could be combined to form bronze.

STEAM-GENERATED POWER

Heat and water were combined to generate steam, and steam engines were developed to convert thermal energy to mechanical energy. Early steam engines drove a piston that was placed between condensing steam and air, as illustrated in Figure 2-1. When steam condenses, it occupies less volume and creates a partial vacuum. The air on the other side of the piston expands and can push the piston. By alternately injecting steam and letting it condense, the piston can be made to move in an oscillating linear motion.

English inventor Thomas Newcomen invented the steam engine in 1705 and built the first practical steam engine in 1712. Newcomen's steam engine was used to pump water from flooded coal mines. Steam condensation was induced in Newcomen's steam engine by spraying cold water into the chamber containing steam. The resulting condensation created the partial vacuum that allowed air to push the piston. A weight attached to the rod used gravity to pull the piston back as steam once again entered the left-hand chamber, as shown in Figure 2-1.

Scottish engineer James Watt improved the efficiency of the steam engine by introducing the use of a separate vessel for collecting and condensing the expelled steam. Watt's assistant, William Murdock, developed a gear system design in 1781 that enabled the steam engine to produce circular motion. The ability to produce circular motion made it possible for steam engines to provide the power needed to turn wheels. Steam engines could be placed on platforms attached to wheels to provide power for transportation. Thus was born the technology needed to develop steam-driven locomotives, paddle wheel boats, and ships with steam-driven propellers. Furthermore, steam engines did not have to be built near a particular fuel source. It was no longer necessary to build manufacturing facilities near

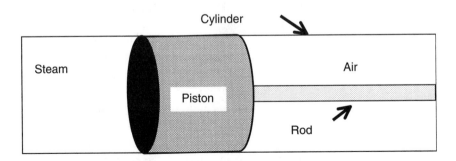

Figure 2-1. Schematic of a simple steam engine.

sources of wind or water, which were formerly used to provide power. Manufacturers had the freedom to build their manufacturing facilities in locations that optimized the success of their enterprise. If they chose, they could build near coal mines to minimize fuel costs, or near markets to minimize the cost of distributing their products.

Steam-generated power was an environmentally dirty source of power. Burning biomass, such as wood, or fossil fuel, such as coal, typically produced the heat needed to generate steam. Biomass and fossil fuels were also used in the home. Attempts to meet energy demand by burning primarily wood in sixteenth-century Britain led to deforestation and the search for a new fuel source [Nef, 1977]. Fossil fuel in the form of coal became the fuel of choice in Britain and other industrialized nations. Coal gas, which is primarily methane, was burned in nineteenth-century homes.

The demand for energy had grown considerably by the nineteenth century. Energy for cooking food and heating and lighting homes was provided by burning wood, oil, or candles. The oil was obtained from such sources as surface seepages or whale blubber. Steam-generated power plants could only serve consumers in the immediate vicinity of the power plant. A source of power was needed that could be transmitted to distant consumers.

By 1882, Thomas Edison was operating a power plant in New York City. Edison's plant generated direct current electricity at a voltage of 110 V. Nations around the world soon adopted the use of electricity. By 1889, a megawatt electric power station was operating in London, England. Industries began to switch from generating their own power to buying power from a power generating company. But a fundamental inefficiency was present in Edison's approach to electric power generation. The inefficiency was not removed until the battle of currents was fought and won.

THE BATTLE OF CURRENTS

The origin of power generation and distribution is a story of the battle of currents, a battle between two titans of business: Thomas Edison and George Westinghouse. The motivation for their confrontation can be reduced to a single, fundamental issue: how to electrify America.

Edison invented the first practical incandescent lamp and was a proponent of electrical power distribution by direct electric current. He displayed his direct current technology at New York City's Pearl Street Station in 1882. One major problem with direct current is that it cannot be transmitted very far without significant degradation.

Unlike Edison, Westinghouse was a proponent of alternating electric current because it can be transmitted over much greater distances than direct electric current. Alternating current could be generated at low voltages, transformed to high voltages for transmission through power lines, and then reduced to lower voltages for delivery to the consumer. Nikolai Tesla (1857–1943), a Serbian-American scientist and inventor who was known for his work with magnetism, worked with Westinghouse to develop alternating current technology. Westinghouse displayed his technology at the 1893 Chicago World's Fair. It was the first time one of the world's great events was illuminated at night, and it showcased the potential of alternating current electricity.

The first large-scale power plant was built at Niagara Falls near Buffalo, New York, in the 1890s. The power plant at Niagara Falls began transmitting power to Buffalo, less than 30 kilometers (20 miles) away, in 1896. The transmission technology used alternating current. The superiority of alternating current technology gave Westinghouse a victory in the battle of currents and Westinghouse became the father of the modern power industry. Westinghouse's success was not based on better business acumen, but on the selection of better technology. The physical principles that led to the adoption of alternating current technology are discussed in the following paragraphs.

A chronology of milestones in the development of electrical power is presented in Table 2-1 [after Brennan, et al., 1996, page 22; and Aubrecht, 1995, Chapter 6]. The milestones refer to the United States, which was the worldwide leader in the development of an electric power industry.

GROWTH OF THE ELECTRIC POWER INDUSTRY

The power industry started out as a set of independently owned power companies. Because of the large amounts of money needed to build an efficient and comprehensive electric power infrastructure, the growth of the power industry required the consolidation of the smaller power companies into a set of fewer but larger power companies. The larger, regulated power-generating companies became public utilities that could afford to build regional electric power transmission grids. The ability to function more effectively at larger scales is an example of an economy of scale. Utility companies were able to generate and distribute more power at lower cost by building larger power plants and transmission grids.

The *load* on a utility is the demand for electrical power. Utilities need to have power plants that can meet three types of loads: base load, intermediate

Table 2-1
Early milestones in the history of the electric power industry in the United States

Year	Event	Comment
1882	Pearl Street Station, New York	Edison launches the "age of electricity" with his DC power station
1893	Chicago World's Fair	Westinghouse displays AC power to the world
1898	Fledgling electric power industry seeks monopoly rights as regulated utility	Chicago Edison's Samuel Insull leads industry to choose regulation over "debilitating competition"
1907	States begin to regulate utilities	Wisconsin and New York are first to pass legislation
1920	Federal government begins to regulate utilities	Federal Power Commission formed

or cycling load, and peak load. The *base load* is the minimum baseline demand that must be met in a 24-hour day. *Intermediate load* is the demand that is required for several hours each day and tends to increase or decrease slowly. *Peak load* is the maximum demand that must be met in a 24-hour day.

Electric power for small towns and rural communities was an expensive extension of the power transmission grid that required special support. The federal government of the United States provided this support when it created the Tennessee Valley Authority (TVA) and Rural Electric Associations (REA).

STATUS OF ELECTRIC POWER GENERATION

The first commercial-scale power plants were hydroelectric plants. The primary energy source (the energy that is used to operate an electricity-generating power plant) for a hydroelectric plant is flowing water. Today, most electricity is generated by one of the following primary energy sources: coal, natural gas, oil, or nuclear power. Table 2-2 presents the consumption of primary energy in the year 1999 as a percentage of total primary energy consumption in the world for a selection of primary energy types. The table is based on statistics maintained at a website by the Energy Information Administration (EIA), United States Department of Energy. The statistics should be considered approximate. They are quoted here because the EIA is a standard source of energy information that is

Table 2-2
Primary energy consumption in 1999 by energy type

Primary energy type	Total world energy consumption
Oil	39.9%
Natural gas	22.8%
Coal	22.2%
Hydroelectric	7.2%
Nuclear	6.6%
Geothermal, solar, wind, and wood	0.7%

Source: EIA website, 2002.

widely referenced. The statistics give us an idea of the relative importance of different primary energy sources. Fossil fuels were clearly the dominant primary energy source at the end of the twentieth century. Electric energy, however, is the most versatile source of energy for running the twenty-first century world, and much of the primary energy is consumed in the generation of electric energy.

2.2   ELECTROMAGNETISM

People have known since ancient times that certain materials can attract other materials. For example, rubbing a hard rubber rod with fur and then placing the rod near small bits of paper in a dry climate will result in the paper bits being picked up by the rod. This is one example of an electric phenomenon. Another is familiar to those of us who have walked across carpets during a cold, dry spell. An electric shock may greet us if we touch a metal doorknob.

Materials, such as magnetite (or lodestone), can attract or even repel other materials without actually coming in contact with them. If you place a small magnet next to the screen of your television, you should observe the television image become distorted in the vicinity of the magnet. This distortion is an example of a magnetic phenomenon. The magnetic field of the magnet is deflecting the electrons that are striking the screen of your television picture tube.

Electric and magnetic phenomena are special cases of a more general class of phenomena known as electromagnetic phenomena. Englishman Michael Faraday performed one of the first and most thorough experimental studies of electromagnetic phenomena. Faraday's work provided much of the experimental data used by other scientists to construct a mathematical theory of electromagnetism.

Scottish mathematician and physicist James Clerk Maxwell was the first to publish what is now considered to be the most complete mathematical formulation of electrodynamics. Maxwell, building on the work of many other researchers, including Faraday, discovered a set of equations that could be used to describe all of the then-known electromagnetic observations. Maxwell's equations synthesized experimental observations in electricity and magnetism, and showed that electrical and magnetic forces are different manifestations of a single electromagnetic force. In addition, Maxwell used his equations to predict a large, but finite, speed for light. His prediction was later confirmed by experiments.

MAXWELL'S EQUATIONS

Maxwell's equations can be written in either integral or differential form. The differential form of Maxwell's equations in SI units follows [Jackson, 1999, Appendix]:

$$\nabla \cdot \vec{D} = \rho \quad \text{(Gauss's Law)} \qquad (2.2.1)$$

$$\nabla \cdot \vec{B} = 0 \quad \text{(Gauss's Law—magnetic)} \qquad (2.2.2)$$

$$\nabla \cdot \vec{H} = \vec{J} + \frac{\partial \vec{D}}{\partial t} \quad \text{(Maxwell-Ampere Law)} \qquad (2.2.3)$$

$$\nabla \cdot \vec{E} = -\frac{\partial \vec{B}}{\partial t} \quad \text{(Faraday's Law)} \qquad (2.2.4)$$

where

$\vec{D}$ = displacement, C/m^2

ρ = charge density, C/m^3

$\vec{B}$ = magnetic induction, Tesla

$\vec{H}$ = magnetic field, A/m

$\vec{J}$ = current density, A/m^2

$\vec{E}$ = electric field, V/m

The displacement $\vec{D}$ and magnetic field $\vec{H}$ depend on the properties of the medium and may be written as

$$\vec{D} = \varepsilon_0 \vec{E} + \vec{P} \tag{2.2.5}$$

$$\vec{H} = \frac{1}{\mu_0} \vec{B} - \vec{M} \tag{2.2.6}$$

where

$\vec{P}$ = polarization of a material medium, C/m^2

$\vec{M}$ = magnetization of a material medium, A/m

μ_0 = permeability of vacuum = $4\pi \times 10^{-7}$ N/A^2

ε_0 = permittivity of vacuum = $\dfrac{1}{\mu_0 c^2}$

c = speed of light in vacuum = 3.0×10^8 m/s

The permeability μ and permittivity ε of an isotropic material medium are obtained by writing Equations (2.2.5) and (2.2.6) as

$$\vec{D} = \varepsilon \vec{E} \tag{2.2.7}$$

$$\vec{B} = \mu \vec{H} \tag{2.2.8}$$

For external sources in vacuum, Equations (2.2.7) and (2.2.8) become

$$\vec{D} = \varepsilon_0 \vec{E} \tag{2.2.9}$$

$$\vec{B} = \mu_0 \vec{H} \tag{2.2.10}$$

The dimensionless ratios $\varepsilon/\varepsilon_0$ and μ/μ_0 are called relative permittivity (or dielectric constant) and relative permeability, respectively. The terms *permeability* and *relative permeability* have an important and entirely different use in describing fluid flow in porous media [Fanchi, 2002].

The total electromagnetic energy density u is

$$u = \frac{1}{2} (\vec{E} \cdot \vec{D} + \vec{B} \cdot \vec{H}) \tag{2.2.11}$$

The Poynting vector gives the flow of electromagnetic energy through a unit area in unit time $\vec{S}$:

$$\vec{S} = \vec{E} \times \vec{H} \tag{2.2.12}$$

The Poynting vector has the dimension of power per unit area, and is named after British physicist John Henry Poynting (1852–1914), the man who first studied its properties. The momentum density of the electromagnetic fields, or momentum per unit volume, is

$$\vec{g} = \frac{\vec{S}}{c^2} \tag{2.2.13}$$

The Poynting vector and momentum density for electromagnetic fields in vacuum are

$$\vec{S}_0 = \frac{\vec{E} \times \vec{B}}{\mu_0} \tag{2.2.14}$$

and

$$\vec{g}_0 = \frac{\vec{S}_0}{c^2} = \frac{\vec{E} \times \vec{B}}{\mu_0 c^2} = \varepsilon_0 \vec{E} \times \vec{B} \tag{2.2.15}$$

where we have used Equation (2.2.10).

The continuity equation

$$\frac{\partial \rho}{\partial t} + \nabla \cdot \vec{J} = 0 \tag{2.2.16}$$

can be derived from Maxwell's equations and is an expression of charge conservation. Maxwell's equations can be used to show that the fields $\vec{E}$ and $\vec{B}$ obey a wave equation in vacuum. The wave equations for $\vec{B}$ and $\vec{E}$ in vacuum are

$$\nabla^2 \vec{B} = -\frac{1}{c^2} \frac{\partial^2 \vec{B}}{\partial t^2} \tag{2.2.17}$$

$$\nabla^2 \vec{E} = -\frac{1}{c^2} \frac{\partial^2 \vec{E}}{\partial t^2} \tag{2.2.18}$$

where ∇^2 is the Laplacian $\nabla^2 = \dfrac{\partial^2}{\partial x^2} + \dfrac{\partial^2}{\partial y^2} + \dfrac{\partial^2}{\partial z^2}$ in Cartesian coordinates.

The speed of the $\vec{B}$ and $\vec{E}$ waves in vacuum is the speed of light $c = 1/\sqrt{\varepsilon_0 \mu_0}$, and the amplitudes of the $\vec{B}$ and $\vec{E}$ waves are related by $E = cB$.

Maxwell's electromagnetic equations are comparable in importance to Newton's laws of mechanics. The combination of Newton's laws and Maxwell's equations forms the theoretical basis of the subject known as classical electrodynamics: the study of the motion of electrically charged objects. A point particle with charge q and velocity $\vec{v}$ moving in the presence of electromagnetic fields is subjected to the force

$$\vec{F} = q\left(\vec{E} + \vec{v} \times \vec{B}\right) \tag{2.2.19}$$

Equation (2.2.19) is called the Lorentz force. Its application is illustrated by the following examples. Classical electrodynamics, and especially the speed of light, played a major role in Einstein's development of relativity and his discovery of the relationship between mass and energy.

Example 2.2.1: Charged Particle in an Electric Field

The force $\vec{F}$ on a particle with mass m and charge q in an electric field $\vec{E}$ is

$$\vec{F} = m\ddot{\vec{r}} = q\vec{E} \tag{2.2.20}$$

This can be expressed in terms of components and position coordinates $\{x, y, z\}$ as

$$\begin{aligned} m\ddot{x} &= qE_x \\ m\ddot{y} &= qE_y \\ m\ddot{z} &= qE_z \end{aligned} \tag{2.2.21}$$

The electric field components can be functions of $\{x, y, z, t\}$. As a simple example, suppose the electric field is constant and oriented along the z-axis so that $\{E_x, E_y, E_z\} = \{0, 0, E_c\}$. The equations of motion become

$$\begin{aligned} m\ddot{x} &= 0 \\ m\ddot{y} &= 0 \\ m\ddot{z} &= qE_c \end{aligned} \tag{2.2.22}$$

The acceleration of a positively charged particle due to the electric field is constant and in the direction of the electric field, thus

$$\ddot{z} = \frac{qE_c}{m} \qquad (2.2.23)$$

The trajectory of the particle is

$$
\begin{aligned}
x &= x_0 + v_{x_0}t \\
y &= y_0 + v_{y_0}t \\
z &= z_0 + v_{z_0}t + \frac{1}{2}\frac{qE_c}{m}t^2
\end{aligned}
\qquad (2.2.24)
$$

where $\{x_0, y_0, z_0\}$ is the initial position of the particle and $\{v_{x_0}, v_{y_0}, v_{z_0}\}$ is the initial velocity of the particle.

Example 2.2.2: Charged Particle in a Magnetic Field

The force $\vec{F}$ on a particle with mass m and charge q moving with velocity $\vec{v}$ in a static magnetic field with magnetic induction $\vec{B}$ is

$$\vec{F} = q\left(\vec{v} \times \vec{B}\right) \qquad (2.2.25)$$

where $\vec{v} \times \vec{B}$ is the cross product of velocity with magnetic induction. Suppose the magnetic induction is uniform and aligned parallel to the y-axis such that $\vec{B} = B_c\hat{k}$ for a unit vector $\hat{k}$ aligned with the z-axis. The equations of motion are

$$m\ddot{\vec{r}} = q\left(\vec{v} \times B_c\hat{k}\right) = qB_c \begin{vmatrix} \hat{i} & \hat{j} & \hat{k} \\ \dot{x} & \dot{y} & \dot{z} \\ 0 & 0 & 1 \end{vmatrix} \qquad (2.2.26)$$

or

$$m\left(\ddot{x}\hat{i} + \ddot{y}\hat{j} + \ddot{z}\hat{k}\right) = qB_c\left(\dot{y}\hat{i} - \dot{x}\hat{j}\right) \qquad (2.2.27)$$

This vector equation is separated into three equations by equating vector components, thus

$$m\ddot{x} = qB_c\dot{y}$$

$$m\ddot{y} = -qB_c\dot{x} \tag{2.2.28}$$

$$m\ddot{z} = 0$$

The particle does not accelerate along the direction of the magnetic field. Particle acceleration is in the plane that is transverse to the direction of motion.

Solution of the equations of motion shows that the particle will follow a helical path if it has a nonzero velocity component aligned with the direction of the magnetic field. If $v_z = \dot{z} = 0$ in our example, the particle will move in a circular path in the x-y plane with a constant radial velocity v_r. We can estimate the radius of the path by equating the magnitude of the magnetic force to the centripetal force:

$$qv_rB_c = \frac{mv_r^2}{r} \tag{2.2.29}$$

or

$$r = \frac{mv_r}{qB_c} \tag{2.2.30}$$

The radius of the path is proportional to the magnitude of the momentum mv_r of the particle. The radial velocity can be written in terms of the radius r and the angular frequency

$$\omega_c = \frac{qB_c}{m} \tag{2.2.31}$$

so that

$$v_r = \omega_c r \tag{2.2.32}$$

The angular frequency ω_c is called the cyclotron frequency. It is used in the design of particle accelerators and the development of some types of nuclear fusion reactors.

SCALAR AND VECTOR POTENTIALS

Electric and magnetic fields may be written in terms of the electric potential Φ and vector potential $\vec{A}$ using the relations

$$\vec{E} = -\frac{\partial \vec{A}}{\partial t} - \nabla \Phi \tag{2.2.33}$$

and

$$\vec{B} = \nabla \times \vec{A} \tag{2.2.34}$$

The electric potential is also called the scalar potential. The scalar and vector potentials have many interesting properties that are beyond the scope of this book. The potentials appear again later in discussions of quantum mechanical equations.

2.3 ELEMENTS OF ALTERNATING CURRENT CIRCUITS

We saw in Section 2.1 that alternating current electricity plays a crucial role in today's electric power industry. Alternating currents are needed for generating electric power and efficiently transmitting electric power over great distances. Before we quantify these statements, we review the basic elements of alternating current circuits in this section. We can then provide a more sophisticated discussion of electric power generation and transmission.

CURRENT IN THE *LCR* SERIES CIRCUIT

Some of the basic elements of an alternating current circuit are sketched in Figure 2-2. They include an alternating electromotive force, an inductor, a capacitor, and a resistor. Our goal here is to solve the *LCR* series circuit equations for the time-dependent current i.[3]

The sum of the electric potential differences across all of the elements in the *LCR* circuit satisfies the equation

$$V_{ha} + V_{bc} + V_{de} + V_{fg} = 0 \tag{2.3.1}$$

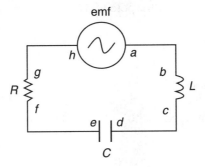

Figure 2-2. An *LCR* series circuit.

An alternating current is generated in the *LCR* series circuit by imposing an oscillating potential difference V_{ha} between points *h* and *a* in Figure 2-2. We impose the oscillating potential difference

$$V_{ha} = \varepsilon_{\text{emf}} \sin \omega t \tag{2.3.2}$$

where the angular frequency ω is the frequency of oscillation of the electromotive force. The frequency of oscillation $\nu = \omega/2\pi$ for commercial alternating current in the United States is typically 60 Hz or $\omega = 377 \text{ s}^{-1}$. The constant voltage ε_{emf} is the initial voltage, that is, voltage at $t = 0$. The potential difference V_{bc} across the inductor is

$$V_{bc} = L\frac{di}{dt} \tag{2.3.3}$$

for an electric current *i* and constant inductance *L*. The potential difference V_{de} across the capacitor is

$$V_{de} = \frac{q}{C} \tag{2.3.4}$$

for a charge *q* and constant capacitance *C*. The electric current is related to charge by the time derivative

$$i = \frac{dq}{dt} \tag{2.3.5}$$

Finally, Ohm's law gives the potential difference V_{fg} across the resistor as

$$V_{fg} = iR \qquad (2.3.6)$$

for a constant resistance R.

The current for the LCR series circuit is obtained by first substituting Equations (2.3.2) through (2.3.6) into (2.3.1) to find

$$\varepsilon_{emf} \sin \omega t + L\frac{di}{dt} + \frac{q}{C} + iR = 0 \qquad (2.3.7)$$

If we differentiate Equation (2.3.7) with respect to time we obtain the following differential equation for current after some rearrangement of terms:

$$L\frac{d^2i}{dt^2} + R\frac{di}{dt} + \frac{1}{C}i = -\varepsilon_{emf}\omega \cos \omega t \qquad (2.3.8)$$

One solution of Equation (2.3.8) is

$$i(t) = a \cos \omega t + b \sin \omega t \qquad (2.3.9)$$

where the coefficients a, b are

$$a = -\frac{\varepsilon_{emf} X}{R^2 + X^2}, \quad b = \frac{\varepsilon_{emf} R}{R^2 + X^2} \qquad (2.3.10)$$

with

$$X \equiv X_C - X_L, \quad X_C = \frac{1}{\omega C}, \quad X_L = L\omega \qquad (2.3.11)$$

The parameters X, X_L, X_C are the reactance, inductive reactance, and capacitive reactance respectively. The denominator in Equation (2.3.10) is the impedance Z. Impedance can be written as

$$Z = \sqrt{R^2 + X^2} = \sqrt{R^2 + (X_C - X_L)^2} = \sqrt{R^2 \left(\frac{1}{\omega C} - L\omega\right)^2} \qquad (2.3.12)$$

Reactances and impedance have the unit of resistance, the ohm in SI units.

Two other solutions to Equation (2.3.8) that appear in the literature are

$$i(t) = i_0 \sin(\omega t - \delta) \tag{2.3.13}$$

and

$$i(t) = i_0 \sin(\omega t - \phi) \tag{2.3.14}$$

The term i_0 is the maximum current

$$i_0 = \frac{\varepsilon_{\text{emf}}}{\sqrt{R^2 + X^2}} \tag{2.3.15}$$

The phase angles δ, ϕ are

$$\delta = \tan^{-1}\frac{X}{R} = \tan^{-1}\left(\frac{X_C - X_L}{R}\right) = \tan^{-1}\left(\frac{1/\omega C - \omega L}{R}\right) = -\phi \tag{2.3.16}$$

A relatively easy way to verify that Equations (2.3.13) and (2.3.14) are solutions to Equation (2.3.8) is to use trigonometric identities to transform Equation (2.3.9) into the form of either Equation (2.3.13) or (2.3.14).

POWER IN THE *LCR* SERIES CIRCUIT

The net flow of energy in the *LCR* series circuit depends on how it is operated. We assume the *LCR* series circuit is operated in a steady-state condition so that the average energy stored in the capacitor and inductor does not change. Electromagnetic energy is transformed into thermal energy in the resistor. The rate of transformation of energy in the resistor is the power

$$P(t) = i^2 R = [i_0 \sin(\omega t + \phi)]^2 R \tag{2.3.17}$$

This flow of energy may be called Joule heating because the power is in the form of Joule's law for electrical heating of a resistor, namely $P = i^2 R$.

From the point of view of a power company, Joule heating is a loss of electromagnetic energy to heat energy.

The average power, or net flow of energy, in the LCR series circuit is

$$P_{av} = (i_{rms})^2 R \qquad (2.3.18)$$

where i_{rms} is the root mean square current, or

$$i_{rms} = \sqrt{\frac{i_0^2}{2}} = \frac{i_0}{\sqrt{2}} \qquad (2.3.19)$$

The factor $\sqrt{2}$ in the denominator comes from the average of the $\sin(\omega t + \phi)$ term in Equation (2.3.17). The root mean square potential difference is

$$\varepsilon_{rms} = \frac{\varepsilon_{emf}}{\sqrt{2}} \qquad (2.3.20)$$

The root mean square current and potential difference i_{rms}, ε_{rms} can be used to write the average power of the alternating current LCR series circuit in the equivalent form

$$P_{av} = \varepsilon_{rms} i_{rms} \cos \phi \qquad (2.3.21)$$

The trigonometric factor $\cos \phi$ is called the power factor. The maximum value of the power factor is 1.

The power factor is a function of impedance and resistance. We show the dependence of power factor on impedance and resistance by using Equation (2.3.12) for impedance to write the power factor as

$$\cos \phi = \frac{R}{Z} = \frac{R}{\sqrt{R^2 + (X_C - X_L)^2}} = \frac{R}{\sqrt{R^2 + (1/\omega C - L\omega)^2}} \qquad (2.3.22)$$

Resonance occurs in an LCR series circuit when capacitive impedance equals inductive impedance. In terms of angular frequency, the condition for resonance in an LCR series circuit is

$$X_C - X_L = \frac{1}{\omega C} - L\omega = 0 \qquad (2.3.23)$$

Solving for angular frequency shows that resonance is achieved when angular frequency satisfies the equation

$$\omega_{\text{resonance}} = \omega = \frac{1}{\sqrt{LC}} \tag{2.3.24}$$

The power factor for an *LCR* series circuit with the angular frequency of the alternating current generator equal to the resonance angular frequency gives a power factor $\cos \phi = 1$. The resulting average power is the maximum average power, namely $P_{\text{av}} = \varepsilon_{\text{rms}} i_{\text{rms}}$ since $\cos \phi = 1$.

Example 2.3.1: Rating of Electrical Devices

The rating of electrical devices is usually expressed in terms of average power and root mean square values for electrical variables. For example, a typical voltage for an electrical appliance in the United States is $\varepsilon_{\text{rms}} = 120$ V. A hair dryer may be rated at an average power of 1500 W for a root mean square voltage of 120 V. If the power factor is 1, the corresponding root mean square current is $i_{\text{rms}} = 12.5$ A.

2.4 ELECTRIC POWER GENERATION

Many of the first commercial electric power plants relied on flowing water as their primary energy source. People have known for some time that falling water can be used to generate electric power. A schematic of a hydroelectric power plant is presented in Figure 2-3. Water flows from an upper elevation to a lower elevation through a pipeline called a penstock. The water current turns a turbine that is connected to a generator. The turbine is called the prime mover because it rotates the generator shaft. The rotating turbine rotates a shaft that is connected to either a magnet adjacent to a coil of wire, or a coil of wire adjacent to a magnet. The mechanical energy of falling water is transformed into the kinetic energy of rotation of the turbine. An alternating current generator converts mechanical energy to electrical energy using a few basic physical principles.[4]

A magnetic field is created by the flow of charged particles through a wire. Figure 2-4 shows the magnetic field created by an electric current in a coil of wire. A simple alternating current generator can be constructed from a loop of wire that can be rotated inside a constant magnetic field. Magnetic flux is a measure of the number of magnetic field lines passing

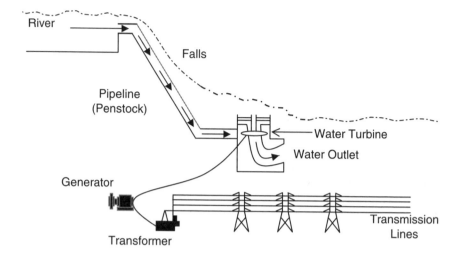

Figure 2-3. Principles of hydroelectric power generation.

through the area of the loop of wire. If the loop of wire is spinning with angular frequency ω inside a uniform magnetic field with magnitude B, it generates a time-varying magnetic flux

$$\Phi_B = BA \cos \omega t \qquad (2.4.1)$$

where A is the area of the loop, and t is time. Faraday's law of induction says that an electromotive force ε_{emf} is induced in a circuit by a time-varying magnetic flux. The electromotive force is proportional to the time rate of change of the magnetic flux through the circuit.

The electromotive force of a simple alternating current generator with N loops of wire is calculated from Faraday's law of induction as

$$\varepsilon_{\text{emf}} = -N\frac{d\Phi_B}{dt} = NBA\omega \sin \omega t \qquad (2.4.2)$$

The minus sign is needed to satisfy Lenz's law, which says that the induced electromotive force must produce a current that generates a magnetic field that opposes the change in magnetic flux through the loop. The maximum electromotive force is $(\varepsilon_{\text{emf}})_{\text{max}} = NBA\omega$. The electromotive force ε_{emf} induces an electric current $I = \varepsilon_{\text{emf}}/R$ in an external circuit with

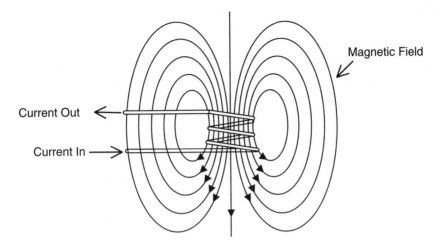

Figure 2-4. Electromagnetism.

resistance R. The induced current is routed into the transmission grid for distribution to consumers.

The efficiency η_{Power} of the power station is the ratio of the power output to the input power, or

$$\eta_{\text{Power}} = \text{Output Power/Input Power} \qquad (2.4.3)$$

A complete assessment of an alternating current generating power station requires knowledge about both the power factor and the power station efficiency.

Example 2.4.1: Power Station Performance

A 100 MW power station uses a 500 kV transmission line to conduct electricity to a consumer. The current in the transmission line estimated from the average power equation $P_{\text{av}} = \varepsilon_{\text{rms}} i_{\text{rms}}$ for the maximum power factor is $i_{\text{rms}} = 200$ A. If the power factor is 0.9 instead of 1, the current calculated from Equation (2.3.21) is $i_{\text{rms}} = 222$ A. If it takes 120 MW of power to generate 100 MW, the efficiency of the power station is $\eta_{\text{Power}} = 100$ MW/120 MW = 0.83 or 83%.

Example 2.4.2: Power Station Rating

Alternating current generators can be rated in terms of volt·amperes, or voltage (volts) times current (amperes). The rating of an alternating current generator that provides a current of 10 A at a voltage of 100 kV is 1,000 kVA.

TRANSFORMERS

Transformers had to be built to use alternating current. A transformer is a device that can convert a small alternating current voltage to a larger alternating current voltage, or vice versa. For example, it is desirable to work with relatively small voltages at the power plant and provide a range of voltages to the consumer. In between, in the transmission lines, a large voltage is required to minimize resistive heating losses in the line. Transformers perform the function of converting, or transforming, voltages. A transformer that increases the voltage is referred to as a *step-up transformer* and is said to step-up the voltage; a transformer that decreases the voltage is referred to as a *step-down transformer* and is said to step-down the voltage. Transformer T$_1$ in Figure 2-5 is a step-up transformer from the low voltage (L.V.) at the power station to a higher voltage (H.V.) in the transmission line. Transformer T$_2$ in the figure is a step-down transformer that converts the relatively high voltage in the transmission line to a lower voltage that is suitable for the consumer. The low voltages shown at opposite ends of the transmission line do not have to be the same. The actual voltages used in the transmission line depend on the properties of the transformer. Typical transmission line voltages can range from under 100,000 V to over 750,000 V [Wiser, 2000, page 197].

A simple alternating current transformer is shown in Figure 2-6. It consists of two coils of wire wrapped around an iron core. The coil of wire that is connected to the alternating current (AC) generator is the primary coil of

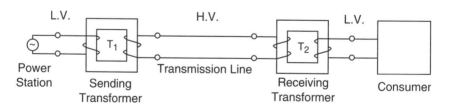

Figure 2-5. Power transmission.

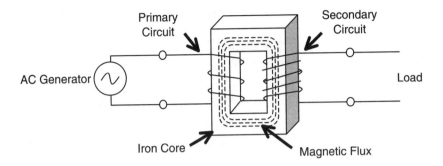

Figure 2-6. A simple voltage transformer.

wire and has N_1 turns of wire; the other coil of wire is the secondary coil and has N_2 turns of wire. The iron core provides a medium for conveying the magnetic flux from the primary coil Φ_B through the secondary coil. The time rate of change of magnetic flux is approximately the same in both coils, so that the induced voltages are given by

$$V_1 = -N_1 \frac{d\Phi_B}{dt} \quad \text{and} \quad V_2 = -N_2 \frac{d\Phi_B}{dt} \tag{2.4.4}$$

Subscript 1 refers to the primary coil in the primary circuit and subscript 2 refers to the secondary coil in the secondary circuit. It is worth noting here that the magnetic flux in a direct current circuit does not change, and the time derivative of magnetic flux is therefore zero. Consequently, transformers do not work with direct current circuits, which was a significant limitation in Edison's plan to produce and sell direct current power.

We can obtain a useful result from Equation (2.4.4) by factoring out the time derivative of magnetic flux $d\Phi_B/dt$. The result is

$$V_2 = V_1 \frac{N_2}{N_1} \tag{2.4.5}$$

If the number of turns in the secondary coil is larger than the number of turns in the primary coil so that $N_2/N_1 > 1$, the voltage in the secondary circuit will be larger than the voltage in the primary circuit. This is the situation in Figure 2-5 when low voltage at the power station is stepped up to high voltage in the transmission line. Conversely, the voltage in the secondary circuit can be made smaller than the voltage in the primary circuit by making the number of turns in the secondary coil less than the number of turns in the primary coil, that is, $N_2 < N_1$ and $N_2/N_1 < 1$. This is the situation in Figure 2-5 when high voltage in the transmission line is being stepped down to a lower voltage for sale to the consumer.

Energy conservation requires that the power input to the primary coil is equal to the power output from the secondary coil. The result is

$$\text{Power} = I_1 V_1 = I_2 V_2 \qquad (2.4.6)$$

where I_1, I_2 are the currents in the primary and secondary coils respectively. Equation (2.4.6) shows that power can be kept constant only if an increase in voltage in the secondary coil relative to the primary coil is accompanied by a decrease in the current in the secondary circuit relative to the current in the primary circuit.

Example 2.4.3: Step-Down Transformer Design

Suppose a step-down transformer is designed to convert 88 kV to 240 V. The voltage in the secondary circuit must be made significantly less than the voltage in the primary circuit. The ratio of turns in the primary coil to the ratio of turns in the secondary coil is

$$\frac{N_1}{N_2} = \frac{V_1}{V_2} = \frac{88000 \text{ V}}{240 \text{ V}} \approx 367 \qquad (2.4.7)$$

2.5   ELECTRIC POWER DISTRIBUTION

Transmission lines are used to distribute electric power. Resistive heat loss in the transmission line can be approximated using Equation (2.3.21). The typical transmission line, such as the power cables lining the highways of America, may be approximated as an LCR series circuit with only a resistor present. In this case, we remove the inductor and capacitor so that the impedance is due to the resistor only, hence $Z \rightarrow R$. The power factor for a resistor in series with an alternating current generator becomes $\cos \phi = 1$ and the average power is $P_{av} = \varepsilon_{rms} i_{rms}$. The resistor in this case is the resistance to the flow of electrons through the conducting wire of the transmission line.

If we know the transmission line resistance as the resistance per unit length Λ_R of line, we can estimate the power loss due to resistive heating for any length ℓ_{TL} of transmission line. In this case, Equation (2.3.19) can be written as

$$P_{av} = (i_{rms})^2 \, \Lambda_R \ell_{TL} \tag{2.5.1}$$

The power loss increases with the length of transmission line and the square of the current i_{rms}. Equation (2.5.1) applies to both alternating current transmission lines and direct current transmission lines.

As a rule, society would like to minimize power loss due to heating to maximize the amount of primary energy reaching the consumer from power plants. We can reduce power loss by reducing the current or by decreasing the distance of transmission. In most cases, it is not a viable option to decrease the transmission distance. It is possible, however. For example, you could choose to build a manufacturing facility near a power station to minimize the cost of transmission of power. One consequence of that decision is that the manufacturer may incur an increase in the cost of transporting goods to market.

A more viable option for reducing power loss is to reduce the current that must be transmitted through transmission lines. Power loss has a greater dependence on current than on transmission distance since Equation (2.5.1) shows that power loss is proportional to the square of the current. This provides a physical explanation of why Edison's direct current concept was not as attractive as Westinghouse's alternating current concept. Transformers do not work with direct current because there is no time-varying magnetic flux, so the transmission of direct current incurs resistive power losses based on the direct current generated at the power plant. The purpose of transformers is to reduce the root mean square current in the transmission line, which is possible with alternating current.

An option for the future is to use superconductors as transmission lines. Superconductors are materials that offer no resistance to electron flow.[5] Dutch physicist Hans Kamerlingh Onnes (1853–1926) discovered the first superconductor in 1911. Onnes produced liquid helium by cooling helium to a temperature below $4.2°$ K. He then observed that the resistivity of mercury vanished when mercury was cooled by liquid helium. Superconducting materials have been developed that operate at higher temperatures than $4.2°$ K, but superconductors can still operate only at temperatures that are far below room temperature. Superconductors are not yet feasible for widespread use in power transmission because they require costly refrigeration.

TRANSMISSION AND DISTRIBUTION SYSTEM

The electric power generating stations are connected to loads (consumers) by means of a transmission system that consists of transmission lines and substations.[6] The substations are nodes in the transmission grid that route electric power to loads at appropriate voltages. Figure 2-7 shows the basic elements of an electric power transmission system. Typical transmission voltages in the United States range from 69 kV to 765 kV, and the alternating current frequency is 60 Hz. The infrastructure for providing electric power to the loads from the substations is the distribution system. A failure of the power transmission system can leave millions of people without power, as it did in the 2003 blackouts in North America and Italy.

Electric power is transmitted and distributed cost-effectively by operating a three-phase system. Three-phase electricity refers to current and voltages that are out of phase with each other by 120 degrees. A three-phase alternating current generator is used to provide three-phase electricity. The equations for power and impedance described in the preceding sections apply to each phase of three-phase electricity. Transmission lines for three-phase electricity are shown in Figure 2-8.

Three-phase electricity can be distributed to three one-phase loads using three separate conductors. The conductors are designed with small areas to minimize the amount of conducting material, such as copper, that must be bought and put into position. Three-phase transmission lines are designed

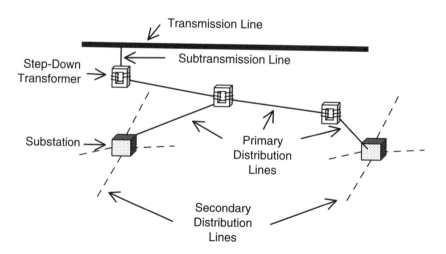

Figure 2-7. Power transmission system.

Figure 2-8. Transmission of three-phase electricity.

Figure 2-9. Pole-mounted transformer.

to operate at high voltages and low currents. High-voltage transformers are used to step-down the voltage for use by consumers.

A typical pole-mounted transformer is shown in Figure 2-9. The transformer can convert three-phase power at a distribution voltage of 13.8 kV to single-phase voltages that are suitable for consumers. Residential consumers in the United States typically use single-phase voltages of 120 V for small appliances and 240 V for larger appliances such as clothes dryers. Large, industrial consumers can use three-phase voltages on the order of 2160 V or higher.

HOUSEHOLD CIRCUITS

Households are among the most common consumers of electricity in the modern world. Electricity is delivered as alternating current to a typical house in the United States using either a two-wire line, or a three-wire line.[7]

The potential difference, or root mean square voltage, between the two wires in the two-wire line in the United States is 120 V, and is 240 V in many parts of Europe. One of the wires in the two-wire line is connected to a ground at the transformer, and the other wire is the "live" wire. The three-wire line has a neutral line, a line at +120 V, and a line at −120 V.

A meter is connected in series to the power lines to measure the amount of electricity consumed by the household. In addition to the meter, a circuit breaker is connected in series to the power lines to provide a safety buffer between the house and the power line. A fuse may be used instead of a circuit breaker in older homes. The fuse contains a metal alloy link, such as lead-tin, with a low melting temperature. If the alloy gets too hot because of resistive heating, the link will melt and break the circuit. Modern circuit breakers are electromagnetic devices that use a bimetallic strip. If the strip gets too hot, the bimetallic strip curls, based on the coefficients of thermal expansion of the two metals that make up the bimetallic strip. The curling strip will break the circuit.

The power line and circuit breaker are designed to handle the current in the circuit, which can be as high as 30 amps, although many household applications require 15 amps of current. Lamps and appliances such as microwaves and toasters operate at 120 V, but electric ranges and clothes dryers operate at 240 V. Each circuit in the house has a circuit breaker or fuse to accommodate different loads.

The circuit in the house is connected in parallel to the power lines. The parallel connection makes it possible to turn on and shut off an electrical device without interfering with the operation of other electrical devices. If a series connection were used, the circuit would be broken whenever one of the electrical devices was turned off. That is why circuit breakers are connected in series between the household circuit and the power lines; the circuit breaker is designed to disconnect electrical devices in the house from the power lines in the event of an overload, such as a power surge. An open circuit occurs when the continuity of the circuit is broken. A short circuit occurs when a low-resistance pathway is created for the current.

Electricity can be harmful if a person touches a live wire while in contact with a ground. Electric shocks can cause burns that may be fatal, and can disrupt the routine functioning of vital organs such as the heart. The extent of biological damage depends on the duration and magnitude of current. A current in excess of 100 mA can be fatal if it passes through a body for a few seconds.

Electrical devices and power lines should be handled with care. Three-pronged power cords for 120-volt outlets provide two prongs that are grounded and one prong that is connected to the live wire. The grounded prongs are provided for additional safety in electrical devices designed to use the three-pronged cords. One of the ground wires is connected to the casing of the appliance and provides a low-resistance pathway for the current if the live wire is short-circuited.

2.6   DISTRIBUTED GENERATION

Practical considerations limit the size of power plants. Most large-scale power plants have a maximum capacity of approximately 1000 megawatts. The size of the power plant is limited by the size of its components, by environmental concerns, and by energy source. For example, the radial speed of the tip of a large wind turbine can approach supersonic speeds. The area occupied by a power plant, called the plant footprint, can have an impact on land use. Conventional power plants that burn fossil fuels such as coal or natural gas can produce on the order of 1000 MW of power. Power plants that depend on nuclear reactors also produce on the order of 1000 MW of power. By contrast, power plants that rely on solar energy presently can produce on the order of 10 MW of power. Power from collections of wind turbines in wind farms can vary from 1 MW to hundreds of MW. If we continue to rely on nuclear or fossil fuels, we can expect power plants to generate on the order of 1000 MW of power. If we switch to power plants that depend on solar energy or wind energy, the power-generating capacity of each plant is less than 1000 MW power, and we must generate and transmit power from more plants to provide existing and future power needs.

In some areas, public pressure is growing to have more power plants with less power-generating capacity and more widespread distribution. The federal government of the United States passed a law in 1978 called the Public Utilities Regulatory Policies Act (PURPA) that allows non-utilities to generate up to 80 MW of power and requires utilities to purchase this power. PURPA was the first law passed in decades to relax the monopoly on power generation held by utilities and reintroduce competition in the power-generating sector of the United States' economy.

Distributed generation of energy is the generation of energy where it is needed and on a scale that is suitable for the consumer.[8] Examples of distributed generation include a campfire, a wood stove, a candle, a battery-powered watch, and a car. Each of these examples generates its own power for its specific application. Historically, distributed generation was the first power-generation technology. The electric power generation and transmission grid that emerged in the twentieth century and is still in use today is a centralized system that relies on large-scale power-generating plants and extensive transmission capability. The transmission grid provides power to distant locations.

Some people believe that the future of energy depends on a renaissance in distributed generation. In this view, a few large-scale power plants in the centralized system will be replaced by many smaller-scale power-generating technologies. Borbely and Kreider define distributed generation as "power generation technologies below 10 MW electrical output that can be sited at or near the load they serve" [Borbely and Kreider, 2001, p. 2]. This definition does not include small-scale power-generating technologies with ideal locations dependent on the locations of their energy source. For example, hydropower and wind-powered generators are not considered distributed generation technologies according to Borbely and Kreider's definition because hydropower and wind-powered generators depend on the availability of flowing water and air, respectively. Consequently, hydropower and wind-powered generators must be located near their energy sources, and these locations are often not near the power consumer. Each of these power generators is discussed in more detail in later chapters.

ENDNOTES

1. The basic principles of the components of power generation and transmission are described in a variety of textbooks. For example, see Serway and Faughn [1985], Cassedy and Grossman [1998], and Young and Freedman [2000].
2. The history of electric power distribution relies primarily on information in Challoner [1993], Burke [1985, Chapter 6], Aubrecht [1995, Chapter 6], Brennan, et al. [1996, Chapter 2], and Bernal [1997, Chapter 10].
3. The analysis of an *LCR* circuit approximately follows the analysis in Halliday and Resnick [1981, Chapters 35 and 36].
4. Electric power is discussed in the references listed in endnote 1 and in Wiser [2000, Chapter 8], Shephard and Shephard [1998, Chapter 3], and Brennan, et al. [1996].
5. The history of superconductivity and a review of high magnetic field strength superconducting magnets are provided by Van Sciver and Marken [2002].
6. Scheinbein and Dagle [2001, Chapter 11] and Shephard and Shephard [1998, Chapter 3] discuss three-phase transmission.

7. For more discussion of household circuits, see Serway and Faughn [1985, Chapter 19], and Young and Freedman [2000, Chapter 27].

8. For more discussion of distributed generation, see Borbely and Kreider [2001].

EXERCISES

2-1. A. An electron with radial velocity v_r is moving in a circular orbit with a 1 m radius due to a constant magnetic field B. If a proton is placed in the same magnetic field B with the same radial velocity v_r, what is the radius of the orbit of the proton?
B. Does the proton orbit in the same direction as the electron?
C. Will the path of a neutron moving through the magnetic field be circular or linear?

2-2. A. Suppose the magnitude of the electric field of an electromagnetic wave is 300 V/m. Calculate the magnitude of the magnetic field.
B. Calculate the energy per unit area per unit time of the electromagnetic wave, that is, calculate the magnitude of the Poynting vector.

2-3. Derive Equation (2.3.8) from Equation (2.3.7).

2-4. Show that Equation (2.3.9) is a solution of Equation (2.3.8). You will have to use Equations (2.3.10) and (2.3.11).

2-5. Show that Equation (2.3.13) is a solution of Equation (2.3.8).

2-6. The root mean square voltage for a household in Britain is 240 V. What is the root mean square current for a hair dryer with an average power of 1500 W. Assume the power factor is one.

2-7. A. A 120 V electric power supply is connected to a load with a resistance of 40 ohms. What is the current in the load?
B. How much power does the load dissipate?

2-8. A. Suppose an alternating current generator produces 10 A of current at 500 V. The voltage is stepped up to 5000 V and transmitted 1000 km through a transmission line that has a resistance of 0.03 ohm per kilometer of line. What is the percentage of power lost by the transmission line?
B. Suppose an alternating current generator produces 10 A of current at 500 V. The voltage is transmitted 1000 km through a transmission

line that has a resistance of 0.03 ohm per kilometer of line. What is the percentage of power lost by the transmission line?
C. Explain the difference between Parts A and B.

2-9. A. What is the maximum induced electromotive force in a simple alternating current generator with 10 turns of wire with area 0.1 m^2 and resistance 5 ohms? Assume the loop rotates in a magnetic field of 0.5 T at a frequency of 60 Hz.
B. What is the maximum induced current?

2-10. A. An LCR series circuit has the following characteristics: resistance is 200 ohms, capacitance is 15 microfarads, inductance is 150 milli-henries, frequency is 60 Hertz, and electromotive force is 80 volts. Calculate capacitive reactance, inductive reactance, and reactance.
B. Calculate impedance.
C. Calculate the power factor and phase angle ϕ.

2-11. A. Use the properties and results of Exercise 2-10 to calculate root mean square current and electromotive force.
B. Calculate the average power of the LCR series circuit in Exercise 2-10.
C. Calculate the resonance frequency of the LCR series circuit in Exercise 2-10.

2-12. How many electrons would enter your body if a current of 100 mA were passing through you for 30 seconds?

2-13. What is the magnitude of the Poynting vector for an electromagnetic wave that has an electric field with an amplitude of 450 V/m? Assume the electromagnetic wave is propagating in vacuum.

2-14. Calculate the current generated by a magnet that is rotating around a loop of wire. The loop of wire contains 100 loops and is connected to an external circuit that has a resistance of 50 ohms. Each loop has an area of 0.01 m^2. The magnet is rotating at an angular frequency of 10 sec^{-1} and has a magnetic field equal to 0.75 T.

2-15. A. What is the power of the primary circuit in a transformer if the current and voltage of the primary circuit are 12 A and 400 V, respectively?
B. Suppose the transformer is used to step-up the voltage of the primary circuit so that the voltage in the secondary circuit is 6000 V. What is the current associated with the higher voltage?
C. What is the advantage of reducing the current in an electrical transmission line?

CHAPTER THREE

Heat Engines and Heat Exchangers

Historically, the concept of heat energy evolved from a study of burning and heating.[1] The English clergyman and gentleman-scientist Joseph Priestly (1733–1804) believed that burning occurred because of the presence of an element called "phlogiston" in air [Wolff, 1967, Chapter 2]. A fellow gentleman-scientist, Frenchman Antoine Lavoisier (1734–1794), identified phlogiston as the element oxygen [Wolff, 1967, Chapter 3]. Today we know that oxygen is consumed during the process of burning in air. Lavoisier also used the concept of caloric to explain the behavior of an object when subjected to heating and cooling. Caloric is a substance that is invisible and "imponderable," or weightless. Lavoisier said that heated objects expanded when caloric filled the space between the particles of the object. Conversely, an object contracted when cooled because it lost caloric. The concept of caloric was eventually replaced by the concept of heat.

American Benjamin Thompson (1753–1814), who became Count Rumford of Bavaria, provided experimental evidence that heat was not a conserved substance. By the nineteenth century, several people independently realized that heat was a form of energy and that energy was conserved [Halliday and Resnick, 1981, Chapter 20]. Early proponents of the principle of conservation of energy included Germans Julius Mayer (1814–1878) and Hermann Ludwig von Helmholtz (1821–1894), James Joule (1818–1889) in England, and L.A. Colding (1815–1888) in Denmark. Wolff [1965, Chapter 8] presented an English version of Helmholtz's seminal paper on energy conservation. The combination of the concepts of heat and conservation of energy with Englishman John Dalton's (1776–1844) concept of atoms [Wolff, 1967, Chapter 4] eventually led to the kinetic theory of atoms. The kinetic theory relates the temperature of an object to the motion of the atoms within the object. According to the kinetic theory, a heated object expands because heating increases the kinetic energy of its atoms. Similarly, a cooled object contracts because cooling decreases the kinetic energy of atoms in the object. The ideas of phlogiston and caloric were at the core of paradigms, or widely held beliefs, in the seventeenth and eighteenth centuries. These paradigms were eventually replaced by the kinetic theory in the nineteenth century.[2]

Thermodynamics is the study of the flow of heat energy. Practical applications of energy require a familiarity with the laws of thermodynamics.[3] We begin by defining fundamental concepts and then describe what we mean by a thermodynamic system. This is followed by a summary of the laws of thermodynamics and an introduction to thermodynamic equilibrium. We end the chapter with a discussion of heat engines, heat transfer mechanisms, and heat exchangers.

3.1 TEMPERATURE AND COMPOSITION

Temperature is a measure of the average kinetic energy of a system. Composition provides quantitative information about the constituents, or components, of a material. Together, temperature and composition are two of the most fundamental concepts in thermodynamics. They are defined in the following sections for ease of reference.

TEMPERATURE

The most commonly used temperature scales are the Fahrenheit and Celsius scales. The relationship between these scales is

$$T_C = \frac{5}{9}(T_F - 32) \tag{3.1.1}$$

where T_C and T_F are temperatures in degrees Celsius and degrees Fahrenheit respectively.

Absolute temperature is needed for many applications, such as the equations of state discussed in the following paragraphs. Absolute temperature is usually expressed in terms of degrees Kelvin or degrees Rankine. The absolute temperature scale in degrees Kelvin is related to the Celsius scale by

$$T_K = T_C + 273 \tag{3.1.2}$$

where T_K is temperature in degrees Kelvin. The Kelvin scale is the basic SI temperature scale, and zero on the Kelvin scale is called absolute zero.

The absolute temperature scale in degrees Rankine is related to the Fahrenheit scale by

$$T_R = T_F + 460 \qquad (3.1.3)$$

where T_R is temperature in degrees Rankine. The Rankine scale is used in many engineering applications.

FLUID COMPOSITION

The components of a material determine the composition of a material. The relative amount of each component in an object is defined as the concentration of the component. Concentration may be expressed in such units as volume fraction, weight fraction, or molar fraction. The unit of concentration should be clearly expressed to avoid errors. Concentration is often expressed as mole fraction. A mole is the SI base unit for the amount of a substance. A mole of a molecular component contains Avogadro's number of molecules. For example, one mole of carbon with atomic mass equal to 12 has a mass of 0.012 kg (12 g), and one mole of carbon dioxide with atomic mass equal to 44 has a mass of 0.044 kg (44 g). The mole fraction y_i of component i in a gas mixture is the number of moles n_i of the component in the gas divided by the total number of moles in the gas, or

$$y_i = \frac{n_i}{\displaystyle\sum_{j=1}^{N_c} n_j} \qquad (3.1.4)$$

where N_c is the number of components in the gas mixture. The apparent molecular weight M_a of the gas mixture is

$$M_a = \sum_{j=1}^{N_c} y_j M_j \qquad (3.1.5)$$

where M_j is the molecular weight of component j. The specific gravity γ_g of the gas mixture is the apparent molecular weight of the gas mixture divided by the molecular weight of air, or

$$\gamma_g = \frac{M_a \, (\text{gas})}{M \, (\text{air})} \approx \frac{M_a \, (\text{gas})}{29} \qquad (3.1.6)$$

Although we are interested in all of the physical states of matter, it is worthwhile to develop the concept of composition for fluids in some detail. Fluids play a key role in society as the dominant energy source: petroleum. Petroleum is a mixture of hydrocarbon molecules, that is, molecules that are primarily composed of hydrogen and oxygen. The types of atoms or molecules that comprise a fluid determine the composition of a fluid. Each type is referred to as a component. A fluid can exist as either a gas or a liquid. A pure fluid consists of a single type of molecule, such as water or methane. A fluid mixture contains several types of molecules. For example, the water in a rock usually contains dissolved solids and may contain dissolved gases. We specify the composition of a fluid by listing the molecular components contained in the fluid and their relative amounts.

The amount of component i in the gas phase relative to the liquid phase can be expressed as the equilibrium K value, which is the ratio

$$K_i = \frac{y_i}{x_i} \tag{3.1.7}$$

where y_i is the gas phase mole fraction and x_i is the liquid phase mole fraction. The liquid phase mole fraction is the number of moles n_i of the component in the liquid divided by the total number of moles in the liquid, or,

$$x_i = \frac{n_i}{\sum\limits_{j=1}^{N_c} n_j} \tag{3.1.8}$$

The allowed range of the equilibrium K value is determined by considering two special cases. If component i is present entirely in the liquid phase, then the gas phase mole fraction y_i is 0 and K_i is 0. Conversely, if component i is present entirely in the gas phase, then the liquid phase mole fraction x_i is 0 and K_i approaches infinity. Thus, the equilibrium K value for component i may range from 0 to infinity. An equilibrium K value can be calculated for each distinct molecular component in a fluid.

3.2 THERMODYNAMIC SYSTEMS AND STATES

A *system* is the part of the universe that we are considering. Everything outside of the system is called the *surroundings*. As an example, suppose we drop a cube of ice in a glass of water. We may view the ice as the system and the water as the surroundings, or we may view the water as the system and the ice as the surroundings. This example demonstrates that the surroundings do not have to surround the system.

An isolated system does not interact with its surroundings. A closed system may exchange energy with its surroundings, but it does not exchange matter. In other words, energy may flow between a closed system and its surroundings, but no matter flows between the closed system and its surroundings. By contrast, energy and matter may flow between an open system and its surroundings.

The *state* of a system is the set of values of the thermodynamic variables that characterize the system. Typical state variables include pressure P, volume V, and temperature T. A *process* is a change from one state of the system to another state. In our example above, we dropped a cube of ice in a glass of water. As the ice melts, some of the water molecules that started in the solid state as ice gain kinetic energy and may enter the liquid state. If we define the system to be ice, the thermodynamic process is the melting of ice. If we define the system to be liquid water, the thermodynamic process is the cooling of water. In either case, the system is an open system because both matter and energy may flow between the open system and its surroundings.

A process is reversible if the system changes from an initial state to a final state by passing through a continuous sequence of equilibrium states. An *equilibrium* state is the state of a closed system that is obtained when the state variables are no longer changing with time. The state of the system can be made to reverse its path at any stage of the reversible process and return to its initial state. If the intermediate states are not equilibrium states, the process is irreversible and a return to the initial state from the final state will not occur by reversing the path of the process. All real processes are irreversible, but in some cases it is possible to approximate a reversible process.

Suppose we subdivide a cell of gas into two halves by inserting a vertical partition. The subdivision is reversible if the partition has the same temperature as the gas in the cell and has negligible volume. Because any real partition will have a finite volume, we have chosen a partition that is small relative to the volume of the cell so that we can neglect the size of the partition and work with an approximately reversible process. If the gas was initially in an equilibrium state, the gas in each half of the cell after inserting the partition should have the same pressure and temperature as it did before the partition was inserted. The mass and volume in each half of the cell will be one half of the original mass and volume, but their ratio, the density, is unchanged. Mass and volume are examples of extensive properties. An *extensive property* is a property of the system that depends on the amount of material. Density, temperature, and pressure are examples of intensive properties. An *intensive property* is a property of the system that is independent of the amount of material.

EQUATIONS OF STATE

An equation of state is used to define the relationship between state variables. The ideal gas law relates P, V, T for an ideal gas through the equation of state

$$PV = nRT \qquad (3.2.1)$$

where R is the universal gas constant and n is the number of moles of the gas. A gas is considered *ideal* if the gas is composed of noninteracting particles. The degree of ideality of a gas can be estimated using the dimensionless gas compressibility factor

$$Z = \frac{PV}{RT} \qquad (3.2.2)$$

for a single mole of gas ($n = 1$ mole). In the case of an ideal gas, the gas compressibility factor satisfies $Z = 1$. If we allow interactions between particles, the ideal gas law must be modified. One modification is the real gas law

$$PV = ZnRT \qquad (3.2.3)$$

The gas compressibility factor Z is not equal to one for real gas. Other equations of state may be constructed that include a set of adjustable parameters $\{a, b\}$ that may be functions of temperature and composition. J.D. van der Waals introduced the parameter b in 1867 to account for the finite size of an atom or molecule.

Table 3-1 presents some equations of state for a single mole of a fluid. The coefficients of the virial expansion in Table 3-1 are called "virial" coefficients: $B(T)$ is the second virial coefficient; $C(T)$ is the third virial coefficient, and so on [Baumann, 1992, page 135]. The Redlich-Kwong, Soave-Redlich-Kwong, Peng-Robinson, and Zudkevitch-Joffe equations

Table 3-1
Examples of equations of state for one mole of fluid

Name	Equation of state
van der Waals	$P = \dfrac{RT}{V-b} - \dfrac{a}{V^2}$
Virial Expansion	$P = \dfrac{RT}{V}\left[1 + B(T)V^{-1} + C(T)V^{-2} + \cdots\right]$
Redlich-Kwong	$P = \dfrac{RT}{V-b} - \dfrac{a/T^{1/2}}{V(V+b)}$
Soave-Redlich-Kwong	$P = \dfrac{RT}{V-b} - \dfrac{a(T)}{V(V+b)}$
Peng-Robinson	$P = \dfrac{RT}{V-b} - \dfrac{a(T)}{V(V+b)+b(V-b)}$
Zudkevitch-Joffe	$P = \dfrac{RT}{V-b(T)} - \dfrac{a(T)/T^{1/2}}{V[V+b(T)]}$

of state in Table 3-1 are called "cubic" because they yield a cubic equation for the compressibility factor Z. The first term in the van der Waals equation of state appears in each of the cubic equations of state. The size parameter b is allowed to depend on temperature in the Zudkevitch-Joffe equation of state. It is often useful to fit more than one equation of state to data because equations of state differ in their ability to match different types of measurements and different types of fluids.

Equations of state are used in a number of disciplines, such as chemical engineering and reservoir engineering. Equation of state parameters such as parameters $\{a, b\}$ in Table 3-1 must be determined before an equation of state can be used as a quantitative model of fluid properties. For a mixture with N_c components, the parameters $\{a, b\}$ have the form

$$a = \sum_{i=1}^{N_c}\sum_{j=1}^{N_c} a_i a_j x_i x_j \left(1 - \delta_{ij}\right)$$

$$b = \sum_{i=1}^{N_c} b_i x_i \tag{3.2.4}$$

where $\{a_i, b_i\}$ refer to equation of state values for the ith component with mole fraction x_i, and δ_{ij} is a symmetric array of numbers called the binary interaction parameters. The binary interaction parameters are determined by fitting equations of state to fluid property measurements on mixtures of pairs of components. It is assumed that the interaction between the component pairs will be the same in a mixture with other components as it was when only the pair of components was present. In practice, this assumption is relaxed and binary interaction parameters are determined for the mixture by fitting the mixture equation of state to laboratory measurements of mixture properties, such as the change in volume that results from a change in pressure.

Several regression techniques exist for fitting the parameters in an equation of state to experimental data. The techniques usually differ in the choice of parameters that are to be varied in an attempt to match laboratory data with the equation of state. The modification of equation of state parameters is called *tuning* the equation of state. One justification for tuning an equation of state is that the parameters are determined for systems with one or two components only, but many fluids of interest are generally a mixture with many components. The equation of state parameter adjustments attempt to match the multicomponent behavior of the fluid system.

3.3 LAWS OF THERMODYNAMICS

The laws of thermodynamics were originally formulated as an attempt to understand observations. They are therefore empirically based. We begin by considering three systems A, B, C with temperatures T_A, T_B, T_C. If two of the systems are in thermodynamic equilibrium with the third system, then they must be in thermodynamic equilibrium with each other. This observation is the 0th law of thermodynamics. The 0th law may be written as follows:

$$\text{If } T_A = T_C \quad \text{and} \quad T_B = T_C, \quad \text{then} \quad T_A = T_B \tag{3.3.1}$$

The first law of thermodynamics recognizes that heat is a form of energy, and that nonrelativistic energy is conserved. The first law of thermodynamics may be written as

$$dU = dQ - dW \tag{3.3.2}$$

where

$$dU = \text{change in internal energy of the system}$$

$dQ =$ heat absorbed by the system
$dW =$ work done by the system on its surroundings

The second law of thermodynamics is a statement about disorder in a closed system. It asserts that the disorder in a system will remain constant or increase, and quantifies this statement using the concept of entropy. An infinitesimal change in entropy dS may be expressed as the ratio of the infinitesimal amount of heat absorbed by the system dQ divided by the temperature T of the system:

$$dS = \frac{dQ}{T} \tag{3.3.3}$$

The second law of thermodynamics says that the change in entropy ΔS must increase or be constant, thus

$$\Delta S \geq 0 \tag{3.3.4}$$

The third law of thermodynamics makes a statement about the behavior of a system as its temperature approaches absolute zero. It says that the entropy of a system approaches a constant value S_0 as the absolute temperature of the system approaches zero. The third law of thermodynamics may be written in the following form:

$$\text{As } T \rightarrow 0_+{}^{\circ}\text{K}, \ S \rightarrow S_0 \tag{3.3.5}$$

The third law of thermodynamics may be considered a corollary of the second law.

HEAT CAPACITY

Heat capacity C is the amount of heat ΔQ required to raise the temperature of an object a small amount ΔT:

$$C = \frac{\Delta Q}{\Delta T} \tag{3.3.6}$$

The specific heat capacity c is heat capacity divided by the mass m of the object, thus

$$c = \frac{1}{m} \frac{\Delta Q}{\Delta T} \tag{3.3.7}$$

If we combine Equations (3.3.7) and (3.3.3), we obtain an expression for entropy in terms of specific heat capacity:

$$dS = \frac{dQ}{T} = \frac{mc\,dT}{T} \qquad (3.3.8)$$

The heat Q absorbed by a material can change its physical state, or phase. The heat required to change the phase of an object with mass m is

$$Q = \pm mL \qquad (3.3.9)$$

where L is the heat required for the phase change. The quantity L is the heat of fusion for a phase change from solid to liquid; L is the heat of vaporization for a phase change from liquid to gas; and L is the heat of sublimation for a phase change from gas to solid. The sign of heat Q depends on the direction of the phase change. An increase in heat absorbed by the material corresponds to $Q > 0$, such as a material melting, and a decrease in heat content corresponds to $Q < 0$, such as a material freezing. Equation (3.3.9) also holds for a change in the physical state of a material that occurs because of a chemical reaction. For example, if the chemical reaction is combustion, the quantity L is the heat of combustion.

Heat capacity depends on the state variables. If we define M as mass per mole, molar heat capacity at constant volume C_V is the product of M and heat capacity at constant volume c_V, thus $C_V = Mc_V$. Molar heat capacity at constant pressure is similarly defined, thus $C_P = Mc_P$. Molar heat capacity at constant volume is related to molar heat capacity at constant pressure for an ideal gas by the equation

$$C_P = C_V + R \qquad (3.3.10)$$

where R is the gas constant. The ratio of heat capacities for an ideal gas is

$$\gamma = \frac{C_P}{C_V} = \frac{c_P}{c_V} \qquad (3.3.11)$$

Typical values of the heat capacity ratio γ range from 1.3 to 1.7.

HEATING VALUE

The heating value of a gas can be estimated from the composition of the gas and the heating values associated with each component of the gas. Heating value of the mixture H_m is defined as

$$H_m = \sum_{i=1}^{N_c} y_i H_i \qquad (3.3.12)$$

where N_c is the number of components, y_i is the mole fraction of component i, and H_i is the heating value of component i. The heating value of a typical natural gas is between 1000 BTU/ft^3 (3.73×10^7 J/m^3) to 1200 BTU/ft^3 (4.48×10^7 J/m^3) at standard conditions. Heating values of molecular components in a mixture are tabulated in reference handbooks.

3.4 EQUILIBRIUM CONDITIONS IN THE ABSENCE OF GRAVITY

We illustrate the thermodynamic principles presented in the preceding section by using them to determine the conditions for thermodynamic equilibrium of a fluid mixture in an open system. We begin the calculation by assuming the mixture is contained in a closed system. This simplifying assumption lets us introduce Gibbs free energy. We then solve the more general problem for an open system.

CLOSED SYSTEM

The first law of thermodynamics was written for a closed system in Section 3.3 as

$$dU = dQ - dW \qquad (3.4.1)$$

where

$dU =$ change in internal energy of the system
$dQ =$ heat absorbed by the system
$dW =$ work done by the system on its surroundings

For an infinitesimal, reversible process, we have

$$dQ = T\,dS \qquad (3.4.2)$$

where $S =$ entropy of the system.

The system is assumed to be in thermal contact with a constant volume heat bath at constant, uniform temperature T. If the only work done by the system is due to expansion or compression, then

$$dW = P\,dV \qquad (3.4.3)$$

where V is the volume of the system and P is a constant, uniform pressure applied to the system. Substituting Equations (3.4.2) and (3.4.3) into Equation (3.4.1) gives

$$dU = T\,dS - P\,dV \qquad (3.4.4)$$

The functional dependence of internal energy U on independent thermodynamic variables is

$$U = U(S, V) \qquad (3.4.5)$$

Another useful thermodynamic quantity is Gibbs free energy. Gibbs free energy is also called the thermodynamic potential. It is given by

$$G = U - TS + PV \qquad (3.4.6)$$

The functional dependence of Gibbs free energy G on independent thermodynamic variables is

$$G = G(T, P) \qquad (3.4.7)$$

The differential of Gibbs free energy is found by taking the differential of Equation (3.4.6) subject to the functional dependence specified in Equation (3.4.7), hence

$$dG = -S\,dT + V\,dP \qquad (3.4.8)$$

We use Gibbs free energy to establish criteria for achieving phase equilibrium.

OPEN SYSTEM

For a single-phase fluid mixture in an open system, the thermodynamic functions U, G also depend on the amount of each component:

$$U = U\left(S, V, n_1, n_2, \ldots, n_{N_c}\right) \tag{3.4.9}$$

$$G = G\left(T, P, n_1, n_2, \ldots, n_{N_c}\right) \tag{3.4.10}$$

for N_c components. The number of moles of component i is n_i. The differential forms of U, G are

$$dU = T\,dS - P\,dV + \sum_{i=1}^{N_c} \mu_i\,dn_i \tag{3.4.11}$$

$$dG = -S\,dT + V\,dP + \sum_{i=1}^{N_c} \mu_i\,dn_i \tag{3.4.12}$$

where chemical potential μ_i is defined by the relations

$$\mu_i = \left(\frac{\partial U}{\partial n_i}\right)_{S,V,n_i} = \left(\frac{\partial G}{\partial n_i}\right)_{T,P,n_i} \tag{3.4.13}$$

Chemical potential is the change in Gibbs free energy with respect to the number of moles of a component. The chemical potential accounts for the movement of mass into and out of the open system.

EQUILIBRIUM AND THE SECOND LAW OF THERMODYNAMICS

The second law of thermodynamics says that all real processes in an isolated system occur with zero or positive entropy change:

$dS > 0$: irreversible process

$dS = 0$: reversible process $\hphantom{xxxxxxxxxxx}$ (3.4.14)

All natural processes occur irreversibly, but if the changes from one equilibrium state to another are small enough, a natural process may be treated as an idealized reversible process.

Table 3-2
Equilibrium conditions

Variable	Extremum	Subject to constant
S	Maximize	U, V, n_i for all i
U	Minimize	S, V, n_i for all i
G	Minimize	T, P, n_i for all i

An isolated system reaches equilibrium when entropy is maximized, that is, when entropy satisfies the relation

$$dS = 0 \tag{3.4.15}$$

subject to the conditions that $U, V, \{n_i\}$ are constant:

$$dU = 0$$
$$dV = 0 \tag{3.4.16}$$
$$dn_i = 0 \quad \forall\, i = 1, \ldots, N_c$$

Equilibrium conditions can be determined for the thermodynamic quantities U, G if U, G are minimized. The equilibrium conditions are summarized in Table 3-2.

3.5   HEAT ENGINES

A *heat engine* is a device that transforms heat into other forms of energy, such as mechanical or electrical energy. For comparison, a *heat pump* is a device that transfers heat from one location to another. Diagrams of a heat engine and a heat pump are presented in Figure 3-1. In a heat engine, the engine transforms heat Q_2 from the hot reservoir to work W and expels heat Q_1 to the cold reservoir. The heat pump, on the other hand, combines heat Q_1 from the cold reservoir and work W to provide heat Q_2 to the hot reservoir. The thermal efficiency of the heat engine is the ratio of the net work done to the heat absorbed:

$$\eta = \frac{W}{Q_2} = \frac{Q_2 - Q_1}{Q_2} = 1 - \frac{Q_1}{Q_2} \tag{3.5.1}$$

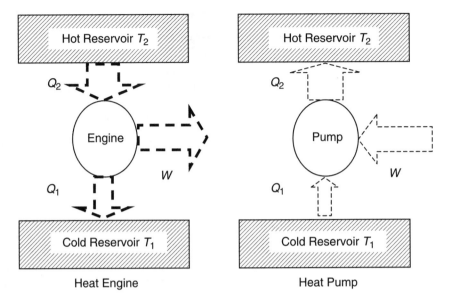

Figure 3-1. Heat engine and heat pump.

The coefficient of performance of a heat pump is the ratio of the heat transferred to the work done by the pump:

$$COP = \frac{Q_2}{W} \tag{3.5.2}$$

CARNOT CYCLE

Frenchman Sadi Carnot (1796–1832) presented a theoretical model of a heat engine in 1824. He used an ideal gas as the working material. The gas was expanded and compressed in four successive, reversible stages. The stages are listed in Table 3-3 and schematically illustrated in Figure 3-2. Stages I and III are isothermal processes, and stages II and IV are adiabatic processes. An *isothermal process* is a process in which the temperature does not change. An *adiabatic process* is a process in which there is no heat exchange—that is, no heat enters or leaves the system.

The efficiency η of an engine was defined in Equation (3.5.1) in terms of heat and work. An alternative definition says that the efficiency η of an engine is the ratio of heat transformed into work divided by the total heat

Table 3-3
Stages of the Carnot cycle

Stage	Process
I	Isothermal expansion at the higher temperature T_2
II	Adiabatic expansion
III	Isothermal compression at the lower temperature T_1
IV	Adiabatic compression

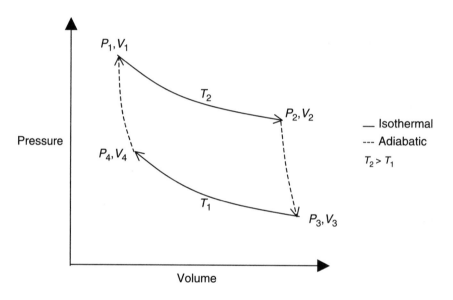

Figure 3-2. Carnot cycle.

absorbed, or

$$\eta = \frac{\text{heat transformed to work}}{\text{total heat absorbed}} = \frac{\text{work output}}{\text{heat input}} \qquad (3.5.3)$$

The efficiency of the Carnot cycle is

$$\eta_{\text{Carnot}} = \frac{T_2 - T_1}{T_2} = 1 - \frac{T_1}{T_2} \qquad (3.5.4)$$

Some aspects of the Carnot cycle are difficult to achieve in practice [Çengel and Boles, 2002, Chapter 9]. For example, if steam is the working fluid, the formation of liquid (water) droplets by condensation can erode turbine blades as the liquid droplets impinge on the turbine blades. It is therefore desirable to maintain a high steam quality. The Rankine cycle resolves many of the difficulties associated with the Carnot cycle.

The Rankine cycle assumes that superheated steam is prepared in a boiler and complete condensation occurs in a condenser. Water is brought into the system by constant entropy (isentropic) compression in a pump. It is transformed to superheated steam by the addition of heat at constant pressure in a boiler. The superheated steam undergoes isentropic expansion in a turbine. Excess heat is then rejected at constant pressure in a condenser. There are no internal irreversibilities in the ideal Rankine cycle. For further discussion of the Rankine cycle, see a thermodynamics text such as Çengel and Boles [2002, Chapter 9].

OTTO CYCLE

The gasoline engine is an internal combustion engine that continues to be the engine of choice in vehicles designed for personal transportation. The Otto cycle is a four-stroke model of the internal combustion engine. The Otto cycle is named after the German inventor Nikolaus August Otto (1832–1891). We can describe the Otto cycle as a five-stage process: four of the five stages are strokes and the fifth stage is ignition. The strokes are intake, compression, power, and exhaust. The stages are listed in Table 3-4 and schematically illustrated in Figure 3-3. Stages I and IV are adiabatic processes, stage II is the ignition stage, and stages III and V are isochoric processes. An *isochoric process* is a process in which the volume does not change. The strokes and stages are described here for a gasoline engine.

Table 3-4
Stages of the Otto cycle

Stage	Process (see Figure 3-3)
I	Point A to Point B: Adiabatic compression
II	Ignition
III	Point B to Point C: Isochoric pressure increase
IV	Point C to Point D: Adiabatic expansion
V	Point D to Point A: Isochoric pressure decrease

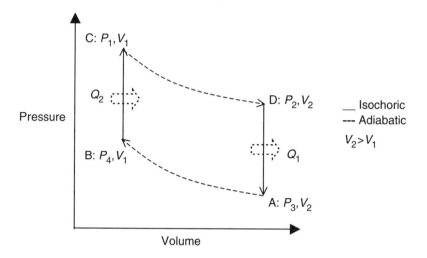

Figure 3-3. Otto cycle.

A gasoline-air mixture enters the system at point A during the intake stroke. The mixture is adiabatically compressed to point B during the compression stroke and then ignited. Heat Q_2 is added to the system when the mixture burns and the system pressure increases isochorically to point C. Adiabatic expansion from point C to point D is the power stroke. The hot gas mixture with heat Q_1 is ejected isochorically, and the pressure decreases to point A during the exhaust stroke.

The efficiency of the Otto cycle is expressed in terms of the compression ratio r and the ratio of heat capacities γ. The compression ratio r is the ratio of the maximum volume of the system V_2 to the minimum volume of the system V_1. Typical values of compression ratio r are 8 to 10. The Otto cycle efficiency is

$$\eta_{\text{Otto}} = 1 - \frac{1}{r^{\gamma-1}}, \quad r = \frac{V_2}{V_1} \quad \text{and} \quad \gamma = \frac{C_P}{C_V} \tag{3.5.5}$$

If $\gamma = 1.4$ and $r = 8$, the Otto cycle efficiency is 0.56 or 56%.

3.6   HEAT TRANSFER

Heat can be transferred from one location to another by convection, conduction, or radiation. *Convection* is the transfer of heat by the movement of a heated substance. *Conduction* is the transfer of heat due to temperature differences between substances that are in contact with each other. *Radiation* is the transfer of heat by the emission and absorption of electromagnetic waves. The three mechanisms of heat transfer can work separately or in combination.

CONVECTION

The heated substance carries the heat to a cooler environment. The heated substance and the cooler environment then achieve thermal equilibrium when the heated substance cools and the cooler environment warms up. Heat exchangers (discussed in a following section) rely on convection to transport heat from a warmer environment to a cooler environment. Conduction is then used to transfer heat between two systems that are separated by a thermal conductor.

CONDUCTION

Kinetic energy can be transferred during conduction by such processes as the collision of molecules or by the movement of atomic electrons. The rate of heat flow P_{Heat} through a slab with cross-sectional area A and length L is

$$P_{Heat} = k_T A \left(\frac{T_{Hot} - T_{Cold}}{L} \right) \qquad (3.6.1)$$

where k_T is the thermal conductivity of the material. The warmer temperature $T_{Hot} > T_{Cold}$ is separated from the cooler temperature T_{Cold} by the distance L. The heat flow rate is the amount of heat conducted per unit time $P_{Heat} = \Delta Q_{conducted} / \Delta t$ and is expressed in the SI unit of power, the watt. In SI units, A is in m^2, L is in m, temperature is in $^\circ K$, and k_T is in $W/m \cdot {}^\circ K$. Thermal conductivities for various substances are shown in Table 3-5.

RADIATION

Our understanding of radiation can be enhanced by reviewing the wave description of light. Let us recall that James Clerk Maxwell developed equations for describing electromagnetic phenomena. Maxwell used his equations to predict the speed of light. These same equations can be

Table 3-5
Thermal conductivities

Substance at 25° C (298.15° K)	Thermal conductivity (W/m·°K)	Substance at 25° C (298.15° K)	Thermal conductivity (W/m·°K)
Air	0.026	Hydrogen	0.14*
Aluminum	237	Iron	13.9
Asbestos (loose fiber)	0.16	Limestone (Bedford, Indiana)	2.20
Concrete (cement, sand, and gravel)	1.82	Mica	0.79
Copper	401	Oxygen	0.023*
Diamond	907	Silver	429
Firebrick	0.75	Steel (carbon)	60.5
Glass (Pyrex)	1.10	Water (saturated liquid)	0.609
Gold	318	Water (saturated vapor)	0.0186
Helium	0.14*	Wood	0.04 to 0.12*

Young and Freedman [2000, page 479].
Source: Cohen, 1996, pages 118–119.

used to show that light has many of the characteristics of water waves. Figure 3-4 shows a single wave. The length of the wave from one point on the wave to an equivalent point is the wavelength. The number of waves passing a particular point, say point B in Figure 3-4, in a specified time interval is the frequency f of the wave in cycles per second, or Hertz (1 Hz = 1 cycle per second). Angular frequency ω is related to frequency by $\omega = 2\pi f$ and has the unit of radians per second. The period of the wave T is given by $T = 1/f$ and has the unit of seconds per cycle.

The mathematical equation for describing the motion of a wave in one space dimension and one time dimension is the wave equation

$$\frac{\partial^2 y}{\partial x^2} = \frac{1}{v^2}\frac{\partial^2 y}{\partial t^2}$$

(3.6.2)

where x is the x-direction coordinate, y is the displacement of the wave along the vertical y-axis, t is time, and v is the speed of the wave in the direction of motion of the wave. Equation (3.6.2) has the same form as the

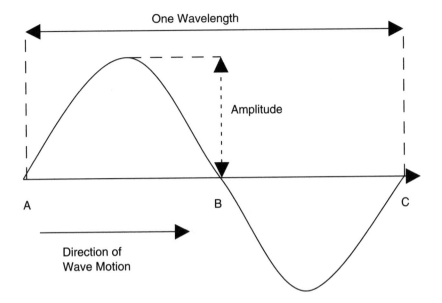

Figure 3-4. A wave (after Fanchi [2002]).

equation for describing the motion of light. Thus light is often thought of as wave motion.

A general solution to Equation (3.6.2) may be written as

$$y(x, t) = y(kx - \omega t) \tag{3.6.3}$$

where $k = 2\pi/\lambda$ is the wave number (m^{-1}) of a wave with wavelength $\lambda \, (\text{m})$ and angular frequency ω. The wave number has the dimension of reciprocal length. Angular frequency (rad/s) is related to frequency of motion f (Hz) by $\omega = 2\pi f$. Particular solutions of the wave equation include

$$y(x, t) = y_0 \cos(kx - \omega t)$$
$$y(x, t) = y_0 \sin(kx - \omega t)$$
$$y(x, t) = y_0 \exp[i(kx - \omega t)] \tag{3.6.4}$$

etc.

where y_0 is the constant amplitude of the wave. The speed of the wave v satisfies the equation

$$v^2 = \left(\frac{\omega}{k}\right)^2 \tag{3.6.5}$$

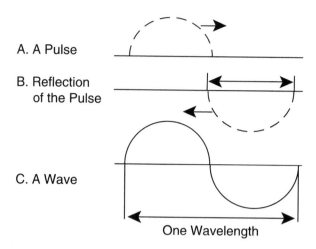

A. A Pulse

B. Reflection
of the Pulse

C. A Wave

One Wavelength

Figure 3-5. Creating a wave (after Fanchi [2002]).

Equation (3.6.5) is called a dispersion relation; it expresses angular frequency as a function of wave number. In our example, the dispersion relation for the wave equation is

$$\omega = kv \quad \text{or} \quad f\lambda = v \tag{3.6.6}$$

Equation (3.6.6) says angular frequency is proportional to wave number, and that the product (frequency times wavelength) gives the speed of the wave.

It is easy to generate wave motion. For example, suppose we tie a rope to the doorknob of a closed door. Hold the rope tight and move it up and then back to its original position. A pulse like that shown in Figure 3-5A should proceed toward the door. When the pulse strikes the door it will be reflected back toward us as shown in Figure 3-5B. This pulse is one half of a wave and it has one half a wavelength. To make a whole wave, move the rope up, back to its original position, down, and then up to its original position. All of these motions should be made smoothly and continuously. The resulting pulse should look something like the complete wave in Figure 3-5C. We can make many of these waves by moving the rope up and down rhythmically. The ensuing series of waves is called a *wavetrain*.

Different colors of light have different wavelengths and different frequencies. The wavelengths of light we see, visible light, represent only a very narrow band of wavelengths of the more general phenomenon known as electromagnetic radiation. Radio waves are an example of electromagnetic radiation with relatively long wavelengths ranging from a fraction of an inch to over a mile in length. Visible light wavelengths are a thousand to a trillion times shorter than the wavelengths of radio waves. Yet electromagnetic radiation with wavelengths smaller than the wavelength of visible light exists. The wavelength of X-rays, for instance, is one hundred to one thousand times smaller than the wavelength of visible light. Gamma rays, products of nuclear detonations, have wavelengths smaller than X-rays. This wide range of wavelengths comprises what is known as the *electromagnetic spectrum.*

Stefan's law expresses the rate of emission of radiant energy as

$$P_{rad} = \frac{\Delta Q_{rad}}{\Delta t} = \sigma A e_T T^4 \tag{3.6.7}$$

where P_{rad} is in watts, A is the surface area of the emitting object in m^2, σ is the Stefan-Boltzmann constant 5.67×10^{-8} W·m^{-2}· $^\circ$K^{-4}, T is the temperature of the object in $^\circ$K, and e_T is the thermal emissivity of the object. An object emits radiant energy to a cooler environment, and it absorbs radiant energy from a warmer environment. The net rate of radiant energy transfer is

$$P_{net\ rad} = \frac{\Delta Q_{net\ rad}}{\Delta t} = \sigma A e_T \left(T^4 - T^4_{environ} \right) \tag{3.6.8}$$

where $P_{net\ rad}$ is in watts and $T_{environ}$ is the temperature of the environment in $^\circ$K. The environment is often called the surroundings. Thermal emissivity e_T depends on the substance and can vary between zero and one. A porous, nonmetallic substance has thermal emissivity $e_T \approx 1$; e_T for a smooth metallic surface is often in the range $0.1 < e_T < 0.3$. Dark surfaces tend to have larger e_T than light surfaces, hence $(e_T)_{dark} > (e_T)_{light}$. Thermal emissivity e_T is approximately equal to 1 for a dull black surface.

BLACK BODY RADIATION

A special case of thermal radiation that arises in several energy contexts is black body radiation. An example of a black body is shown in Figure 3-6. It is just a hollow metal cube, or cavity, with a small hole in one side. The walls of the cavity are coated with a thin layer of black pigment. If we heat the walls of the cavity, its temperature will rise and the material of the walls

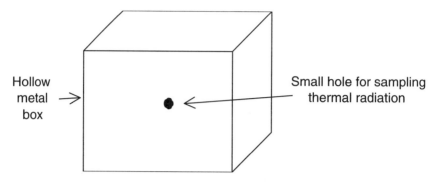

Figure 3-6. Black body.

will begin to emit electromagnetic radiation. We can study this black body radiation by measuring the wavelength of the radiation that escapes from the small hole in the side of the cavity.

The energy density of the observed radiation emitted by a black body follows the curve labeled "Planck" in Figure 3-7. Energy density u is the energy of black body radiation per unit volume and is given by the distribution [Weinberg, 1988, page 173]

$$du = \frac{\dfrac{8\pi hc}{\lambda^5}}{\left[\exp\left(\dfrac{hc}{kT\lambda}\right) - 1\right]} d\lambda \tag{3.6.9}$$

for the narrow range of wavelengths λ to $\lambda + d\lambda$. In addition to wavelength and temperature T, Equation (3.6.9) depends on three fundamental physical constants: Planck's constant h; Boltzmann's constant k; and the speed of light in vacuum c. Equation (3.6.9) is called the Planck distribution after German physicist Max Planck, the first person to derive the equation.

Energy density is plotted as a function of wavelength in Figure 3-7. Also sketched in the figure is the calculated description of the black body phenomenon based on the classical theory of physics (labeled "Rayleigh-Jeans" in the figure). From Figure 3-7 it is clear that the Rayleigh-Jeans theoretical description does not account for the observed behavior of the experimental system (the "Planck" curve) when the wavelength of the emitted radiation gets very small. The discrepancy between classical theory and experiment is known as the *ultraviolet catastrophe*. Theory based on classical physics predicts that an extremely large intensity of black body radiation should be emitted at very short wavelengths. Instead, experiments show

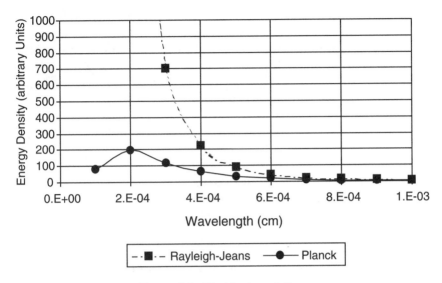

Figure 3-7. Blackbody radiation.

the intensity becomes small and eventually vanishes when the wavelength gets smaller. What went wrong with the classical theory?

Planck provided the answer to this question in 1900. He could derive Equation (3.6.9), the equation that matched experimental results, only by assuming energy was a discrete variable and obeyed the condition

$$E = nh\nu, \quad n = 0, 1, 2, \ldots \tag{3.6.10}$$

where ν is the frequency of oscillation of the thermal radiation emitted by the black body and h is Planck's constant. Before Planck's work, physicists believed energy was continuous. In the classical limit, Planck's constant $h \to 0$ and we can make the first order approximation

$$\exp\left(\frac{hc}{kT\lambda}\right) - 1 \approx \frac{hc}{kT\lambda} \tag{3.6.11}$$

in Equation (3.6.9). The result is the Rayleigh-Jeans formula:

$$du = \frac{\dfrac{8\pi hc}{\lambda^5}}{\left[\dfrac{hc}{kT\lambda}\right]} d\lambda = \frac{8\pi kT}{\lambda^4} d\lambda \tag{3.6.12}$$

Equation (3.6.12) is the formula for the Rayleigh-Jeans curve shown in Figure 3-7. If we integrate over wavelength, we find that the total rate of emission of energy from a black body obeys the Stefan-Boltzmann law with thermal emissivity equal to 1. The Stefan-Boltzmann law gives the total energy flux, or energy passing through a unit area per unit time, emitted by a black body U_b as

$$U_b = \sigma T^4 \tag{3.6.13}$$

where σ is the Stefan-Boltzmann constant.

The Rayleigh-Jeans formula corresponds to the equation we would obtain if we made the classical assumption that energy was a continuous variable. Planck was forced to reject this concept and replace it with his new and somewhat bold concept: energy comes in discrete packets. Planck's hypothesis that energy comes in discrete packets has been substantiated by many experimental tests and applications. These discrete energy packets are called *quanta* and the theory of which quanta are a part is called the *quantum theory*.

3.7   HEAT EXCHANGERS

Heat exchangers[4] use a thermally conducting material such as stainless steel to separate two fluids. Heat from one fluid can be transferred to the other fluid through the thermally conducting material. An example of a heat exchanger is a coil heat exchanger. A coil heat exchanger consists of a small-diameter tube placed concentrically inside a larger-diameter tube. The combined coaxial tubes are wound into a helix shape. The simple heat exchanger on the left-hand side of Figure 3-8 uses energy from hot water to heat air.

Fluid flow inside a heat exchanger is either laminar or turbulent. Fluid flow is *laminar* when there is no fluid motion transverse to the direction of bulk flow. Fluid flow is *turbulent* when the velocity components of fluid flow fluctuate in all directions relative to the direction of bulk flow. The dimensionless Reynolds number N_{Re} characterizes the flow regime, or type of fluid flow. It is defined as

$$N_{Re} = \frac{\rho v D}{\mu} \tag{3.7.1}$$

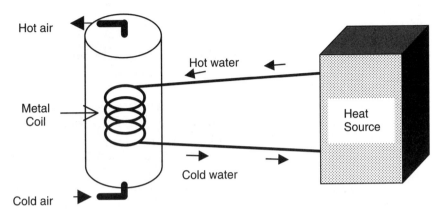

Figure 3-8. Schematic of heat exchanger.

where ρ is fluid density, v is bulk flow velocity, D is tube diameter for flow in a tube, and μ is the dynamic viscosity of the fluid. Fluid flow is laminar when $N_{Re} \leq 2000$, and fluid flow is fully turbulent when $N_{Re} > 6000$ [Bartlett, 1996, page 19].

The type of flow provides information about the degree of mixing and the pressure gradient in the heat exchanger. Turbulent flow facilitates heat transfer because the eddy currents in turbulent flow mix the fluid. The pressure drop in the heat exchanger determines pumping power requirements. Small pressure drops associated with laminar flow require less pumping power than large pressure drops associated with turbulent flow.

The heat transfer rate of a heat exchanger depends on the design of the heat exchanger and the two fluids flowing in the heat exchanger. Energy conservation requires that the heat transferred to the cooler fluid must equal the heat lost by the warmer fluid. The heat transfer rate $\dot{Q}$ in energy per unit time is

$$\dot{Q} = [\dot{m}c\,(T_{\text{out}} - T_{\text{in}})]_{\text{cool}} = - [\dot{m}c\,(T_{\text{out}} - T_{\text{in}})]_{\text{hot}} \qquad (3.7.2)$$

where $\dot{m}$ is mass flow rate in mass per unit time, c is specific heat, and T is temperature. The subscripts "in" and "out" refer to the entry and exit points, respectively, of the heat exchanger. The effectiveness of a heat exchanger is the ratio of the actual heat transferred from the warmer to the cooler fluid divided by the heat that could be transferred by a heat exchanger of infinite size.

ENDNOTES

1. References include Wolff [1965, 1967], Halliday and Resnick [1981], Serway and Faughn [1985], Greiner, et al. [1995], and Young and Freedman [2000].
2. The change in paradigm is an example of a paradigm shift. Several authors have discussed paradigm shifts, including Kuhn [1970], Lakatos [1970], and Root-Bernstein [1989].
3. Some references of note for the thermodynamic concepts discussed in this chapter include Baumann [1992], Greiner, et al. [1995], Young and Freedman [2000], Çengel and Boles [2002], Bernstein, et al. [2000], Serway, et al. [1997].
4. Supplemental references on heat exchangers and conduction include Bartlett [1996], Çengel and Boles [2002, Chapter 4], and Young and Freedman [2000, Chapter 15].

EXERCISES

3-1. Express the temperature $T = 2.7°$ K in degrees Centigrade and degrees Fahrenheit.

3-2. A. What is the molar composition of a gas with the following mass distribution?

Component	Mass (kg)	Molecular weight (kg/kg mole)
Methane (CH_4)	20	16
Ethane (C_2H_6)	6	30
Propane (C_3H_8)	3	44
Carbon dioxide (CO_2)	4	44

Hint: Complete the following table:

Component	Mass (kg)	Molecular wt. (kg/kg mole)	# of moles	Mole fraction
CH_4	20	16		
C_2H_6	6	30		
C_3H_8	3	44		
CO_2	4	44		

B. What is the apparent molecular weight of the gas?

C. What is the specific gravity of the gas?

3-3. Standard temperature and pressure (STP) for a volume of gas in the SI system of units are $0°$ C and 1 atmosphere (1 atm $= 1.01 \times 10^5$ Pa). The standard volume of gas for many engineering applications is calculated at the standard conditions $60°$ F and 1 atm. Use the ideal gas law to estimate the ratio of gas volume at standard conditions V_{SC} to the gas volume at STP (V_{STP}).

3-4. A black, spherical rock exposed to sunlight on a warm day is at a temperature of $110°$ F while the surrounding air has a temperature of $90°$ F. If half of the rock is exposed, and the radius of the rock is 2 m, estimate the amount of radiant energy emitted by the rock during the afternoon between noon and 6 P.M. Express your answer in J and BTU.

3-5. A. Ideal gross heating values for the components of a gas can be found in the literature, such as the *Gas Processors Association Manual of Petroleum Measurements*. Use the ideal gross heating values and gas component mole fractions to complete the following table.

Gas component	Ideal gross heating value (H_{ideal}) (BTU/ft^3)	(J/m^3)	Gas component mole fraction (y_i)	($H_{ideal} \times y_i$) (BTU/ft^3)	(J/m^3)
Nitrogen	0.0		0.0030		
Carbon dioxide	0.0		0.0042		
Methane	1009.7		0.9115		
Ethane	1768.8		0.0510		
Propane	2517.5		0.0157		
iso-Butane	3252.7		0.0038		
n-Butane	3262.1		0.0049		
iso-Pentane	4000.3		0.0020		
n-Pentane	4009.6		0.0020		
Hexane	4756.2		0.0019		

B. Does the sum of the gas component mole fractions add up to one?

C. What is the ideal gross heating value of the gas with the composition shown in Part A?

3-6. Verify that each of the three functions in Equation (3.6.4) is a solution of the wave equation given in Equation (3.6.2).

3-7. Suppose a laser emits a quantum of light called a photon with wavelength $\lambda = 10.6$ microns $= 10.6 \times 10^{-6}$ m. Calculate the frequency and energy of the photon.

3-8. The surface temperature of a star like the sun is approximately $6,000°$ K. How much energy is emitted per second by a black body with a temperature of $6,000°$ K? Express your answer in watts.

3-9. A. Suppose water is the fluid in a heat exchanger. Assume water has density $\rho = 1$ g/cc and dynamic viscosity $\mu = 1$ cp. If the tube diameter of the heat exchanger is $D = 1$ cm, calculate the flow velocity for laminar flow at $N_{Re} = 2000$ and the flow velocity for turbulent flow at $N_{Re} = 6000$. Express velocities in m/s. Hint: Convert all physical parameters to SI units first.
B. Calculate the heat transfer rate for laminar flow of water at $N_{Re} = 2000$ and the heat transfer rate for turbulent flow of water at $N_{Re} = 6000$. Assume the temperature of the water increases by $10°$ C and the specific heat is 4190 J/(kg $\cdot °$K). Express heat transfer rate in J/s.

3-10. A. Calculate the rate of heat flow through an aluminum cylinder that is 2 m long and has a radius of 0.1 m. One end of the bar is heated to $200°$ F and the other end is at the ambient temperature of $60°$ F.
B. Calculate the rate of heat flow through a Pyrex glass cylinder that is 2 m long and has a radius of 0.1 m. One end of the bar is heated to $200°$ F and the other end is at the ambient temperature of $60°$ F. Compare your answer to Part A.

3-11. A. Calculate the rate of heat loss through a single-paned window with a 1/4-inch pane of glass. The area of the window is 3 ft by 4 ft. Assume the outside temperature is $-6.7°$ C and the inside temperature is $21.1°$ C.
B. Calculate the rate of heat loss through a double-paned window with a 1/8-inch air-filled separation between 1/4-inch panes of glass. The area of the window is 3 ft by 4 ft. Assume the outside temperature is $-6.7°$ C and the inside temperature is $21.1°$ C.

CHAPTER FOUR

The Earth and Geothermal Energy

To us, the earth is the solid planet beneath our feet. The earth is also a celestial object. Scientists generally believe that celestial objects such as gaseous nebulae, galaxies, stars, and planets formed from the atomic gases permeating the universe after the Big Bang. These celestial objects may serve as significant sources of energy in the twenty-first century. One possible scenario for their formation is described in the following section.[1] We then consider the heat energy stored inside the earth: geothermal energy.

4.1 FORMATION OF CELESTIAL OBJECTS

Relatively dense accumulations of matter in space exert a stronger gravitational attraction than less-dense accumulations of matter in neighboring regions of space. Gradients in the attractive forces facilitate the local accumulation of matter and provide a mechanism for enlarging the local accumulation. Figure 4-1 illustrates this effect by showing the increase in mass density from initial to final conditions around two points A and B. The gravitational capture of matter further increases the total gravitational pull of the growing accumulation. This snowballing effect results in a universe with vast distances separating regions containing substantial quantities of matter.

Permeating the greatly cooled universe, according to cosmological theory, is a relic of the Big Bang: the microwave background radiation. The microwave background radiation is an example of black body radiation. The microwave background radiation is a microwave radiation with a temperature of approximately $2.7°$ K. Arno Penzias and Robert Wilson's discovery of this background radiation in 1965 is a significant experimental verification of the Big Bang model of cosmology.

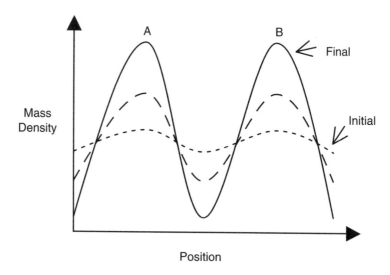

Figure 4-1. Local accumulation of matter.

Regions of space having relatively dense concentrations of matter appear at first to be gaseous clouds relative to adjacent regions with less-dense concentrations of matter. Gaseous nebulae, such as the Eta Carinae and Horsehead Nebulae, are modern examples of these regions. In time, local disturbances of the matter distribution within the gaseous clouds cause further coalescing of matter to form stars. All primordial matter does not necessarily coalesce into stars. Some locally dense gaseous clouds are neither massive enough nor energetic enough to initiate the nuclear forces necessary to balance the gravitational forces present in the coalescing cloud. Those clouds that can establish a balance between nuclear repulsion and gravitational attraction become stars.

The most important constituent of an infant star is hydrogen, the simplest and most plentiful atom in the universe. Helium is a distant second, followed by such common elements as oxygen, nitrogen, and carbon. Hydrogen is the primary fuel for nuclear fusion reactions: two atoms of hydrogen are fused together in the dense, hot core of the coalescing cloud to form helium. Not all of the mass-energy of the two hydrogen atoms is needed to form helium. The excess is converted to electromagnetic energy and is a major source of solar radiation. A star burns up its atomic hydrogen fuel as it ages. In the process of burning hydrogen, the core of a star grows and its pressure and temperature increase. Eventually the pressure and temperature of the core are high enough to begin burning helium. These nuclear reactions—the burning of hydrogen and helium—are important processes in the construction of larger atoms such as iron and magnesium. Nuclear reactions and their role in nuclear energy technology are discussed in more detail later.

4.2   KANT–LAPLACE HYPOTHESIS

The preceding discussion provides a sampling of the rich variety of fates befalling primordial matter as the universe matured. It is a scenario stemming from ideas first suggested by the German philosopher Immanuel Kant in 1755 and, independently, by the French mathematician Pierre Laplace in 1796. The Kant–Laplace hypothesis is the idea that the solar system formed out of a rotating cloud of gas and dust. It is the basis of modern theories of nebular, galactic, stellar, and planetary formation.

Pieces of evidence supporting the Kant–Laplace hypothesis include observations by the Infrared Astronomical Satellite (IRAS). IRAS was launched on a voyage to scan the sky in search of infrared (heat) radiation on January 25, 1983. According to a Jet Propulsion Laboratory Fact Sheet, IRAS found "...numerous small clouds of molecular gas and dust that are sites of formation for stars like our Sun.... The survey found that many nearby dark clouds, some within 650 light-years of Earth, harbor newly formed stars in a stage of evolution much like that of the Sun when it formed 4.6 billion years ago" [Jet Propulsion Laboratory, 1984]. IRAS findings were not limited to stellar activity: "Two of the most significant findings during the 10-month IRAS mission were the discoveries of solid material around the stars Vega and Fomalhaut (26 and 22 light-years from Earth, respectively).... The findings provide the first direct evidence that solid objects of substantial size exist around a star other than the Sun" [Jet Propulsion Laboratory, 1984].

We now know that the interstellar medium contains atomic nuclei, molecules, and grains of "interstellar dust" [Hester, et al., 2002, Section 14.2; and Taylor, 2001, Chapter 4]. Approximately 90% of the nuclei are hydrogen nuclei (H), and 9% are helium nuclei (He). The rest of the atomic nuclei include nuclei of carbon (C), oxygen (O), neon (Ne), sodium (Na), magnesium (Mg), aluminum (Al), silicon (Si), sulfur (S), calcium (Ca), iron (Fe), nickel (Ni), etc. Molecules in the interstellar medium include hydrogen (H_2), carbon monoxide (CO), methane (CH_4), ammonia (NH_3), hydrochloric acid (HCl), silicon oxide (SiO), sodium chloride (NaCl), water (H_2O), sulfur dioxide (SO_2), hydrogen sulfide (H_2S), and potassium chloride (KCl). More complex molecules in the interstellar medium include methanol (CH_3OH) and acetone [$(CH_3)_2CO$]. Notice that both inorganic and organic molecules are present in the interstellar medium. The presence of organic molecules in space has encouraged the belief by some in the extraterrestrial origin of life, as discussed further in Section 5.5. Interstellar dust includes grains of material similar to soot from a candle. The material in interstellar dust consists of inorganic substances such as carbon and silicates, as well as icy and organic components. The source of much of the material found in space is discussed in Chapter 10 when we discuss nucleosynthesis.

The planets in our solar system provide two additional pieces of evidence that support the Kant–Laplace hypothesis of planetary formation. The first piece of evidence comes from observations of planetary orbits. Most planetary orbits lie in the ecliptic plane. The *ecliptic plane* is a plane that intersects the sun and is illustrated in Figure 4-2. The line of intersection between the orbital plane and the ecliptic plane is the *line of nodes*. The orbits of the outermost planet Pluto and the innermost planet Mercury are tilted at a slight angle relative to the ecliptic plane. The angles of inclination

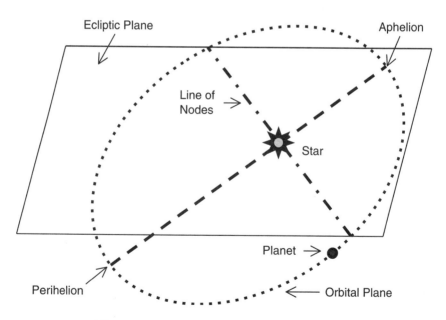

Figure 4-2. Planetary orbit and the ecliptic plane.

of the orbits of Mercury and Pluto relative to the ecliptic plane are about 7 degrees for Mercury, and 17 degrees for Pluto.

The second piece of evidence comes from observations of the direction of rotation of planetary orbits relative to the direction of rotation of the sun. Our sun rotates about its axis much like a spinning top. All nine planets of the solar system orbit in the same direction as the direction of rotation of the sun. These two observational facts—the direction of orbital rotation, and orbits confined to the ecliptic plane—suggest that the sun and planets were formed from the dust and debris of a rotating gas cloud.

Johannes Kepler first quantified the motion of planets in 1619 using Tycho Brahe's observations of the motion of the planet Mars. From empirical evidence, Kepler formulated three laws of planetary motion.

1. The orbit of a planet is an ellipse with the sun at one focus of the ellipse.
2. The orbit of a planet lies in a plane that passes through the sun, and the area swept out by a line joining the sun and planet is proportional to the elapsed time.
3. The time for one revolution, or period T, of the planet around the sun is proportional to $a^{3/2}$ where a is the semi-major axis of the elliptical planetary orbit.

Isaac Newton's classical mechanics can be used to derive Kepler's laws. Doing so, we find the period T of a planetary orbit given by

$$T = \frac{2\pi a^{3/2}}{\sqrt{Gm_s}}$$

(4.2.1)

where a is the semi-major axis, G is Newton's gravitational constant, and m_s is the mass of the sun. Equation (4.2.1) is an expression of Kepler's third law.

CELESTIAL COLLISIONS

The Kant–Laplace hypothesis is considered a single-body theory because it does not require a collision between celestial objects. Alternatives to the single-body Kant–Laplace hypothesis are second-body theories. Second-body theories employ the near approach of two celestial objects as the primary mechanism for forming a third celestial object. For example, in 1750 the French naturalist Georges de Buffon suggested that planets were formed from the sun by the passing of a large object such as a comet. Two University of Chicago professors advanced a more recent second-body theory of planetary formation in 1905. T.C. Chamberlain and F.R. Moulton argued that the second body was not a comet, as Buffon suggested, but a star. As two stars approach one another, strong gravitational forces pull stellar matter from each star toward the other. It is conceivable that large amounts of matter are essentially ripped from their solar homes. In the cold interstellar space the matter cools and condenses by gravitational contraction into planetary size "droplets."

Collisions or close approaches between two astronomical objects used to be considered too unlikely to be the primary mechanism of celestial body formation. Distances between celestial objects are so large that the probability of close approach is small. Modern studies of the moon, including the study of lunar rocks brought back by the Apollo astronauts, support the theory that the moon is actually a piece of the earth that was broken off after the earth collided with a massive celestial object.[2] There are other known examples of collisions between celestial objects, such as the collision of the two galaxies NGC 4038 and NGC 4039. The names of these galaxies, which are about 50 million light-years from the earth, refer to their catalogue number in the New General Catalogue. The New General Catalogue got its start as a compilation of galaxies by the German-born English astronomer William Herschel. Herschel's son John continued the work and published the General Catalogue of Nebulae in 1864. J.L.E. Dreyer published an expanded version in 1888.

Scientists do not know what the precise mechanism of celestial body formation is. It is possible that many mechanisms have played an important role in various regions of the universe. Most people believe some variation of the Kant–Laplace hypothesis was the primary mechanism. Despite our ignorance, we can say with some confidence what the formation sequence was. This sequence is summarized in Table 4-1. Estimates of the time when the events occurred are included. In general, the formation of a nebula is

Table 4-1
Timetable of planetary formation

Years ago	Epoch	Event
14 to 15 billion years	Decoupling Era	Galaxies begin to form and cluster
10 billion years		First stars form
4.5 billion years		Earth's interstellar cloud forms
3.9 billion years	Archeozoic Era	Oldest terrestrial rocks form

a necessary precursor to the formation of stars and the transformation of a nebula into a galaxy. Planetary formation either succeeds or is concurrent with star formation, depending on the formation mechanism. We have reached the point of universal evolution when it is time to address the physical evolution of planets, with special attention focused on the earth.

4.3   EVOLUTION OF THE PRIMORDIAL EARTH

Until the beginning of the twentieth century, people could only study the earth from its surface. A broader view of our planet became possible with the advent of flight. We are now able to see our planet from the perspective of space. We have improved our understanding of the planet we live on with each technological advance. Today we can use satellites and sensitive measurements of slight gravitational variations on the earth's surface to effectively see through the oceans and construct an unimpeded view of the hard outer crust of the earth. Special seismic detectors let us measure the speed of propagation of wavelike vibrations caused by earthquakes and nuclear detonations. These seismic measurements give us an indirect picture of the earth's interior. In addition to an arsenal of powerful devices for extending our senses, we can also look into the earth's past by simply looking at the Grand Canyon in Arizona, examining fossil remains from Siberia, or studying drill cuttings from oil and gas wells drilled in the crust of the earth. A wealth of observations obtained from a combination of methods practiced by an army of geoscientists over several decades is the raw material underlying the modern picture of the earth.[3]

Imagine a cloud of rotating gas and dust. Gravitational attraction between the constituents of the cloud is greatest near the center. As the cloud collapses, increases in pressure and temperature at the core of the cloud are eventually high enough to ignite nuclear fusion processes: the sun is born. The accretion, or aggregation, of matter creates protoplanets. The shape of the sun continues to define itself as perturbations in the rotating cloud surrounding the infant star coalesce into protoplanets that eventually become the large, gaseous planets Jupiter and Saturn. Dust grains, some made of such relatively heavy elements as iron and magnesium, are aggregating in other parts of the cloud engulfing the sun. These heavy aggregates are called *planetesimals* and are the precursors of the planets Mercury, Venus, Earth, and Mars. Planetesimal growth is greatest near the sun where nebular debris is most abundant, yet not so close that the debris is pulled into the sun.

The dense core of each planetesimal acquires an exterior of less-dense material from the dwindling molecular gas cloud. Gravitational pressure from the planetesimal exterior generates a large interior pressure. The temperature also rises, in part due to crushing gravitational pressure, and in part due to nuclear reactions. Eventually the pressure and temperature of the planetesimal interior becomes great enough to melt a portion of the once solid core. Moving outward from the interior we find the pressure and temperature decreasing as the amount of confining material decreases and we approach the cold of space. Heavier radioactive elements migrate toward the core, while lighter elements migrate toward the planetesimal surface. The planetesimals are large enough to retain a gaseous atmosphere of light elements, including hydrogen and helium. Atmospheres of the planetesimals vary from a trace of helium in the case of Mercury to an atmosphere of molecules such as ammonia and methane. Lighter molecules, especially hydrogen, escape the gravitational pull of the planetesimals.

Embodied in this scenario of planetary formation is the explanation of some observational facts. The cross-section of the earth's interior is subdivided into an inner core, outer core, mantle, and crust (Figure 4-3).

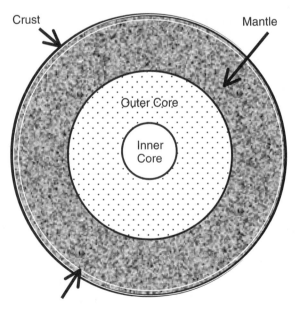

Crust

Mantle

Outer Core

Inner Core

Crust–Mantle Boundary (Moho)

Figure 4-3. The interior of the earth.

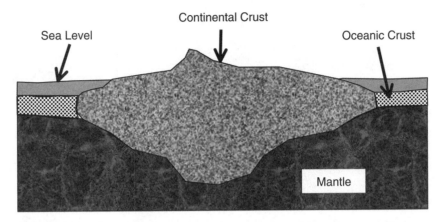

Figure 4-4. The crust of the earth.

Seismic measurements of earthquake waves have shown the core to consist of molten metal enveloping a solid core. The earth's electric and magnetic properties, and the density of the core, provide evidence for identifying the metal as an alloy of iron and nickel. Iron is the dominant constituent. Existence of a solid inner core implies the planetesimal seed did not entirely melt. The core has a radius of approximately 3500 kilometers (2200 miles) and consists of a crystalline inner core and a molten outer core.

Exterior to the core of the earth is a rock mantle that is 2900 kilometers (1800 miles) thick. The mantle is thought to be primarily basalt, a dark volcanic rock. Basalt is composed of magnesium and iron silicates. Basalt at the surface of the mantle exists in a semi-molten state. This layer of semi-molten basalt is called the *asthenosphere*. As we descend through the mantle, the rigidity of the basalt increases until the mantle acquires rigidity in excess of the rigidity of steel.

The ratio of iron to magnesium and silicon in the earth is about the same as the ratio observed in stars like our sun. This observation lends support to models of planetary formation that are analogous to models of stellar formation. The materials used to form the sun and the planets appear to be from the same source.

Above the mantle is approximately 30 kilometers (19 miles) of crust. Figure 4-4 is a sketch of the crust of the earth. The granitic continental crust is composed principally of silica and aluminum; the oceanic crust contains silica and magnesia. Underlying the continental and oceanic crusts is a layer of solid basalt. The boundary between the crustal basalt and the semi-molten basalt of the mantle is the Mohorovicic discontinuity, or

Moho for short. The combination of crust and solid basalt above the Moho is known as the lithosphere.

The *lithosphere* is the mobile part of the upper mantle and crust. Lithospheric plates drift on a denser, partially molten material called the asthenosphere. As the earth cooled from its hot, gaseous state, the surface of the earth was subjected to forces that caused great changes in its topography, including the formation of continents and the uplift of mountain ranges. Pressure from the earth's interior could crack the sea floor and allow less dense molten material to flow onto the sea floor. Such cracks in the Earth's crust are called *subsea ridges.*

Studies of material extruded from a subsea ridge show that the material spread laterally on each side of the ridge. The symmetry of the material spread on each side of the ridge supports the contention that the material was in a molten state as it gradually moved outward from the ridge. As the material cooled, magnetic constituents within the molten material aligned themselves in conjunction with the polarity of the earth's magnetic field at the time that the material solidified. Several periods of polarity have been identified and dated.

Measurements made by satellites of the earth's gravitational field have identified boundaries between continents. The shapes of the boundaries are indicative of vast plates. These plates are referred to as *tectonic plates,* and their behavior is the subject of *plate tectonics.*

We live on the lithosphere. Its history is intertwined with the history and distribution of life. As with human history, a chronology of events must be developed. In the case of lithospheric history, the chronology relies heavily on a technique known as radioactive dating.[4] Radioactive dating depends on the detection of radioactivity from the decay of atomic nuclei. In addition to providing a means of dating rock, radioactivity provides energy to heat the interior of the earth. Radioactivity is the subject of the following section.

4.4   RADIOACTIVITY

The number of protons in the nucleus of an element is used to classify the element in terms of its positive electric charge. Isotopes are obtained when electrically neutral neutrons are added to or subtracted from the nucleus of an element. Changes in the number of neutrons occur by several mechanisms. Nucleus changing mechanisms include nuclear emission of helium nuclei (alpha particles), electrons (beta particles), or highly ener-getic photons (gamma rays). An isotope is said to decay radioactively when the number of protons in its nucleus changes. Elements produced by radioactive decay are called *decay products*.

PARTICLE DECAY

Experimentally, the rate of decay of a collection of unstable particles is proportional to the number of particles. It may be empirically described using an exponential (Poisson) probability distribution. Let $N(t')$ be the number of particles observed at time t' and write the total number of par-ticles at $t' = 0$ as $N(0)$. Then the number of particles decaying in the time duration t' to $t' + dt'$ is

$$dN(t') = -\lambda N(t') \, dt' = -A \, dt' \tag{4.4.1}$$

where λ is the constant probability of particle decay per unit time. The product $A = \lambda N(t')$ is the activity of the radioactive sample.

Integrating Equation (4.4.1) from $t' = 0$ to $t' = t$ gives

$$N(t) = N(0) \, e^{-\lambda t} \tag{4.4.2}$$

Equation (4.4.2) represents the exponential decay of an unstable object. The decay rate is

$$R = \frac{dN(t)}{dt} = -\lambda N(t) \tag{4.4.3}$$

The probability of observing a particle at time t is

$$P_{obs} = \frac{N(t)}{N(0)} = e^{-\lambda t}. \tag{4.4.4}$$

Equation (4.4.4) is commonly used for describing phenomenological results. It should be noted that an unstable particle does not conserve prob-ability within the context of conventional quantum theories that normalize probability over spatial volume only.[5]

MEASURES OF RADIOACTIVITY

Several measures of radioactivity have been introduced.[6] The activity of a radioactive sample was introduced in Equation (4.4.1). Activity is the number of decays per second of the sample, and is measured in Curies. The Curie, with symbol Ci, is named after pioneers in the study of radioactivity, French physicists Pierre and Marie Curie. One Curie equals 3.7×10^{10} decays per second. The Curie only contains information about the number of decays that are occurring.

The Roentgen is used to measure the amount of ionizing charge produced per unit mass of radioactive substance. The Roentgen, with symbol R, is named after Wilhelm Conrad Roentgen, the discoverer of X-rays. One Roentgen equals 2.58×10^{-4} Coulombs of ionizing charge per kilogram of radioactive substance.

The radiation unit that measures the amount of radiation energy being absorbed per gram of absorbing material is called the rad, or radiation absorbed dose. One rad equals 100 erg of radiation energy absorbed by a gram of absorbing material, or 1 rad = 0.01 J/kg in SI units. The rad is the radiation dose unit in the cgs system. The cgs system is a system of units based on the centimeter, the gram, and the second. The SI unit of radiation dose is the Gray (Gy), and 1 Gy = 1 J/kg, or 1 Gy = 100 rads.

A measure of radiation is needed to monitor the biological effects of radiation for different types of radiation. The measure of radiation is the dose equivalent H. Dose equivalent is the product of the radiation dose D times a qualifying factor QF, thus $H = D \times QF$. The qualifying factor is a dimensionless number that indicates how much energy is produced in a material as it is traversed by a given radiation. Table 4-2 illustrates several qualifying factors. The alpha particle referred to in Table 4-2 is the helium nucleus.

Dose equivalent is measured in sieverts (Sv) if the dose is measured in Grays. If the dose is measured in rads in cgs units, dose equivalent is

Table 4-2
Typical qualifying factors

X-rays, gamma rays	1
Thermal neutrons (0.025 eV)	2
High-energy protons	10
Heavy ions, including alpha particles	20

Source: After Murray, 2001, pg. 213, Table 16.1.

measured in rems. The sievert is the SI unit of dose equivalent, and the rad is the older cgs unit. Murray [2001, pg 214] reports that a single, sudden dose of 400 rems (4 Sv) can be fatal, while the typical annual exposure to natural and manmade (e.g. medical and dental) radiation is 360 mrems (3.6 mSv).

ISOTOPES

Most elements have stable isotopes. For example, the carbon atom has six protons in its nucleus. Carbon-12 is the isotope of carbon with six neutrons in its nucleus. The number 12 following carbon is the sum of six protons and six neutrons. Carbon-12 does not undergo radioactive decay and is therefore stable. An important isotope of carbon, carbon-14, has six protons and eight neutrons in its nucleus. It is a radioactively decaying isotope of carbon. Lifetimes of radioactive isotopes are expressed in terms of half-lives. Half-life $t_{1/2}$ is defined as the length of time it takes half of the original quantity of an isotope to decay. Using the distribution in Equation (4.4.2), we can express half-life as

$$t_{1/2} = \frac{\ln 2}{\lambda} = \tau \ln 2 \qquad (4.4.5)$$

where τ is the mean life, or average lifetime, of the radioactive nucleus. The age of a fossil or rock can be estimated by combining the half-life of an isotope with knowledge of its abundance and the abundance of its decay products in a source rock. Determining the age of a rock or fossil using radioactive decay techniques is called *radioactive dating*.

The common occurrence of carbon-14 in living things, combined with a relatively long half-life (about 5600 years), makes carbon-14 decay a valuable process for dating fossils. Another important decay process is uranium decay. Uranium isotopes decay by alpha emission and have half-lives ranging from 4.5 to 13.9 billion years. Decay products of uranium are lead and helium. The appearance of uranium in many rocks makes uranium decay an important process for estimating the age of rocks. Rock dating, in turn, is crucial to determining the age of the earth[7] and developing a theory of plate tectonics.

4.5 PLATE TECTONICS

In the theory of plate tectonics, the entire crust of the earth is seen as a giant, ever-shifting jigsaw puzzle. The pieces of the puzzle are tectonic plates. The movement of the plates tells us the history of the lithosphere. We can begin to understand the history of the lithosphere by first recognizing that a dynamic relationship exists between the lithosphere and the asthenosphere. The lithosphere consists of a set of rigid plates floating on the semi-molten asthenosphere. The tectonic plates are often associated with continental land masses.

Satellite measurements of the gravitational field of the earth provide pictures of the earth's surface devoid of both oceans and vegetation. Satellite pictures clearly show boundaries between continents, and tremendous mountain ranges rising from ocean floors. The shapes of the boundaries are suggestive of vast plates, as depicted in Figure 4-5. Only the largest of the known plates are depicted in the figure. Many of these plates are associated with continental land masses.

Molten material in the asthenosphere can enter the lithosphere through cracks between plates. This transfer of material can be by violent volcanic eruptions, or through the gradual extrusion of basaltic lavas at the boundaries between plates. Volcano examples are well known. Extrusion examples are not as well known because until recently the major locations of basaltic extrusions were not accessible.

Marine scientists have discovered that oceanic mountain ranges are sources of basaltic extrusion and seafloor spreading. The continental plates are forced to move when seafloors spread. Movement of continental plates is known as continental drift, and was first proposed by the German

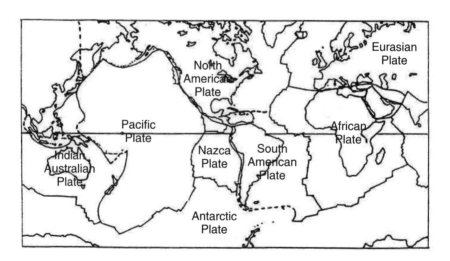

Figure 4-5. Tectonic plates.

geophysicist Alfred Wegener[8] in 1950. Regions of some moving continental plates may collide. A collision of two plates can form great mountain ranges, such as the Himalayas. Alternatively, a collision can deflect one plate beneath another. Material in the deflected region, or subduction zone, may be forced down through the Moho and into the semi-molten asthenosphere. Thus *subduction zones* are locations where crustal material is returned to the mantle. Together, seafloor spreading and subduction zones are the primary mechanisms for transferring material between the crust and the upper mantle. The theory describing the movement of lithospheric plates is known as plate tectonics. Using radioactive dating and similarities in geologic structures, a reconstruction of lithologic history is sketched in Figure 4-6. The acronym MYA in Figure 4-6 refers to millions of years ago.

Figure 4-6 shows the hypothesized movement of tectonic plates during the past 225 million years. The first map shown in Figure 4-6 begins when all surface land masses were coalesced into a single land mass

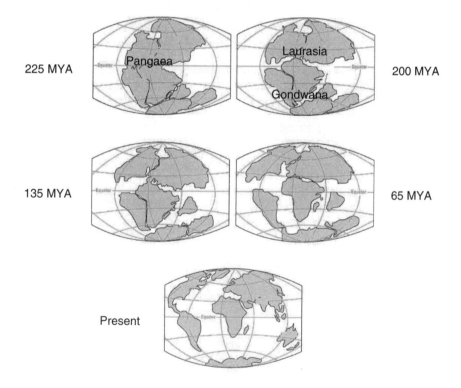

Figure 4-6. Tectonic plate movement [after USGS website, 2001].

called Pangaea. Geoscientists believe that Pangaea was formed by the movement of tectonic plates, and the continued movement of plates led to the break-up of the single land mass into the surface features we see today. Forces that originate in the earth's interior drive the movement of tectonic plates. As the plates pull apart or collide, they can cause such geologic activities as volcanic eruptions, earthquakes, and mountain range formation. Plate movement provides an explanation for the geographic distribution of organisms around the world, and is responsible for much of the geologic heterogeneity that can be found in hydrocarbon-bearing reservoirs.

Along with the occasional impact of a meteor or asteroid, the movement and position of tectonic plates have been theorized to cause extensive environmental changes. These environmental changes include global sea level and atmospheric changes. Plate tectonics can lower the sea level, creating a period of vast erosion and deposition that can affect the biosphere. Plants and animals may thrive in one set of conditions, and readily become extinct when the conditions change. Based on these changes, geologists have found that the geologic history of the earth can be subdivided into convenient periods.

Table 4-3 is an abridged version of the geologic time scale, beginning with the formation of the earth. The most encompassing period of time is the eon, which is subdivided into eras, and further subdivided into periods.

Table 4-3
Abridged geologic time scale

Eon	Era	Period	Approximate start of interval (MYPB)
Phanerozoic	Cenozoic	Quaternary	0.01
		Tertiary	5
	Mesozoic	Cretaceous	144
		Jurassic	208–213
		Triassic	245–248
	Paleozoic	Permian	286
		Carboniferous	320
		Devonian	408
		Silurian	438
		Ordovician	505
		Cambrian	570–590
Precambrian	Proterozoic		2500
	Archean		3800
	Hadean		

The acronym MYPB in Table 4-3 stands for millions of years before the present. The starting time of each interval is reported from two references [Levin, 1991; Ridley, 1996] and is considered approximate because there is still uncertainty in the actual chronology of the earth.

The solid crust of the earth first appeared approximately four billion years ago in the Precambrian. The earth's atmosphere and oceans appeared shortly thereafter, geologically speaking. Life began to flourish during the Paleozoic Era. The Mesozoic Era was the age of dinosaurs. Fragments of the Pangaean supercontinent acquired the shapes of modern continents by the Cretaceous period. Mammals did not begin to flourish until relatively recently, during the Cenozoic Era.

According to plate tectonics, the land masses of the earth have been moving for millions of years. As shown in Figure 4-6, the land masses have usually been separated by vast bodies of water. At one point in geologic history all modern continents combined to form a single, great continent called Pangaea. The supercontinent Pangaea existed some 220 to 240 million years ago in the Triassic Period. This date has a special significance in the context of the origin of life. Scientists have found traces of life in fossils as old as three billion years. The existence of a single land mass, Pangaea, millions of years after the origin of life suggests that life could have originated at a single region on earth, and then spread onto all of the continents without having to cross great bodies of water. Indeed, fossil evidence has been found showing the existence of such relatively advanced life forms as mammal-like reptiles and dinosaurs at the same time that Pangaea existed.

Pangaea obviously did not retain its unity. Basalt extrusions from the mantle split the Pangaean plates. Gradually Pangaea began to break up. The Hawaiian Islands are a good example of the dynamic character of plate tectonics.[9]

HAWAIIAN HOTSPOT

Mount Kilauea on the "Big Island" of Hawaii is an active volcano over a hotspot in the earth's mantle (Figure 4-7). Scientists believe that magma, or melted rock, flows from the mantle and up through the volcano's vents. The magma becomes lava when it flows onto the crust of the earth. When the magma contacts the ocean water, it cools and solidifies. The result is mountain building and, if the mountain gets high enough above the seabed, it becomes an island.

The hotspot in the mantle below Kilauea has created the highest mountains on earth when measured from the seabed, and it has created the

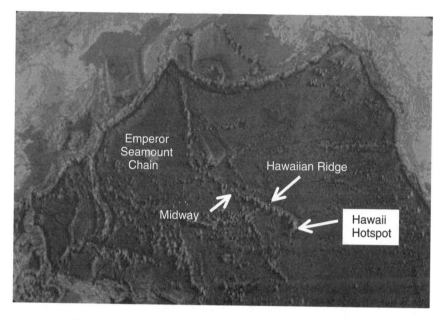

Figure 4-7. Hawaiian Hotspot [after USGS website, 2002].

Hawaiian Island chain, a chain of islands that stretches from the Kure atoll west of the Midway Islands to Hawaii. Many of these volcanic islands have eroded and become atolls. The hotspot is in the process of creating another Hawaiian island, which has been named Loihi. Loihi is an undersea volcano east of Hawaii that is expected to grow and rise above the surface of the Pacific Ocean in a few thousand years. How did one hotspot in the mantle accomplish all this? Plate tectonics can provide the answer.

The Hawaiian Islands are part of the crust of the earth called the Pacific Plate. Plate tectonics tells us that the crustal plates have been moving relative to the mantle for millions of years. The Hawaiian Ridge, including the Hawaiian Islands, was created as the Pacific Plate moved over a relatively stable hotspot in the mantle that is now erupting through Kilauea. Radioactive dating provides evidence to support this idea because the data show that the age of the islands increases as you travel west from Hawaii along the Hawaiian Ridge to Midway.

CRUSTAL ROCK FORMATION

The movement of tectonic plates across the surface of the earth generated forces that can cause rocks to form. We can think of the process of rock creation as a cycle. The beginning of the cycle occurs with the cooling of molten magma and subsequent hardening into rock. Typically, the formation of new rock occurs at plate boundaries, but it can also occur over hotspots within the earth's mantle, as in the Hawaiian example. When plates collide, pressure and heat can cause part of the plate to melt, and result in molten rock being thrust to the surface. After cooling, surface rock is subjected to atmospheric phenomena.

Chemical and physical processes cause exposed rock to break into smaller and smaller particles. Wind and water transport these particles from their source location in a process called *erosion*. The particles continually become finer and finer as they collide with other objects during the transport process. The particle is deposited along with other particles when the energy of the wind or water dissipates to the point where there is not enough energy to transport the particle. The accumulation of particles becomes thicker and thicker.

Slowly, over millions of years, tectonic plates move up and down relative to sea level, alternately causing erosion and deposition. Deposition can range from thousands of feet of sediment in an area to none at all. Erosion can carve canyons, level towering mountains, or remove all traces of a formation that was once hundreds of feet thick. High pressure and temperature can cause rocks to change character in a process called *metamorphism*. Particles may become fused together to form considerably larger objects. Given enough time, pressure, and heat, rocks will melt and start the cycle again.

Based on this rock cycle, geologists recognize three primary types of rocks: igneous, sedimentary, and metamorphic. The cooling of molten material called magma forms igneous rocks. Sedimentary rocks are formed from mineral grains. A *mineral grain* is an inorganic crystal with a specific chemical composition. Mineral grains occur naturally. They can be weathered, transported, and deposited at sites of accumulation. If mineral grains accumulate and are cemented together, they form sedimentary rocks. Metamorphic rocks are formed from rocks that have had their original texture or their mineral or chemical composition changed by heat, pressure, or chemical activity. Sedimentary rocks are usually the most interesting to professionals working to characterize commercially important reservoirs such as petroleum reservoirs.

Weathering processes at the surface of the earth create the grains that form sedimentary rocks. Weathering creates particles that can be practically any size, shape, or composition. A glacier may create and transport a particle the size of a house, and a desert wind might create a uniform bed of very fine sand. The particles, also known as sediments, are transported to the site of deposition, usually by aqueous processes. Sometimes the particles are transported very far. In these cases, only the most durable particles survive the transport. The grains of sand roll and bump along the transport pathway. Grains that started out as angular chunks of rock slowly become smaller and more rounded. A grain of quartz, for example, is a relatively hard mineral. It may be able to withstand multiple cycles of deposition and erosion. This leaves a grain that is very rounded. The minerals that make up a sedimentary rock will depend on many factors. The source of the minerals, the rate of mineral breakdown, and the environment of deposition are important factors to consider in characterizing the geologic environment.

4.6   FLUIDS IN POROUS MEDIA

Subsurface reservoirs in the earth are examples of porous media.[10] A *porous medium* is a medium that contains rock grains and pore space. Figure 4-8 is a sketch of a block of rock with grains of sand filling the block. *Bulk volume* is the volume of the block and includes both grain volume and the volume of space, or pore volume. Bulk volume V_B of the porous medium is the product of area A in the horizontal plane times gross thickness H:

$$V_B = AH \tag{4.6.1}$$

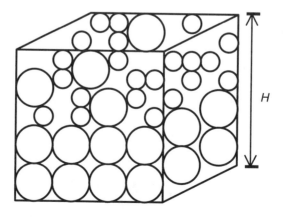

Figure 4-8. Porous medium.

The volume that is not occupied by grains of sand is the pore space available for occupation by fluids such as oil, gas, and water. The gas and liquid captured in the pore spaces of rock typically consist of a mixture of atoms and molecules.

Porosity ϕ is defined as the ratio of pore volume to bulk volume. Pore volume V_P is the volume remaining when the volume of grains V_G is subtracted from the bulk volume, thus

$$\phi = \frac{V_P}{V_B} = \frac{V_B - V_G}{V_B} \tag{4.6.2}$$

There are different kinds of porosity, but porosity can simply be thought of as the void space in a rock. Upon rearrangement, we see from Equation (4.6.2) that pore volume is the product of bulk volume and porosity:

$$V_P = \phi V_B \tag{4.6.3}$$

When sedimentary rocks are being deposited, the pore space is filled with water. The pores of a rock can be filled at a later time with commercially important fluids ranging from potable water to oil and gas. As a rule, we are interested in void spaces that are connected with other void spaces. Connected pore spaces form a conduit for fluid flow. *Permeability* is a measure of the connectivity of pore spaces. A rock is considered impermeable if there is no connectivity between pore spaces. Sedimentary rock such as sandstone tends to be permeable when compared to relatively impermeable shale.

The preceding definitions of bulk volume and pore volume are measures of the gross volume in a system. To determine the volume of the system that is commercially significant, the gross volume must be adjusted by introducing the concept of *net thickness*.

Net thickness h is the thickness of the commercially significant formation. For example, if the gross thickness includes 5 m of impermeable shale and 15 m of permeable sandstone, the gross thickness is 20 m and the net thickness is 15 m. If all of the permeable sandstone is not connected with a production well, then the value of net thickness is reduced further.

The net to gross ratio η_{NTG} is the ratio of net thickness h to gross thickness H:

$$\eta_{NTG} = h/H, \quad 0 \le \eta_{NTG} \le 1 \tag{4.6.4}$$

The inequality highlights the fact that net thickness is always less than or equal to gross thickness. The volume of net pay, or commercially significant zone, is the product of pore volume and net to gross ratio:

$$V_{\text{pay}} = \eta_{\text{NTG}} V_{\text{P}} = \eta_{\text{NTG}} H A \phi = h A \phi \qquad (4.6.5)$$

The saturation S_ℓ of phase ℓ is the fraction of the volume of pore space occupied by phase ℓ. Oil, water, and gas are the most common fluid phases. The volume V_ℓ of phase ℓ in the pay interval is the product of net pay volume and phase saturation:

$$V_\ell = S_\ell V_{\text{pay}} = S_\ell h A \phi \qquad (4.6.6)$$

The sum of the saturations in the pay interval must equal 1. If the system has N_ℓ phases, the saturation constraint is

$$1 = \sum_{\ell=1}^{N_\ell} S_\ell \qquad (4.6.7)$$

For an oil-water-gas system, the saturation constraint is $S_o + S_w + S_g = 1$ where the subscripts $\{o, w, g\}$ refer to oil, water, and gas respectively.

When fluids are produced from a reservoir, they travel through the reservoir rock from a place of higher energy to a place of lower energy. The wellbore of a producing well is a point of lower energy. The route that the fluid takes can be straight or very circuitous. The property of the rock that measures the length of the path from one point A to another point B relative to a straight line is called *tortuosity*. If the path is a straight line, which can occur in a fracture, the tortuosity is 1. In most cases, the flow path between points A and B will be longer than a straight line, so that tortuosity is greater than 1. Figure 4-9 illustrates the concept of tortuosity.

DARCY'S LAW

The basic equation describing fluid flow in porous media is called Darcy's law. Darcy's equation for calculating the volumetric flow rate q for linear, horizontal, single-phase flow is

$$q = -0.001127 \frac{K A_\perp}{\mu} \frac{\Delta P}{\Delta x} \qquad (4.6.8)$$

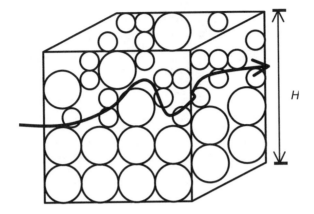

Figure 4-9. Tortuosity.

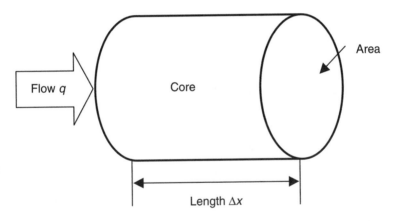

Figure 4-10. Darcy's law.

Figure 4-10 illustrates the terms in Darcy's law for a cylindrical core of rock. The movement of a single-phase fluid through a porous medium depends on cross-sectional area $A_\perp$ that is normal to the direction of fluid flow, pressure difference ΔP across the length Δx of the flow path, and viscosity μ of the flowing fluid. The minus sign indicates that the direction of fluid flow is opposite to the direction of increasing pressure; the fluid flows from high pressure to low pressure in a horizontal (gravity-free) system. The proportionality constant K in Equation (4.6.8) is called permeability.

Table 4-4
Oilfield units in Darcy's law

Variable	Oilfield unit	Conversion factor
Flow rate	barrel per day = bbl/day	1 bbl/day = 0.1589 m^3/day
Permeability	millidarcies = md	1 md = 0.986923 × 10^{-15}m^2
Area	square feet = ft^2	1 ft^2 = 0.0929 m^2
Pressure	pound-force per square inch = psi	1 psi = 6894.8 Pa
Fluid viscosity	centipoise = cp	1 cp = 0.001 Pa · s
Length	feet = ft	1 ft = 0.3048 m

The units of the physical variables determine the value of the constant (0.001127) in Equation (4.6.8). The constant 0.001127 corresponds to variables expressed in the following oilfield units:

q = volumetric flow rate, bbl/day
K = permeability, md
A = cross-sectional area, ft^2
P = pressure, psi
μ = fluid viscosity, cp
Δx = length, ft

The oilfield units for the variables in Equation (4.6.8) are related to SI units in Table 4-4.

If we rearrange Equation (4.6.8) and perform a dimensional analysis, we see that permeability has dimensions of L^2 (area) where L is a unit of length:

$$K = \frac{\text{rate} \times \text{viscosity} \times \text{length}}{\text{area} \times \text{pressure}} = \frac{\left(\dfrac{L^3}{\text{time}}\right)\left(\dfrac{\text{force} \times \text{time}}{L^2}\right)L}{L^2\left(\dfrac{\text{force}}{L^2}\right)} = L^2$$

(4.6.9)

The area unit (L^2) is physically related to the cross-sectional area of pore throats in rock. A *pore throat* is the opening that connects two pores. The size of a pore throat depends on grain size and distribution. For a given grain distribution, the cross-sectional area of a pore throat will increase as grain size increases. Relatively large pore throats imply relatively large values of L^2 and correspond to relatively large values of permeability. Permeability typically ranges from 1 md (1.0 × 10^{-15} m^2) to 1 Darcy (1000 md or 1.0 × 10^{-12} m^2) for commercially successful oil and gas fields.

Darcy's law shows that flow rate and pressure difference are linearly related. The pressure gradient from the point of fluid injection to the point of fluid withdrawal is found by rearranging Equation (4.6.8):

$$\frac{\Delta P}{\Delta x} = -\left(\frac{q}{0.001127 A_\perp}\right)\frac{\mu}{K} \tag{4.6.10}$$

Superficial velocity is the volumetric flow rate q in Darcy's law divided by the cross-sectional area $A_\perp$ normal to flow [Bear, 1972; Lake, 1989], thus $u = q/A_\perp$ in appropriate units. The interstitial, or "front," velocity v of the fluid through the porous rock is the actual velocity of a fluid element as the fluid moves through the tortuous pore space. Interstitial velocity v is the superficial velocity u divided by porosity ϕ, or $v = u/\phi = q/\phi A_\perp$. Because porosity is a fraction between 0 and 1, interstitial velocity is usually larger than superficial velocity.

4.7 EQUILIBRIUM CONDITIONS IN THE PRESENCE OF GRAVITY

In Chapter 3, we introduced the problem of determining equilibrium conditions for a fluid mixture in the absence of gravity. Here we extend the problem to calculating the vertical distribution of components of a fluid mixture in the presence of gravity. This distribution is of interest in the study of atmospheric gases, a column of fluid in a tank or vertical pipe, and fluid flow in a porous medium. Our focus here is on the compositional gradient associated with a fluid in a porous medium. Once again, we consider the problem in a closed system and then allow mass transfer by extending the problem to an open system.

CLOSED SYSTEMS

Suppose a single component mass m is undergoing a change in volume and a change in position with respect to a gravitational field with constant acceleration g. Figure 4-11 illustrates the rock–fluid system. The rock–fluid system is called the *reservoir*. The reservoir is bounded above and below by impermeable rocks. The bounding rocks form a seal that is called an unconformity in this example. We can specify the location of reservoir point A in the vertical direction by measuring elevation from the lower bound to point A, or by measuring depth from the sea level. The elevation

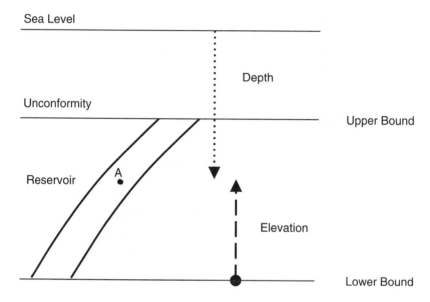

Figure 4-11. The rock–fluid system.

measurement increases as we move up in the reservoir, while the depth measurement increases as we move down in the reservoir. Either approach is acceptable as long as it is clear which measurement is being made.

We choose to measure elevation so that the work done to raise m from an elevation z to a higher elevation $z + dz$ through an infinitesimal distance dz is $(mg)\,dz$. The work done by the system is

$$dW = P\,dV - mg\,dz \qquad (4.7.1)$$

The first law of thermodynamics becomes

$$dU = T\,dS = P\,dV + mg\,dz \qquad (4.7.2)$$

with elevation z now included in the functional relationship for internal energy, thus

$$U = U(S, V, z) \qquad (4.7.3)$$

OPEN SYSTEMS

The change in internal energy of a multicomponent, open system in the absence of gravity is Equation (3.6.11). We write it here in the form

$$dU_0 = T\,dS - P\,dV + \sum_{i=1}^{N_c} \mu_i\,dn_i; \mu_i = \left(\frac{\partial U_0}{\partial n_i}\right)_{S,V,n_i} \tag{4.7.4}$$

In the presence of gravity we add the potential energy term mgz to get

$$dU = dU_0 + d\,(mgz) = T\,dS - P\,dV + \sum_{i=1}^{N_c} \mu_i\,dn_i + d\left[\left(\sum_{i=1}^{N_c} M_i n_i\right)gz\right] \tag{4.7.5}$$

where the mass m of the system is

$$m = \sum_{i=1}^{N_c} M_i n_i \tag{4.7.6}$$

and M_i is the molecular weight of component i. Expanding the differential of the gravity term and rearranging Equation (4.7.5) gives

$$dU = T\,dS - P\,dV + \sum_{i=1}^{N_c} (\mu_i + M_i gz)\,dn_i + mg\,dz \tag{4.7.7}$$

We determine equilibrium criteria by first calculating the differential of the Gibbs free energy. Remembering that S, V are constants in the Gibbs free energy, we obtain

$$dG = dU - S\,dT + V\,dP = -S\,dT + V\,dP + \sum_{i=1}^{N_c} (\mu_i + M_i gz)\,dn_i + mg\,dz \tag{4.7.8}$$

Equilibrium criteria are found by requiring $dG = 0$ subject to the constraints:

1. The process is isothermal (or very nearly so)
2. The system is isolated ($dn_i = 0 \;\forall\, i = 1,\dots,N_c$).

The first constraint says the system reaches equilibrium in a constant temperature environment. Entropy drops out of dG because of the isothermal ($dT = 0$) assumption. The second constraint requires that the number of molecules is constant. The constraint for an isolated system implies that the coefficient of each dn_i term must be constant, thus

$$m_i + M_i gz = \text{constant} \quad \forall\, i = 1, \ldots, N_c \tag{4.7.9}$$

Combining Equation (4.7.9) with the remaining terms in dG gives the additional criterion

$$V\, dP + mg\, dz = 0 \tag{4.7.10}$$

Solving for dP gives

$$dP = -\frac{m}{V} g\, dz = -\rho g\, dz \tag{4.7.11}$$

where ρ is mass density. The pressure difference between elevations z_1 and z_2 is

$$\int_{P_1}^{P_2} dP = P_2 - P_1 = \int_{z_1}^{z_2} (-\rho g)\, dz = -\rho g\,(z_2 - z_1) \tag{4.7.12}$$

or

$$P_2 = P_1 - \rho g\,(z_2 - z_1) \tag{4.7.13}$$

If elevation z_2 is greater than elevation z_1, pressure P_2 at elevation z_2 is less than pressure P_1 at elevation z_1. This demonstrates an observation that is generally true: pressure increases with depth as we drill deeper into the crust of the earth where the gravitational acceleration of the earth is approximately the constant g.

Equation (4.7.9) may be expressed in differential form for an isothermal process as

$$d\mu_i + M_i g\, dz = 0 \quad \forall\, i = 1, \ldots, N_c \tag{4.7.14}$$

The integral of Equation (4.7.14) between elevations z_1 and z_2 is

$$\int_{\mu_{i1}}^{\mu_{i2}} dm_i = -M_i g \int_{z_1}^{z_2} dz \tag{4.7.15}$$

with the result

$$\mu_{i2} - \mu_{i1} = -M_i g (z_2 - z_1) \tag{4.7.16}$$

Equation (4.7.16) shows that the chemical potential of each component in a fluid varies with elevation. This variation represents a compositional gradient.

4.8   GEOTHERMAL ENERGY

We pointed out previously that the earth's interior is subdivided into a crystalline inner core, molten outer core, mantle, and crust. Basalt, a dark volcanic rock, exists in a semi-molten state at the surface of the mantle just beneath the crust. Drilling in the earth's crust has shown that the temperature of the crust tends to increase linearly with depth. The interior of the earth is much hotter than the crust. The source of heat energy is radioactive decay, and the crust of the earth acts as a thermal insulator to prevent heat from escaping into space.

Geothermal energy[11] can be obtained from temperature gradients between the shallow ground and surface, subsurface hot water, hot rock several kilometers below the earth's surface, and magma. Magma is molten rock in the mantle and crust that is heated by the large heat reservoir in the interior of the earth. In some parts of the crust, magma is close enough to the surface of the earth to heat rock or water in the pore spaces of rock. The heat energy acquired from geological sources is called *geothermal energy*. Magma, hot water, and steam are carriers of energy.

The heat carried to the surface from a geothermal reservoir depends on the heat capacity and phase of the produced fluid. We illustrate this dependence by considering an example. Suppose the pore space of the geothermal reservoir is occupied by hot water. If the temperature of the produced water is at the temperature T_{res} of the geothermal reservoir, the heat produced with the produced water is

$$\Delta H_w = m_w c_w \Delta T \tag{4.8.1}$$

where ΔT is the temperature difference $T_{res} - T_{ref}$, T_{ref} is a reference temperature such as surface temperature, m_w is the mass of produced water, and c_w is the specific heat capacity of water. The mass of produced water can be expressed in terms of the volumetric flow rate q_w, the period of flow Δt and the density of water ρ_w, thus

$$m_w = \rho_w q_w \Delta t \tag{4.8.2}$$

Substituting Equation (4.8.2) into (4.8.1) gives

$$\Delta H_w = (\rho_w q_w \Delta t)\, c_w \Delta T = (\rho_w q_w \Delta t)\, c_w\, (T_{res} - T_{ref}) \tag{4.8.3}$$

The heat produced from a geothermal reservoir in time Δt is the geothermal power, or

$$P_{geo} = \frac{\Delta H_w}{\Delta t} = (\rho_w q_w)\, c_w \Delta T = \rho_w q_w c_w\, (T_{res} - T_{ref}) \tag{4.8.4}$$

The electrical power that can be generated from geothermal power depends on the efficiency η_{geo} of conversion of geothermal power P_{geo} to electrical power, thus

$$P_{out} = \eta_{geo} P_{geo} = \eta_{geo} \rho_w q_w c_w\, (T_{res} - T_{ref}) \tag{4.8.5}$$

If steam is produced instead of hot water or in addition to hot water, the heat produced must account for the latent heat of vaporization.

Some of the largest geothermal production facilities in the world are at the geysers in California, and in Iceland. These areas are determined by the proximity of geothermal energy sources. The technology for converting geothermal energy into useful heat and electricity can be categorized as geothermal heat pumps, direct-use applications, and geothermal power plants.[3] Each of these technologies is discussed in the following sections.

GEOTHERMAL HEAT PUMPS

A geothermal heat pump uses energy near the surface of the earth to heat and cool buildings. The temperature of the upper three meters of the earth's crust remains in the relatively constant range of 10° C to 16° C. A geothermal heat pump for a building consists of ductwork in the building connected through a heat exchanger to pipes buried in the shallow ground nearby. The building can be heated during the winter by pumping water through the geothermal heat pump. The water is warmed when it passes through the pipes in the ground. The resulting heat is carried to the heat exchanger where it is used to warm air in the ductwork. During the summer, the direction of heat flow is reversed. The heat exchanger uses heat from hot air in the building to warm water that carries the heat through the pipe system into the cooler shallow ground. In the winter, heat is added to the building from the earth, and in the summer heat is removed from the building.

DIRECT-USE APPLICATIONS

A direct-use application of geothermal energy uses heat from a geothermal source directly in an application. Hot water from the geothermal reservoir is used without an intermediate step such as the heat exchanger in the geothermal heat pump. Hot water from a geothermal reservoir may be piped directly into a facility and used as a heating source. A direct-use application for a city in a cold climate with access to a geothermal reservoir is to pipe the hot water from the geothermal reservoir under roads and sidewalks to melt snow.

Minerals that are present in the geothermal water will be transported with the hot water into the pipe system of the direct-use application. Some of the minerals will precipitate out of the water when the temperature of the water decreases. The precipitate will form a scale in the pipes and reduce the flow capacity of the pipes. Filtering the hot water or adding a scale retardant can reduce the effect of scale. In either case, the operating costs will increase.

GEOTHERMAL HEATING SYSTEMS

An example of a geothermal application with a heat exchanger is shown in the system sketched in Figure 4-12. The geothermal reservoir is an aquifer with hot water or steam. The production well is used to withdraw

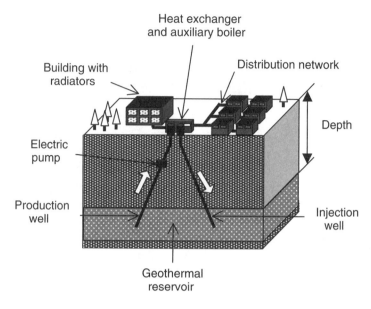

Figure 4-12. Geothermal heating system (after Shepherd and Shepherd [1998, page 149]; they refer to "World Energy Resources: 1985–2020," World Energy Conference, London, United Kingdom).

hot water from the geothermal reservoir and the injection well is used to recycle the water. Recycling helps maintain reservoir pressure. If the geothermal reservoir is relatively small, the recycled, cooler water can lower the temperature of the aquifer. The electric pump in the figure is needed to help withdraw water because the reservoir pressure in this case is not high enough to push the water to the surface. Heat from the geothermal reservoir passes through a heat exchanger and is routed to a distribution network.

GEOTHERMAL POWER PLANTS

Geothermal power plants use steam or hot water from geothermal reservoirs to turn turbines and generate electricity. Dry-steam power plants use steam directly from a geothermal reservoir to turn turbines. Flash steam power plants allow high-pressure hot water from a geothermal reservoir to flash to steam in lower-pressure tanks. The resulting steam is used to turn turbines. A third type of plant called a binary-cycle plant uses heat from moderately hot geothermal water to flash a second fluid to the vapor phase.

The second fluid must have a lower boiling point than water so that it will be vaporized at the lower temperature associated with the moderately hot geothermal water. There must be enough heat in the geothermal water to supply the latent heat of vaporization needed by the secondary fluid to make the phase change from liquid to vapor. The vaporized secondary fluid is then used to turn turbines.

MANAGING GEOTHERMAL RESERVOIRS

Like oil and gas reservoirs, the hot water or steam in a geothermal reservoir can be depleted by production. The phase of the water in a geothermal reservoir depends on the pressure and temperature of the reservoir. Single-phase steam will be found in low-pressure, high-temperature reservoirs. In high-pressure reservoirs, the water may exist in the liquid phase or in both the liquid and gas phases, depending on the temperature of the reservoir. When water is produced from the geothermal reservoir, both the pressure and temperature in the reservoir can decline. In this sense, geothermal energy is a nonrenewable, finite resource unless the produced hot water or steam is replaced. A new supply of water can be used to replace the produced fluid or the produced fluid can be recycled after heat transfer at the surface. If the rate of heat transfer from the heat reservoir to the geothermal reservoir is slower than the rate of heat extracted from the geothermal reservoir, the temperature in the geothermal reservoir will decline during production. To optimize the performance of the geothermal reservoir, it must be understood and managed in much the same way that petroleum reservoirs are managed. Petroleum reservoir management is discussed in Section 6.6.

HOT, DRY ROCK

Another source of geothermal energy is hot, dry rock several kilometers deep inside the earth. These rocks are heated by magma directly below them and have elevated temperatures, but they do not have a means of transporting the heat to the surface. In this case, it is technically possible to inject water into the rock, let it heat up, and then produce the hot water. Figure 4-13 illustrates a hot, dry rock facility that is designed to recycle the energy-carrying fluid. Water is injected into fissures in the hot, dry rock through the injector and then produced through the producer. The power plant at the surface uses the produced heat energy to drive turbines

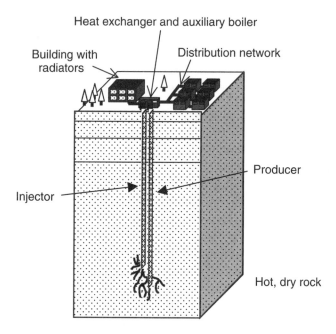

Figure 4-13. Geothermal energy from hot, dry rock.

in a generator. After the hot, produced fluid transfers its heat to the power plant, the cooler fluid can be injected again into the hot, dry rock.

ENDNOTES

1. For further discussion of the theories of galaxy formation, see Hoyle [1963], Silk [1987], Abell, et al. [1991], Börner [1993], van den Bergh and Hesser [1993], Peebles [1993], Mather and Boslough [1996], and Hester, et al. [2002]. Hoyle [1963], Stahler [1991], Abell, et al. [1991], and Hester, et al. [2002] discuss the birth of stars. Sargent and Beckwith [1993] describe the search for forming planetary systems. Binzel, et al. [1991] describe the origin of asteroids.
2. Ringwood [1986] reviewed theories of the terrestrial origin of the moon, and Taylor [1994] summarized evidence from the Apollo program for the terrestrial origin of the moon. Muller [1988] has

suggested that our sun is actually one of two stars in a binary star system, and the orbit of the companion star around the sun is responsible for periodically showering the earth with comets.

3. Geological and geophysical sources include Stokes [1960], Press and Siever [1982; 2001], Siever [1983], Maxwell [1985], Skinner [1986], Montgomery [1990], Levin [1991], Ahrens [1994], Lowrie [1997], Jeanloz and Romanowicz [1997], and Gurnis [2001]. Skinner [1986] reviewed evidence for the age of the earth and plate tectonics. Herbert [1986] discussed Charles Darwin's role as a geologist.

4. Discussions of radioactive dating are provided in a variety of sources, including Montgomery [1990], Levin [1991], York [1993], Lowrie [1997], and Press and Siever [2001] in geology and geophysics; Gould [1993] and Ridley [1996] in biology; and Williams [1991], Serway, et al. [1997], Bernstein, et al. [2000], and Lilley [2001] in physics.

5. For further discussion, see Fanchi [1993, Chapter 13].

6. For more discussion of measures of radioactivity, see Bernstein, et al. [2000, Section 15-6], or Murray [2001, Section 16.2].

7. The scientific age of the earth was not determined without dispute. Gjertsen [1984] documented the development of a geological theory by Charles Lyell, and Hellman [1998] described the debate between physicist Lord Kelvin and members of the geology and biology communities.

8. Proponents of the normal science of his time did not readily accept Wegener's ideas. For a discussion of the hurdles Wegener encountered, see Hellman [1998].

9. For technical descriptions, see Decker and Decker [1998] and Lowrie [1997]. Orr and Cook [2000] present a simple and insightful view of Hawaiian development. Kane [1996] described the view of volcanic eruption from the point of view of Hawaiian mythology and the goddess Pele.

10. Some references of note for the discussion of fluid flow in porous media include Collins [1961], Bear [1972], Selley [1998], Ahmed [2000], and Fanchi [2002]. There are many other references in these sources.

11. Some references of note for the discussion of geothermal energy include Sørensen [2000], Shepherd and Shepherd [1998], and Brown [1996].

EXERCISES

4-1. Estimate the semi-major axis of Earth's orbit using Kepler's third law.

4-2. Tectonic plates move relative to one another at the rate of up to 4 inches per year. How far apart would two plates move in 135 million years? Express your answer in meters and kilometers.

4-3. South America and Africa are about 4500 miles apart. If they began to separate from Pangaea about 150 million years ago, what is their rate of separation? Express your answer in meters per year and inches per year. How does your answer compare to the rate of separation given in Exercise 4-2?

4-4. The temperature in some parts of the earth's crust increases by about 1° F for every 100 ft of depth. Estimate the temperature of the earth at a depth of two miles. Assume the temperature at the surface is 60° F. Express your answer in °C.

4-5. A. A man ingests some radioactive material and receives 6×10^{-5} J of energy. If we assume all of this energy is absorbed in the man's gastrointestinal tract, which weighs 2 kg, what is the dose received by the man? Express your answer in Grays.
B. Suppose the radioactive material in Part A was alpha radiation. What is the dose equivalent received by the man? Express your answer in sieverts.

4-6. Use the definition of half-life and Equation (4.4.2) to derive Equation (4.4.5).

4-7. Suppose the rate of erosion is 3 mm per century. How long would it take to completely erode a mountain that is 1 mile high? Express your answer in years.

4-8. A. A formation consists of 24 ft of impermeable shale and 76 ft of permeable sandstone. What is the gross thickness of the formation?
B. What is the net-to-gross ratio of the formation?

4-9. A. A reservoir block has a length of 1000 ft, a width of 2000 ft, and a gross thickness of 15 ft. What is the bulk volume of the block? Express your answer in ft^3, bbl, and m^3.
B. If the reservoir block porosity is 0.2 and the net-to-gross ratio is 0.8, what is the pore volume of the block? Express your answer in ft^3, bbl, and m^3.

C. If the reservoir block has a gas saturation of 0.7, what is the volume of gas in the block? Express your answer in ft^3, bbl, and m^3.

4-10. The pressure at an injection well is 3000 psi and the pressure at a production well is 1500 psi. The injection well and production well are separated by a distance of 1000 ft. The mobile fluid in the reservoir between the injection well and the production well has a viscosity of 0.9 cp. The net thickness of the reservoir is 15 ft and the effective width of the reservoir is 500 ft. Use Darcy's law to fill in the following table.

Permeability	Flow rate from injector to producer		
(md)	(bbl/day)	(ft^3/day)	(m^3/day)
1			
10			
100			
1000			

4-11. The pressure in a column of water is 1000 psi at a depth of 2300 ft. What is the pressure at a depth of 2200 ft. Assume the density of water is 1 g/cc, the acceleration of gravity is 9.8 m/s^2. Express your answer in psi and kPa.

4-12. A. A geothermal power plant was able to provide 2000 MWe power when it began production. Twenty years later the plant is only able to provide 850 MWe from the geothermal source. Assuming the decline in power production is approximately linear, estimate the average annual decline in power output (in MWe/yr).
B. Suppose the plant operator has decided to close the plant when the electric power output declines to 10 MWe. How many more years will the plant operate if the decline in power output calculated in Part A continues?

4-13. A sandstone core sample is cleanly cut and carefully measured in a laboratory. The cylindrical core has a length of 3 inches and a diameter of 0.75 inch. The core is dried and weighed. The dried core weighs 125.00 grams. The core is then saturated with fresh water. The water-saturated core weighs 127.95 grams. Determine the porosity of the sandstone core. Neglect the weight of air in the dried core and assume the density of water is 1 g/cc.

CHAPTER FIVE

Origin of Fossil Fuels

Fossil energy comes from the combustion of material that was formed by the death, decay, and transformation of organisms over a long period of time. This material is called fossil fuel. The origin and composition of fossil fuels depend fundamentally on the origin and biochemical composition of life. If we want to understand the origin of fossil fuels, we must understand the molecular basis of life.[1]

Cells are the basic units of life. The word *cell* originated with the English physicist Robert Hooke in 1665. Using a microscope, he observed regularly shaped structures in thin sections of cork. These structures he called cells. Today we know that Hooke's "cells" were only cellulose walls. Modern usage of the word *cell* refers to the basic unit of an organism. This connotation is a consequence of the work of the German botanist Matthias J. Schleiden and his countryman Theodor Schwann, a zoologist.

Schleiden and Schwann independently postulated in 1839 [Wallace, 1990, page 94] that all living systems, from microscopic organisms to man, are composed of cells. Their postulate has been borne out by observation. Every living system is constructed from cells; the chemical contents of every cell are contained within a boundary called the cell wall or cell membrane; and every cell contains chemicals of great size and complexity known as proteins and nucleic acids. The material presented here is designed to facilitate our discussion of fossil energy and biomass in later chapters.

5.1 MODELS OF THE ATOM

Chemists in the eighteenth and nineteenth centuries made a distinction between chemicals of inanimate material and chemicals of living organisms. They called the former inorganic, and the latter organic. A distinction was considered necessary because post-Renaissance chemists were unable to make organic chemicals from inorganic materials. Although the distinction is still made today, it is used as a matter of convenience rather than necessity. Friedrich Wohler, a German chemist, rendered the distinction unnecessary in 1832. At that time Wohler reported the synthesis of an organic compound from an inorganic compound. He was able to make urea by heating ammonium cyanate. Wohler's work is a milestone of chemistry and a precursor of molecular biology. To properly understand the relationship between these branches of science, we must acquire some basic knowledge about chemistry and biochemistry.

Chemistry is the study of relationships between atoms and molecules. Molecules are groupings of atoms. Atoms consist of an electron cloud enveloping a nucleus. The particles in a nucleus are positively charged protons and electrically neutral neutrons. These particles occupy a small volume of space relative to the volume occupied by the atom. A problem of great importance in the first half of the twentieth century was the determination of the mechanism that kept the nucleus from flying apart because of the electrical repulsion between protons. This problem was solved by the discovery that nuclei are bound by the strong interaction.

THE STRONG INTERACTION AND THE YUKAWA POTENTIAL

Japanese physicist Hideki Yukawa introduced the idea of a strong interaction potential in 1934. The Yukawa potential is an analog of the Coulomb potential. Yukawa viewed the strong interaction as a force between two nuclei. The nuclear force was mediated by the exchange of a particle with mass m_μ. The electromagnetic interaction is a force between two charged particles that is mediated by the exchange of a photon. A photon is a massless particle of light. The Coulomb potential energy between two particles with charges q_1, q_2 and separation r is

$$V_{\text{Coulomb}} = \frac{q_1 q_2}{4\pi \varepsilon_0} \frac{1}{r} \tag{5.1.1}$$

The Yukawa potential energy between two nucleons with separation r is

$$V_{\text{Yukawa}} = -g^2 \frac{\exp\left(\dfrac{-m_\mu cr}{\hbar}\right)}{r} \tag{5.1.2}$$

where g is a coupling constant analogous to electric charge, c is the speed of light in vacuum, and $\hbar$ is Planck's constant h divided by 2π.

The dependence of the Yukawa potential on separation r reduces to the dependence of the Coulomb potential on separation when the mass of the mediating particle goes to zero, that is, when $m_\mu \to 0$. The exponential factor causes a decline in the magnitude of the nuclear force as the distance r between the nucleons increases. The range r_0 of the potential is the value of r that corresponds to the exponential factor equaling e^{-1}, or

$$e^{-1} = \exp\left(\frac{-m_\mu c r_0}{\hbar}\right) \Rightarrow 1 = \frac{m_\mu c r_0}{\hbar} \tag{5.1.3}$$

so that

$$r_0 = \frac{\hbar}{m_\mu c} \tag{5.1.4}$$

The range r_0 of the strong interaction is inversely proportional to the mass m_μ of the exchanged particle. The range goes to infinity as the exchanged particle mass goes to zero.

The binding force of the strong interaction is greater than the repulsive force of the electromagnetic interaction between protons for volumes the size of nuclei. As the spatial volume increases, the dominant interaction switches from the strong interaction to the electromagnetic interaction. The electron cloud associated with electrically neutral atoms is bound to the atomic nucleus by the electromagnetic interaction.

BOHR MODEL

Danish physicist Niels Bohr (1885–1962) proposed the first modern model of the atom in 1913. Although Bohr spent most of his professional career as a Danish physicist in Copenhagen, he received his post-doctoral education in physics at the Cavendish Laboratory in Cambridge, England with Sir J.J. Thomson. Thomson is credited with discovering the electron. Bohr then worked with Sir Ernest Rutherford in Manchester, England. Rutherford pioneered the modern concept of a small nucleus at the core of the atom. Rutherford was an experimentalist who sought to explain the results of scattering experiments in which alpha particles (positively charged helium ions) were scattered off a thin metal foil. Rutherford observed sharp scattering angles that were best explained by an atom with a small, positively charged, massive nucleus. Following Rutherford, Bohr envisioned electrons orbiting nuclei in direct analogy with planets orbiting the sun, but Bohr added new ideas from quantum theory to quantize atomic energy levels.

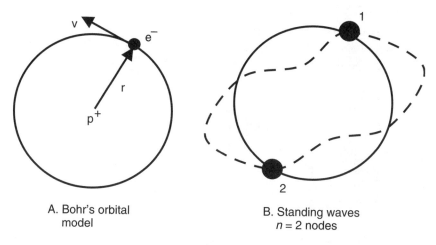

A. Bohr's orbital
model

B. Standing waves
$n = 2$ nodes

Figure 5-1. Bohr's model of the hydrogen atom.

A sketch of Bohr's model of the hydrogen atom is shown in Figure 5-1A. Newton's force law for the orbiting electron expresses equality between the centripetal motion of the electron and the Coulomb force acting on the much more massive, positively charged proton at the center of the atom. Writing m_e as mass of the electron, q as the electrical charge, v_e as the speed of the electron, and r as the distance from the electron orbit to the center of the atom, we obtain

$$m_e \frac{v_e^2}{r} = \frac{1}{4\pi\varepsilon_0} \frac{q^2}{r^2} \tag{5.1.5}$$

The kinetic energy of the electron is found from Equation (5.1.5) to be

$$K_e = \frac{1}{2} m_e v_e^2 = \frac{1}{4\pi\varepsilon_0} \frac{q^2}{2r} \tag{5.1.6}$$

Summing the kinetic energy of the orbiting electron and the potential energy of the Coulomb interaction gives the total energy as

$$E = K_e + V_{\text{Coulomb}} = \frac{1}{4\pi\varepsilon_0} \frac{q^2}{2r} - \frac{1}{4\pi\varepsilon_0} \frac{q^2}{r} = \frac{1}{4\pi\varepsilon_0} \frac{q^2}{r} \left(\frac{1}{2} - 1 \right)$$

$$= -\frac{1}{4\pi\varepsilon_0} \frac{q^2}{2r} \tag{5.1.7}$$

When the total energy is negative, the system is in a bound state. Equations (5.1.5) to (5.1.7) are relationships from classical physics. We now apply concepts from quantum theory.

We assume the electron has wavelength

$$\lambda = \frac{h}{m_e v_e} = \frac{h}{p_e} \qquad (5.1.8)$$

where h is Planck's constant and p_e is the magnitude of electron momentum. We further assume that the atom does not radiate electromagnetic waves while it is in a stationary state. A stationary state is an orbit that satisfies the quantum condition

$$n\lambda = 2\pi r_n, \quad n = 0, 1, 2, 3, \ldots \qquad (5.1.9)$$

where n is called the principal quantum number. Equation (5.1.9) says that there are an integral number of wavelengths, or standing waves, along the circumference of the orbit. Figure 5-1B illustrates a stationary state corresponding to two standing waves for $n = 2$. Substituting Equation (5.1.9) into (5.1.8) yields

$$n\frac{h}{m_e v_e} = 2\pi r_n \qquad (5.1.10)$$

where r_n is the allowed orbital radii, or

$$n\frac{h}{2\pi} = n\hbar = m_e v_e r_n \qquad (5.1.11)$$

The right hand side of Equation (5.1.11) is the angular momentum of the electron. We use Equation (5.1.11) to find the allowed values of electron speed:

$$(v_e)_n = \frac{n\hbar}{m_e r_n} \qquad (5.1.12)$$

Substituting Equation (5.1.12) into (5.1.6) gives

$$\frac{1}{2}m_e \left(\frac{n\hbar}{m_e r_n}\right)^2 = \frac{1}{4\pi\varepsilon_0}\frac{q^2}{2r_n} \qquad (5.1.13)$$

Another expression for allowed orbital radius is obtained from Equation (5.1.13). It is

$$r_n = (4\pi\varepsilon_0)\frac{n^2\hbar^2}{m_e q^2} \tag{5.1.14}$$

The energy of each stationary state, or orbital, is found by combining Equations (5.1.7) and (5.1.14) to find

$$E_n = -\frac{1}{4\pi\varepsilon_0}\frac{q^2}{2r_n} = -\frac{1}{4\pi\varepsilon_0}\frac{q^2}{2}\left(\frac{1}{4\pi\varepsilon_0}\frac{m_e q^2}{n^2\hbar^2}\right) \tag{5.1.15}$$

or

$$E_n = -\frac{1}{(4\pi\varepsilon_0)^2}\frac{m_e q^4}{2n^2\hbar^2}, \quad n = 0, 1, 2, 3, \ldots \tag{5.1.16}$$

Bohr used his model to explain hydrogen spectra. The spectrum of an element such as hydrogen arises from the release of electromagnetic radiation when an electron moves from a higher energy state to a lower energy state. The energy of a photon emitted when an electron makes a transition from state i to state j is calculated from Bohr's model as

$$h\nu_{ij} = E_i - E_j = -\frac{1}{(4\pi\varepsilon_0)^2}\frac{m_e q^4}{2\hbar^2}\left(\frac{1}{i^2} - \frac{1}{j^2}\right) \tag{5.1.17}$$

Equation (5.1.17) agrees with experiment.

The idea of a transition between energy states is an important concept in such applications as laser technology. A laser emits light with a specific wavelength by exciting electrons in atoms to a more energetic state, and then focusing the light that is emitted when the electron makes a transition to a less energetic state. The relationship between energy E, wavelength λ, and frequency ν of light is given by

$$E = h\nu = \frac{hc}{\lambda} \tag{5.1.18}$$

where h is Planck's constant. In addition to potential military applications, lasers are being used in the development of nuclear fusion. The ability of

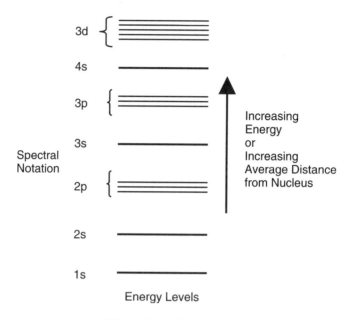

Figure 5-2. Energy levels.

a laser beam to penetrate matter has motivated research in the application of lasers to drilling.

Equation (5.1.16) shows that the difference in energy between allowed states is quantized. Electron energies are quantized in both Bohr's model and modern atomic models. In retrospect, the success of Bohr's atomic model was due in large part to the incorporation of quantized electron energy levels in his model. Some energy levels are sketched as horizontal lines in Figure 5-2, along with their spectral notation. These levels represent the only energies an electron can have in an atom. No other energy values are allowed by quantum theory. Experimental measurements support this theoretical requirement.

Bohr's model was in accord with many experimental results, particularly those of atomic spectroscopy, but it contained assumptions that violated then-accepted theoretical principles. Bohr's work on atomic structure came at a time when the flaws of classical physics could no longer be ignored. Indeed, Bohr's atomic model was based on an ad hoc synthesis of classical and early quantum concepts. It was a transitory model.

Modern models of atomic structure are based on the tenets of quantum theory. The modern models have many similarities to Bohr's model. Electrons still occupy most of the volume of the atom, and nuclei are

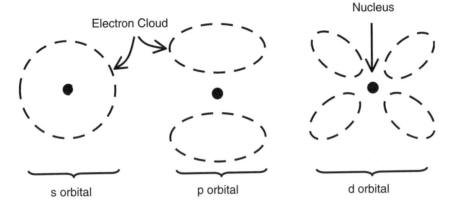

Figure 5-3. Electron orbitals in atoms.

still surrounded by electrons. The positions of electrons are not known with certainty, however, in accord with Heisenberg's uncertainty principle. Bohr's deterministic electron orbits are now viewed as probabilistic electron orbitals or electron clouds to denote the lack of definiteness in the location of the electron. We can determine the probability of locating an electron at various positions around the nucleus. The shape of the probability distribution is the shape of the electron cloud. Typical shapes of electron orbitals, or clouds, are depicted in Figure 5-3.

ATOMIC IDENTITIES

The identities of atoms are based on the number of protons in the nucleus. An electrically neutral atom must have as many negatively charged electrons in its electron cloud as it does positively charged protons in its nucleus. The atomic number is the number of protons in the nucleus of an atom. The number of electrons equals the number of protons when the atom is electrically neutral. The number of neutrons in the nucleus of the atom is usually equal to or greater than the number of protons of the element.

The atomic symbol for an element may be written as $^A_Z X$ where X is the symbol of the element, A is the mass number, and Z is the atomic number. The mass number of an atom is the total number of protons and neutrons in the nucleus. For example, carbon-12 is a stable isotope of carbon with six protons and six neutrons. It has atomic symbol $^{12}_6 C$, mass number 12, and atomic number 6. The number of neutrons is the difference between the mass number and the atomic number.

Atoms of an element may have different numbers of neutrons in the nucleus. They are called *isotopes* of the element. Carbon-14 is an unstable isotope of carbon with six protons and eight neutrons. It has atomic symbol $^{14}_6C$, mass number 14, and atomic number 6. The atomic mass of the element is the average of the masses of the isotopes found in nature.

Atomic electrons fill energy levels with the lowest energy first. Only two electrons are allowed in each energy level. The restriction to two electrons is quantum mechanical in origin, and is related to the spin of the particle.

Electrons behave in the presence of magnetic fields as if they were spinning like tops. The U.S. physicist Otto Stern suggested the idea of electron spin in 1921. A year later Stern and his countryman, Walther Gerlach, experimentally demonstrated the existence of electron spin. They observed a beam of silver atoms split into two beams by an inhomogeneous magnetic field. Their results were explained in 1925 when the Dutch-American physicists S.A. Goudsmit and G.E. Uhlenbeck introduced the concept of electron spin. Only two values of electron spin are possible. The two values are referred to by many names: spin up and spin down, clockwise and counterclockwise spin, and helicity. Helicity is the most general concept. Helicity defines the direction of a particle's spin relative to the direction of motion of the particle. Electron spin is one of four quantum numbers characterizing the behavior of an electron in an atom. The other three quantum numbers define the spatial configuration of the electron cloud.

The most important quantum number, the *principal quantum number,* represents the average distance of an electron from the atomic nucleus. The secondary quantum number characterizes the shape of the electron orbital. Three examples of secondary quantum numbers are shown in Figure 5-3. Letters are used to represent the secondary quantum numbers. This practice is a relic of early spectroscopic notation. The first few letters are s, p, d, and f. They signify secondary quantum numbers with absolute values of 0, 1, 2, and 3, respectively. Figure 5-2 shows the principal and secondary quantum numbers of the lowermost electron energy levels. For example, the spectral notation 2p denotes a principal quantum number of 2 and a secondary quantum number with an absolute value of 1.

As the principal quantum number increases, the average distance of the electron from the nucleus increases. The secondary quantum number defines the shape of the electron orbital. A third quantum number, called the magnetic quantum number, defines the orientation of the electron cloud in space. The s orbital, for example, is a spherical distribution. By contrast, the p orbital is split into three energy levels because it can be aligned along the x-axis, y-axis, or z-axis. Five unique orientations are possible for the

	1s	2s	2p$_x$	2p$_y$	2p$_z$
Hydrogen	↑				
Helium	↑↓				
Lithium	↑↓	↑			
Berylium	↑↓	↑↓			
Boron	↑↓	↑↓	↑		
Carbon	↑↓	↑↓	↑	↑	
Nitrogen	↑↓	↑↓	↑	↑	↑
Oxygen	↑↓	↑↓	↑↓	↑	↑
Fluorine	↑↓	↑↓	↑↓	↑↓	↑
Neon	↑↓	↑↓	↑↓	↑↓	↑↓

↑ Electron spin up
↓ Electron spin down

Figure 5-4. Simple atoms.

d orbital, and are depicted by the five energy levels shown in Figure 5-2 for spectral notation 3d. Figure 5-4 adds the magnetic quantum number to the spectral notation of the 2p orbital.

Austrian physicist Wolfgang Pauli recognized in 1924 that he could recreate the Periodic Table only if he assumed that every electron in an atom must have a unique set of quantum numbers. Every electron orbital could contain only up to two electrons, and the two electrons in each orbital must have different values of the spin quantum number. Pauli's assumption is known as the *Pauli exclusion principle*. The exclusion principle is profoundly related to the statistical behavior of many identical particles. A discussion of this point would lead us too far afield for our purposes.

Electron orbitals for the first few elements are shown in Figure 5-4. The electron configuration of helium, for example, is the same as the configuration of hydrogen, plus one electron added to complete the 1s orbital. Electron configurations of each succeeding atom are constructed by adding an electron to the lowest-energy, unoccupied electron orbital. The electron configuration of every known atom can be understood using this prescription.

5.2   MOLECULAR BIOLOGY

Molecules, or collections of atoms, are held together by interactions between electron orbitals. The modern theory of molecular bonding is

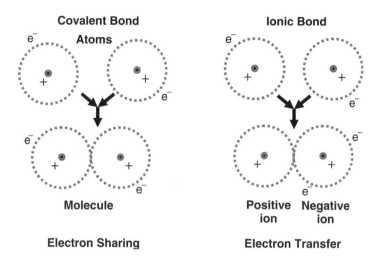

Figure 5-5. Molecular bonds.

known as the *molecular orbital theory*. According to molecular orbital theory, the energy of an atom is at a minimum when all of its outermost electron orbitals are filled. Helium and neon in Figure 5-4 are examples of filled electron orbitals. Both of these elements are chemically inert. To achieve this state, atoms with partially filled outer electron orbitals will share electrons, give up electrons, or take electrons from other atoms. For example, hydrogen will give up its single electron to fluorine in an explosive synthesis of the hydrogen fluoride molecule. Two hydrogen atoms and one oxygen atom can combine to form a water molecule. Table salt, sodium chloride, is a molecule with one atom of sodium and one atom of chlorine. Water and table salt represent two important types of bonds between atoms in a molecule: ionic bonds and covalent bonds (Figure 5-5).

The bonds between oxygen and hydrogen atoms in water are covalent: they are a sharing of electrons. Electrons are shared when their orbitals overlap in space. This overlap represents a probability of finding a shared electron in the orbital of its host atom or in the orbital of the bound atom.

Sodium chloride bonds are ionic: sodium gives up one electron and becomes positively charged while chlorine accepts the sodium electron and becomes negatively charged. The opposite charges of the ionized sodium and chlorine atoms result in an electromagnetic attraction. This attraction is the ionic bond. The ionic bond can be broken when the ionic molecule dissolves in solution. A simple solution consists of a lesser amount of a substance called the *solute* dissolved in a greater amount of a substance

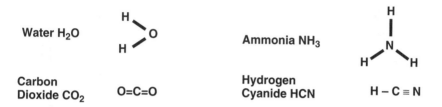

ATOMS

Hydrogen	H
Oxygen	O
Nitrogen	N
Carbon	C

DIATOMIC MOLECULES

Hydrogen H_2	H–H	Single Bond
Oxygen O_2	O=O	Double Bond
Nitrogen N_2	N≡N	Triple Bond

SIMPLE MOLECULES

Water H_2O

Ammonia NH_3

Carbon Dioxide CO_2 O=C=O

Hydrogen Cyanide HCN   H – C ≡ N

Figure 5-6. Simple molecules.

called the *solvent*. Water is a good solvent for dissolution of ionic molecules. The resulting ionic solution can respond to externally applied electric and magnetic fields. This observation has several applications in fields ranging from electrochemistry to reservoir engineering. For example, resistivity measurements can be used to determine the saturation of brine, a salt-bearing aqueous solution, in rock. In electrochemistry, the Voltaic cell is an ionic solution that reacts with a metal strip to generate electrical energy. Voltaic cells are commonly found in batteries.

Examples of molecules are shown in Figure 5-6. A molecular bond is denoted by a dash between the two bound atoms. Each dash represents the sharing of two electrons. The sharing of two or three electron pairs can form a molecular bond. These multiple bonds are referred to as double and triple bonds, and are depicted by two and three dashes, respectively. The oxygen molecule and the nitrogen molecule are examples of molecules with double and triple bonds.

Figure 5-7 includes sketches of three simple organic molecules. The basic difference between inorganic and organic chemicals is that organic chemicals are compounds of carbon. There are so many molecules with carbon that it has turned out to be worthwhile to differentiate the chemistry of carbon compounds from the chemistry of everything else. Organic chemistry is the chemistry of carbon compounds. Biochemistry, the chemistry

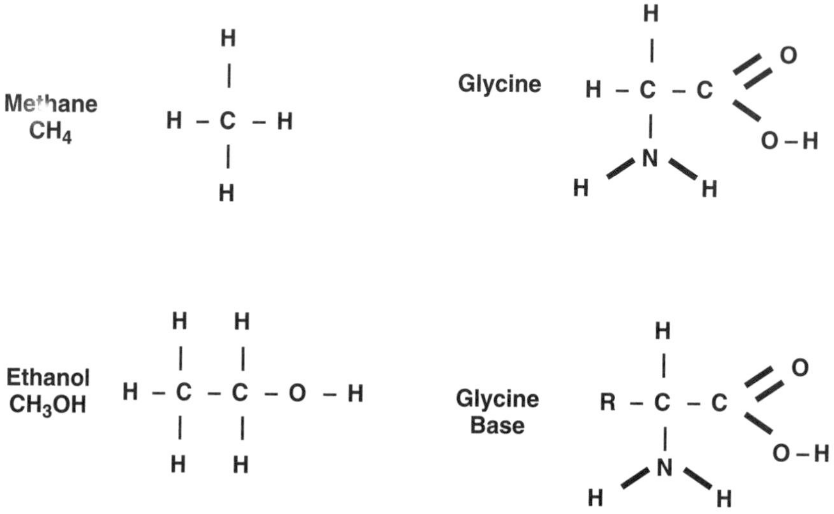

Figure 5-7. Organic molecules.

of life, can be considered a branch of organic chemistry. Molecular biology is a study of the chemistry of life on the molecular level.

Two of the most important classes of biochemicals are proteins and nucleic acids. Proteins are built up from a set of twenty smaller molecules called amino acids. Glycine, an example of an amino acid, is shown in Figure 5-7. Except for proline, every naturally occurring amino acid has the glycine base. The symbol R in the glycine base represents groups of atoms that make up different amino acids. Glycine is formed by replacing R with a hydrogen atom H. Substitution of a glycine hydrogen with other organic groups generates various amino acids. Some example substitutions and the resulting amino acid are listed in Table 5-1.

Table 5-1
Typical amino acids

R-Group	Amino acid
Hydrogen H	Glycine
Methyl group CH_3	Alanine
CH_2OH	Cerine
$C_4H_8NH_2$	Lycine

We can form a bond between two amino acids by combining a hydroxyl (OH) group from one amino acid with a hydrogen atom from another amino acid. The result is the formation of water and a bond between the two amino acids. Repetition of this process leads to the formation of long chains of amino acids. These chains, or polymers, of amino acids are proteins. Enzymes are an important class of proteins. They function as catalysts in the cell. As such, enzymes can speed up a chemical process without being changed by the process. Enzymes are necessary for the replication of nucleic acids.

There are basically two types of nucleic acids: deoxyribonucleic acid (DNA), and ribonucleic acid (RNA). DNA is a large molecule formed from combinations of four smaller molecules attached to a sugar (deoxyribose phosphate) base. A sketch of a DNA molecule is presented in Figure 5-8. DNA exists as a double helix. Each helical strand is a chain of deoxyribose phosphate molecules. Attached to the helical strand are organic molecules

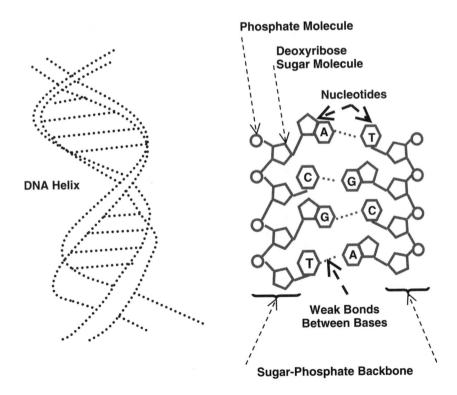

Figure 5-8. DNA.

known as purines (adenine and guanine) and pyrimidines (thymine and cytosine). A purine can pair with a pyrimidine to form a bond between two helical strands to form the DNA double helix. The letters A and G are used to denote the purines adenine and guanine in Figure 5-8, and the letters T and C are used to denote the pyrimidines thymine and cytosine.

Nucleic acid is the genetic material in the cell. The nucleic acid codes the function of the cell and controls reproduction. Ribosomes in the cell translate the genetic code carried by nucleic acids into proteins. The rules governing nucleic acid bonding are known as the Watson-Crick rules after their discoverers, American biophysicists Francis Crick and James Watson. Crick and Watson were the first to deduce the molecular structure of the DNA molecule. They found that adenine attached only to thymine, and guanine paired only with cytosine. The spatial ordering of adenine-thymine and guanine-cytosine pairs along the DNA structure is the identifying characteristic of a DNA molecule. The RNA molecule is the template by which DNA replication occurs.

RNA is a single helical structure. Unlike DNA, which is found primarily in the nucleus of a cell, RNA exists principally in the protein synthesizing ribosomes. Figure 5-9 illustrates the basic components of a cell and

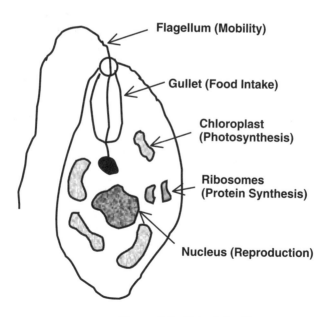

Figure 5-9. Unicellular life.

their functions. The cell is a biochemical factory encased in a porous membrane. The chemical content of RNA is similar to that of DNA. In addition to assisting the replication of DNA during cell division, RNA also plays an important role in the synthesis of cellular proteins. Cellular proteins are synthesized according to instructions in the DNA molecule. The DNA information cannot be used directly in protein synthesis; it must be communicated using an intermediary. RNA is the intermediary.

Reproduction of a unicellular organism, such as the ancestral flagellate sketched in Figure 5-9, requires accurate replication of the cellular DNA. Mutations, or mistakes in the replication of DNA, can occur. These mistakes make changes in a unicellular species possible. More complex, sexually active organisms are constantly making new combinations of genetic material. These new combinations, and mistakes in the combinations, provide mechanisms for the evolution of sexual organisms. The same mechanisms do not exist for asexual organisms. Mistakes in DNA replication are essential to a viable theory of evolution from simple unicellular organisms to complex multicellular organisms.

5.3   WHAT IS LIFE?

The need for a definition of life has acquired new urgency as humanity has learned more about the reproductive process and as we extend our reach to other planets. How will we know if life exists elsewhere in the universe if we do not know how to define life? How can we have confidence in laws that govern the reproductive rights of human beings when we do not know what life is?

Life is difficult to define.[2] Leslie Orgel defined a living entity as a CITROENS, a Complex, Information-Transforming Reproducing Object that Evolves by Natural Selection. A structure is considered living if it satisfies the following criteria: the object is complex and yet well defined by a substance like DNA; and the object can reproduce or is descended from reproducing objects.

In his search for life in the solar system as part of the National Aeronautics and Space Administration, Norman Horowitz presented a definition of life based on genetics. Horowitz says that life "is synonymous with the possession of genetic properties" [Horowitz, 1986, page 13]. He recognized two important genetic properties: self-replication, and mutation. In response to the questions of who or what is the designer of the living world, Horowitz says the designer is "the cumulative effects of natural selection acting on spontaneous mutations over long periods of time" [Horowitz, 1986, page 13].

Noam Lahav reviewed several historical definitions of life and adopted a "scientific" view. He said that living entities "are complex, far-from-equilibrium structures maintained by the flow of energy from sources to sinks" [Lahav, 1999, page 113]. According to Lahav, living entities can replicate, mutate, exchange matter and energy with their environment, and evolve.

The criteria for life should be characteristics that are common to all living systems. To be considered living, an entity must be able to reproduce. If the entity is sexually sterile, as mules are, the entity must be descended from other entities that are able to reproduce. Mistakes or genetic variations associated with the reproductive process are the mechanisms by which mutations occur. Mutations, in turn, are necessary for evolution. Another behavioral characteristic of a living system is the ability to acquire and process food. Food in this sense can be as simple as molecules or as complex as other organisms. A necessary part of the metabolic process, in addition to the acquisition and processing of food, is the removal of waste products. Every biological system can excrete waste products. A final trait is the ability of biological systems to respond to external stimuli. The degree of sophistication of actual responses depends on the type of external stimulus, and the complexity of the responding organism.

5.4   SPONTANEOUS GENERATION

Throughout most of history people believed that inanimate systems were somehow different from living organisms. They thought that living organisms contained a quality, such as soul or karma, that set them apart from both inanimate objects and even the remains of once living organisms. Is life endowed with a special quality beyond our grasp, or can we create living systems from inanimate materials?

The concept of spontaneous generation—the creation of life from inanimate materials—has existed since before the age of the ancient Greek philosopher Aristotle. Aristotle was a major advocate of spontaneous generation. He believed, for example, that frogs formed from mud. His views were widely accepted for two thousand years. Italian physician Francesco Redi mounted the first serious attack on the concept of spontaneous generation in 1668.

Prior to Redi's work, people believed that maggots formed from decaying red meat. Redi laid a piece of fine meshed gauze on the exposed surface of a piece of red meat. The gauze prevented the eggs of insects from settling on the meat. As long as the eggs of insects were isolated from the meat, no maggots formed. Redi's experiment challenged the validity of the concept of spontaneous generation. His work might have been more influential if Dutch naturalist Anton van Leeuwenhoek had not made a new discovery: the microscope. Using a microscope of his own design, Leeuwenhoek made careful drawings and descriptions of microbes. He reported his work to the British Royal Society in a series of letters dated as early as 1674. The observation of microbes led some to suggest that these tiny creatures, which appeared suddenly in decaying food, were generated spontaneously. Although spontaneous generation was in retreat, it had not yet been defeated.

French biochemist and microbiologist Louis Pasteur delivered what was once considered to be the final blow to the Aristotelian version of spontaneous generation. Through a series of clever experiments, Pasteur proved that microbes would not appear in decaying food unless the food was accessible to them or their parents. Pasteur reported his results in April 1864. The spontaneous generation of life, as understood by the ancient Greeks, ceased to be a viable theory.

Spontaneous generation remained dormant for sixty years. In 1923, the Russian biochemist A.I. Oparin revived the concept of spontaneous generation, but in a different form. He conceded that spontaneous generation does not occur on the modern earth, but hypothesized that conditions on the primitive earth were different from what they are today. He suggested that life could have emerged from inanimate material under the prevailing conditions of the primitive earth. British geneticist J.D.S. Haldane independently proposed a similar hypothesis. Is it possible that life arose from inanimate material on the primitive earth?

5.5   THE MILLER–UREY EXPERIMENT

One of the most important pieces of evidence in support of Oparin and Haldane's primordial spontaneous generation hypothesis was provided in 1953 by two American chemists, Stanley Miller and Harold Urey. Miller was a graduate student of Urey's when he performed an experiment of great significance to the development of a believable scientific theory of evolution. Miller's experiment used an apparatus similar to the one

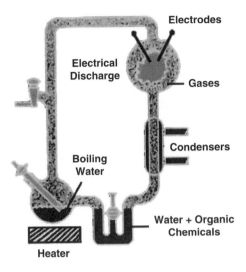

Figure 5-10. Miller–Urey experiment.

sketched in Figure 5-10. At the bottom left-hand side of the figure is a flask containing liquid water and three gases: hydrogen, ammonia, and methane. The fluids in the flask comprise a mixture known as the prebiotic soup. The prebiotic soup consists of inanimate materials that Oparin and Haldane suggested were the components of the earth's atmosphere prior to the appearance of life.

All of the prebiotic constituents—water, hydrogen, methane, and ammonia—are relatively simple compounds. In addition to hydrogen atoms, water contains oxygen, ammonia contains nitrogen, and methane contains carbon. Gases such as methane and ammonia could have been expelled into the atmosphere of the primitive earth by volcanic activity or by outgassing from the earth's crust. Water was present in the earth's oceans, and hydrogen existed in abundance throughout the universe.

In addition to the prebiotic soup, ultraviolet radiation from the sun and volcanic activity were some major heat sources present in the earth's atmosphere. The heater under the flask containing the prebiotic soup represents these heat sources. As the water boils, it vaporizes. The resulting steam mixes with the other gases and warms them. These heated gases are forced to flow through the path bounded by the glass tubes. Once the gases are thoroughly mixed, they are subjected to electrical discharges (sparks). Electrical discharges are supposed to represent energy sources such as lightning flashes, geothermal heat, ionizing radiation such as cosmic rays, or ultraviolet light irradiating the primitive earth. The products of this activity are condensed in the water-cooled condenser and recirculated. After running the experiment for a week, the spark was turned off and the end products analyzed. Among the chemical products were four naturally occurring amino acids: glutamic acid, glycine, alanine, and aspartic acid. Since the publication of the Miller–Urey results in 1953, other researchers have performed similar experiments. All twenty naturally occurring amino acids have been produced under prebiotic conditions. Another product of biological significance has been discovered: adenine. Adenine is one of the four basic constituents of nucleic acids.

Besides producing important biochemicals from simple molecules under prebiotic conditions, people have learned through experimental tests that it is unlikely that amino acids can be produced from an atmosphere such as the one we have today—that is, an atmosphere composed of oxygen, nitrogen, and carbon dioxide. This experimental observation provides evidence in support of Oparin and Haldane's conception of the atmosphere of the primitive earth. Development of an oxygen-rich atmosphere did not occur until oxygen-producing organisms had evolved.[3] Oxygen began to accumulate in the atmosphere about two billion years ago.

Attempts to synthesize the molecules of life from inorganic constituents on a primordial earth have had some success.[4] Cyril Ponnamperuma of the University of Maryland has performed experiments under primitive earth conditions in which amino acids are produced and link together to form polymers. Sydney Fox of the University of Miami has shown that these polymers acquire a spherical shape (called microspheres) when exposed to a watery environment similar to an ancient tidal pool. Despite these successes, there are many holes in our understanding of the evolutionary mechanisms leading from simple molecules to unicellular organisms. Chemist Robert Shapiro presented a critique of the molecular origin of life paradigm. People do not know, for example, how the simplest cell membranes first evolved. Nor has anyone been able to synthesize even the simplest living organism in the laboratory. This has led some to consider yet another possible source of life on earth: an extraterrestrial source.

The Oparin–Haldane–Miller–Urey model assumes the atmosphere of the primordial earth was reducing—that is, the atmosphere was hydrogen-rich. If the earth's atmosphere was not reducing, and the approach by Miller and Urey is wrong, then where did life come from? One possibility is the theory of Panspermia.[5]

PANSPERMIA

The theory of Panspermia is based on the hypothesis that life was brought to Earth from outer space. Swedish chemist Svante Arrhenius (1859–1927) introduced the theory of Panspermia in 1908 because of the inability of science to explain the origin of life at the beginning of the twentieth century. By the middle of the twentieth century, people thought the Miller–Urey experiment provided the necessary mechanism for the origin of life, but questions about its validity have been raised in the context of a few plane formation models. Crick and Orgel took the concept of Panspermia one step further and suggested that life on earth was seeded by a space traveling civilization. Although the reducing atmosphere hypothesis assumed by Miller and Urey has not been disproved, other alternatives, like Panspermia, are being considered while the evidence is being gathered. Is there any evidence in support of Panspermia?

On September 28, 1969, a meteorite landed near the town of Murchison, Australia. Pieces of the meteorite were carefully collected and analyzed. The analysis found relatively large amounts of the same natural amino acids produced in the Miller–Urey experiment. This discovery, based as it was on very careful analyses designed to minimize the possibility of earthly contamination, demonstrated that some of the building blocks of life exist elsewhere in the universe. The Murchison meteorite, as it is called, provided encouragement to proponents of the belief that life on earth may have originated somewhere in outer space, and that extraterrestrial life exists. Another rock found in the Antarctic was identified as a meteorite from Mars in 1996. Study of the rock revealed mineralogy and structures that some researchers attributed to biological activity, presumably while still on Mars. Alternative explanations are possible that do not require such an extrapolation of normal science, and other researchers have not been convinced. The question of the existence of extraterrestrial life is still unresolved.

THE DRAKE EQUATION

Frank Drake of Cornell University developed an equation in the 1950s for estimating the probability of existence of extraterrestrial life. It is worth considering the possibility of extraterrestrial life further by examining the Drake equation in more detail. The effort is justified in a study of energy because many of the energy sources that are discussed in later chapters have their origin in life forms or the decay of life forms. If life exists or

Table 5-2
Variables in the Drake equation

Variable	Meaning
N	Number of intelligent civilizations
R	Rate of star formation per year
f_P	Fraction of stars with planets
n_E	Number of planets in the habitable zone of a star
f_L	Fraction of habitable planets where life does arise
f_i	Fraction of planets with intelligent life
f_C	Fraction of the lifetime of a planet that includes civilizations that can and will communicate
L	Length of time (in years) that a civilization communicates

has existed on other planets, then the possibility of finding energy sources associated with life also exists.

The Drake equation is [Trefil, 1985, page 171; Shermer, 2002]

$$N = R f_P\, n_E\, f_L\, f_i\, f_C L \tag{5.5.1}$$

The meanings of the variables in Equation (5.5.1) are presented in Table 5-2. Depending on your particular bias, you can use Drake's equation to show that extraterrestrial life is both likely and unlikely. Another Cornell University professor, astronomer Carl Sagan, used a variation of Drake's equation to show that extraterrestrial life is probable [Sagan, 1980, pp. 299–302]. Several other authors have used variations of Drake's equation to argue that life on earth is rare.[6]

Peter Ward and Donald Brownlee have argued that the probability of finding microbial life elsewhere in the universe may be high, but the probability of finding more complex life forms will be low. Ward and Brownlee proposed an alternative to the Drake equation [2000, page 270] that we write in the form

$$N = N_{MW} f_S f_P\, n_E\, f_L f_c f_i \tag{5.5.2}$$

The meanings of the variables in Equation (5.5.2) are presented in Table 5-3. Notice that Ward and Brownlee have taken a step back from estimating the number of planets with intelligent civilizations, and believe

Table 5-3
Variables in the Ward–Brownlee equation

Variable	Meaning
N	Number of planets with complex, multicellular organisms
N_{MW}	Number of stars in the Milky Way galaxy
f_S	Fraction of stars suitable for supporting life (Sun-like)
f_P	Fraction of stars with planets
n_E	Number of planets in the habitable zone of a star
f_L	Fraction of habitable planets where life does arise
f_C	Fraction of planets with complex multicellular organisms
f_l	Fraction of the lifetime of a planet that includes multicellular organisms

that it is more meaningful to ask how many planets are suitable for the evolution of complex, multicellular organisms they call metazoans.

The terms in Equations (5.5.1) and (5.5.2) are conceptually easy to understand, but they require making estimates of quantities that are poorly known. To illustrate the range of uncertainty, Sagan [1980, pg. 299] said the number of stars in the Milky Way galaxy was approximately 4×10^{11}. Twenty years later Ward and Brownlee [2000, pg. 267] said the number of stars in the Milky Way galaxy was approximately 3×10^8. The estimates differ by a factor of 1000. Presumably the more recent estimate is also the more correct estimate. If it is, it represents a significant decrease in the possibility of complex life forms evolving elsewhere in the galaxy. The other variables in Drake's equation are even more difficult to quantify with certainty. These uncertainties explain why estimates of N vary from millions to a handful. What we can say is that the basic building blocks of life, such as simple organic molecules like methane, do exist in space and may support the existence of some form of life on other celestial bodies besides earth.

Even if extraterrestrial life exists, it does not prove life on earth was brought here by a spore on a meteorite or by alien space travelers. Arguing that life on earth originated somewhere in outer space does not solve the problem of the origin of life; it merely shifts it from earth to outer space. Since scant evidence exists for the extraterrestrial origin of life, the most plausible scenario is still the origin of life from a prebiotic soup.

TRANSITIONS TO LIFE

Richard Cowen identified several conditions that he considered necessary for the origin of life.[7] Energy is needed to form complex

Table 5-4
Summary of transition to life

Step	Entity
1	Inorganic chemicals
2	Organic chemicals
3	Nucleic acids and proteins
4	Prokaryotes (no nucleus; e.g. bacteria)
5	Eukaryotes (nucleic acid in nucleus; e.g. amoeba)

organic molecules. Once formed, complex organic molecules need protection from strong radiation to prevent the disassociation of the molecule. Several media, such as water, ice, or rock, can provide the protection. The chemicals needed for life need to be concentrated. This can occur, for example, in a pool of water that evaporates after the chemicals of life have formed. Finally, catalysts are needed to facilitate the chemical reactions needed for the origin of life. Enzymes are organic catalysts.

Table 5-4 summarizes a plausible evolutionary sequence for the development of life from inorganic chemicals. It depicts an increasing level of chemical complexity as we move from inorganic chemicals to the simplest unicellular organisms. Bacteria are generally thought to be the simplest forms of life because bacteria contain both DNA and RNA. As we have seen, DNA replication requires RNA as an intermediary. The presence of both DNA and RNA in bacteria means that bacteria can reproduce without any external assistance. Viruses cannot reproduce without the help of a host organism because they contain only DNA or RNA. It is not clear whether viruses played a role in the development of the first unicellular organisms. Biologists generally believe that the evolutionary sequence depicted in Table 5-4 is essentially correct.

The set of steps shown in Table 5-4 can easily be enlarged as details are added and we consider more complex organisms. For example, Smith and Szathmáry [1999, pages 16–19] identified eight major transitions that are necessary to move from Step 3 in Table 5-4 to human society. Their transitions are summarized as follows:

- Transition 1: Replicating molecules combine to form collections of molecules in a compartment. The compartment containing replicating molecules becomes a protocell.
- Transition 2: Independent replicating molecules link together to form chains that become chromosomes.

- Transition 3: Nucleic acids (DNA and RNA) now store and transmit genetic information, while proteins catalyze chemical reactions. This is a complex process that must have evolved from a simpler process in which a molecule like RNA performed the functions of both a gene and an enzyme. At some point in the evolutionary sequence a division of labor must have occurred.
- Transition 4: Relatively simple bacterial cells evolved into cells with nuclei and intracellular structures called organelles. Mitochondria and chloroplasts are examples of organelles. This transition is a transition from Step 4 to Step 5 in Table 5-4—that is, prokaryotes evolved into eukaryotes.
- Transition 5: Asexual clones evolve into populations of sexually active cells.
- Transition 6: Single-celled organisms evolve into multicellular animals, plants, and fungi.
- Transition 7: Individual organisms form colonies.
- Transition 8: Primate societies evolve into human societies and language originates.

This set of transitions is not unique, which is why Smith and Szathmáry qualified the transitions as "major." Many other transitions are required to achieve the major transitions, and the term *evolve* allows us to move from one step to another without actually knowing all of the details of the transition process.

5.6 PHOTOSYNTHESIS

Organisms require two things from their environment: energy, and chemicals [Mihelcic, 1999, Section 5.3]. Energy is used to provide power and chemicals provide substance, such as food to fuel the organism. Chemical elements are cycled through an ecosystem, but energy is used and eventually dissipated as heat.

The energy needed for biochemical processes is provided by the chemical decomposition of relatively simple organic molecules such as sugars and fats. On a more fundamental level, the primary source of energy for the biosphere as a whole is the sun. The sun is often considered the ultimate source of energy used by all organisms on earth. The sun may be either the direct or indirect source of energy for life on the surface of the earth. Gold [1999] has suggested that a biosphere exists within the earth that receives energy from radioactivity and heat inside the earth.

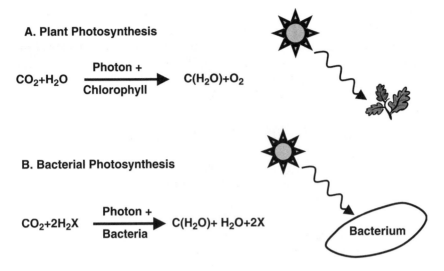

A. Plant Photosynthesis

$$CO_2+H_2O \xrightarrow[\text{Chlorophyll}]{\text{Photon +}} C(H_2O)+O_2$$

B. Bacterial Photosynthesis

$$CO_2+2H_2X \xrightarrow[\text{Bacteria}]{\text{Photon +}} C(H_2O)+ H_2O+2X$$

Bacterium

Figure 5-11. Photosynthesis.

Sunlight, or electromagnetic energy from the sun, is converted to more biologically useful sources of energy, such as chemical energy, by the process of photosynthesis. We consider two types of photosynthesis: green plant photosynthesis, and bacterial photosynthesis. Figure 5-11 illustrates the two photosynthetic processes.

Photosynthesis occurs in chloroplast membranes. Plant pigments, especially chlorophyll, convert light energy to chemical energy through the process of plant photosynthesis shown in Figure 5-11A. Chlorophyll is a mixture of two compounds known as chlorophyll a and chlorophyll b. The chlorophyll compounds are examples of photosensitive chemicals. Plant cells containing chlorophyll are green because the chlorophyll mixture reflects green light. When a photosensitive molecule absorbs light, electrons in the molecule are boosted to a higher energy state where they are more loosely bound and can be captured by other molecules. If chlorophyll loses these energetic electrons, it replaces them by dissociating water molecules. The oxygen atoms from two dissociated water molecules combine to form diatomic oxygen as the gas O_2. The photosynthetic reaction turns water and carbon dioxide into sugars and oxygen. The term $C(H_2O)$ represents organic molecules with the general formula $C_nH_{2n}O_n$ such as the sugar glucose. Glucose has the formula $C_6H_{12}O_6$ and corresponds to $6\times C(H_2O)$.

The efficiency of the conversion of light energy to chemical energy by photosynthesis is about 14% [Sørenson, 2000, pages 300–302]. This overall efficiency accounts for the fraction of frequencies in the solar spectrum that can be used in photosynthesis, the efficiency of collecting the incident electromagnetic radiation from the sun, and the efficiency of the chemical process that converts light energy to chemical energy.

RESPIRATION AND REDOX REACTIONS

The process of photosynthesis converts light energy to stored chemical energy. The process of respiration converts stored chemical energy to energy that can be used by an organism. The efficiency of the respiration process ranges from 5% to 50% [Mihelcic, 1999, page 238].

Respiration is a redox, or oxidation-reduction, reaction. In a redox reaction, an electron donor transfers one or more electrons to an electron acceptor. The electron donor is oxidized and its valence state becomes more positive. The electron acceptor is reduced and its valence state becomes more negative. During the process of respiration, carbon atoms lose electrons and are oxidized while oxygen atoms accept electrons and are reduced.

The redox reaction in the respiration process may be considered a two-step process. The first step is the oxidation reaction

$$C(H_2O) + H_2O \rightarrow CO_2 + 4H^+ + 4e^- \tag{5.6.1}$$

The four electrons are released when the valence state of carbon changes from 0 in the hydrocarbon molecule to 4+ in carbon dioxide. The second step in the redox reaction is the reduction reaction

$$O_2 + 4H^+ + 4e^- \rightarrow 2H_2O \tag{5.6.2}$$

In this step, the four electrons from the oxidation reaction change the valence state of oxygen from 0 in the oxygen molecule to 2− in the water molecule. The combined redox reaction may be written as

$$C(H_2O) + O_2 \rightarrow CO_2 + H_2O + \varepsilon_{redox} \tag{5.6.3}$$

where ε_{redox} is the energy released during the reaction. Although the number of electrons is unchanged, the electron distribution is changed.

BACTERIAL PHOTOSYNTHESIS

Plants are not the only organisms that can convert sunlight to useful chemical energy. Some types of bacteria can use sunlight to dissociate compounds of the form H_2X with the reaction shown in Figure 5-11B. Sulphur bacteria can dissociate hydrogen sulfide H_2S, and fermentation bacteria can dissociate ethanol C_2H_5OH. Most photosynthetic bacteria can absorb light in the infrared region [Sørenson, 2000, page 302]. The infrared region has wavelengths in the range $800-1000 \times 10^{-9}$ m and is associated with heat energy.

An important difference between plant photosynthesis and bacterial photosynthesis is the way converted light energy is used. Plant photosynthesis converts light energy to stored chemical energy for later use in respiration. By contrast, photosynthetic bacteria consume most of the light energy and do not store much of the converted energy as chemical energy.

Two respiration processes have been identified in bacterial photosynthesis. The aerobic respiration process consumes molecular oxygen O_2. The anaerobic respiration process does not use oxygen. Anaerobic bacteria can use electron acceptors such as nitrate (NO_3^-), manganese (Mn^{4+}), or ferric ion (Fe^{3+}) [Mihelcic, 1999, page 239] in place of oxygen.

The role of anaerobic bacteria in the energy industry may only be partially understood. Anaerobic bacteria can digest polymers—long chain organic molecules—that are useful for producing oil. This is a negative effect because of the expense associated with polymer injection. On the other hand, the leading theory of oil and gas formation is a theory of biological decay. Anaerobic bacteria may contribute to a hypothesis that is challenging this theory: the hypothesis that a deep, hot biosphere exists in the earth [Gold, 1999]. These ideas are discussed in the next section.

5.7   ORIGIN OF FOSSIL FUELS

Fossil fuels are sources of energy that were formed by the death, decay, and transformation, or diagenesis, of life.[7] The term *diagenesis* encompasses physical and chemical changes that are associated with lithification and compaction. Sediment can be *lithified*, or made rocklike, by the movement of minerals into sedimentary pore spaces. The minerals can form cement that binds grains of sediments together into a rocklike structure that has less porosity than the original sediment in a process called *cementation*. *Compaction* is the process of applying pressure to compress the rocklike structure.

Much of the scenario of the evolution of life presented thus far is based on empirical observations. Though the observations are firmly established, their interpretation is not. It is the interpretation of observations and the intellectual leaps across gaps in the observational evidence that are a principal source of controversy among competing scientific theories. We illustrate a source of controversy in this section by considering two theories of the origin of fossil fuels: the biogenic theory, and the abiogenic theory.

BIOGENIC THEORY

The biogenic theory is the mainstream scientific view of the origin of fossil fuels. In the biogenic theory, a type of biochemical precipitation called organic sedimentation forms coal, oil, and gas. When vegetation dies and decays in aqueous environments such as swamps, it can form a carbon-rich organic material called peat. If peat is buried by subsequent geological activity, the buried peat is subjected to increasing temperature and pressure. Peat can eventually be transformed into coal by the process of diagenesis. A similar diagenetic process is thought to be the origin of oil and gas.

Oil and gas are petroleum fluids. A petroleum fluid is a mixture of hydrocarbon molecules and inorganic impurities, such as nitrogen, carbon dioxide, and hydrogen sulfide. Petroleum can exist in solid, liquid, or gas form, depending on its composition and the temperature and pressure of its surroundings. Natural gas is typically methane with lesser amounts of heavier hydrocarbon molecules such as ethane and propane. The elemental mass content of petroleum fluids ranges from approximately 84% to 87% carbon and 11% to 14% hydrogen, which is comparable to the carbon and hydrogen content of life. This is one piece of evidence for the origin of petroleum from biological sources.

The biochemical process for the formation of petroleum is illustrated in Figure 5-12. It begins with the death of microscopic organisms such as algae and bacteria. The remains of the organisms settle into the sediments at the base of an aqueous environment as organic debris. Lakebeds and seabeds are examples of favorable sedimentary environments. Subsequent sedimentation buries the organic debris. As burial continues, the organic material is subjected to increasing temperature and pressure, and is transformed by bacterial action into oil and gas. Petroleum fluids are usually less dense than water and will migrate upwards until they encounter impermeable barriers and are collected in traps. The accumulation of hydrocarbon in a geologic trap becomes a petroleum reservoir.

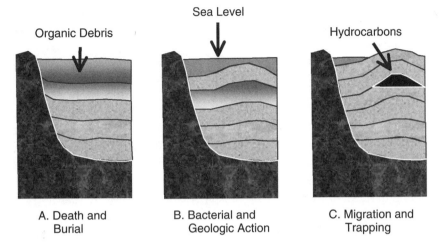

Figure 5-12. Biogenic origin of oil and gas.

ABIOGENIC THEORY

In the biogenic theory, the origin of oil and gas begins with the death of organisms that live on or near the surface of the earth. An alternative hypothesis called the abiogenic theory says that processes deep inside the earth, in the earth's mantle, form petroleum. Thomas Gold, an advocate for the abiogenic theory, pointed out that the biogenic theory was adopted in the 1870s. At the time, scientists thought that the earth was formed from molten rock that was originally part of the sun. In 1846, Lord Kelvin, (William Thomson) estimated the age of the earth from the rate of cooling of molten rock to be about 100 million years old. French physicist Antoine Henri Becquerel's discovery of radioactivity in 1896 provided a means of estimating the age of the earth from the concentration of long-lived radioactive materials in rock. In 1905, Ernest Rutherford proposed using radioactivity to measure the age of the earth. The earth is now believed to be over four billion years old.

Scientists now believe that the earth was formed by the accumulation and compression of cold nebular material, including simple organic molecules. Gold argues that simple inorganic and organic molecules in the accreting earth were subjected to increasing heat and pressure, and eventually formed more complex molecules. The interior of the earth is viewed by proponents of the abiogenic theory as the crucible for forming life. According to this point of view, the Miller–Urey experiment can be considered a model of the conditions that existed in the earth's mantle. The electrical discharge that represents lightning in the Miller–Urey experiment would be considered a heat source in the abiogenic theory.

Gold [1999] refutes challenges to the abiogenic theory and presents several pieces of evidence in support of the abiogenic theory, and the possible existence of a biological community deep inside the earth. Some of Gold's evidence includes the existence of microbial populations that can thrive in extreme heat. These microbes, notably bacteria and archaea, grow at hot, deep ocean vents and can feed on hydrogen, hydrogen sulfide, and methane. Gold considers life forms at deep ocean vents transitional life forms that exist at the interface between two biospheres.

One biosphere is the surface biosphere and includes life that lives on the continents and in the seas on the crust of the earth. Gold postulates that a second biosphere exists in the mantle of the earth. He calls the second biosphere the deep biosphere. The surface biosphere uses chemical energy extracted from solar energy, but the deep biosphere feeds directly on chemical energy. Oxygen is a requirement in both biospheres. Gold's deep biosphere is the source of life that eventually forms hydrocarbon mixtures, petroleum, in the earth's mantle. Crustal oil and gas reservoirs are formed by the upward migration of petroleum fluid until the fluid is stopped by impermeable barriers and accumulates in geological traps. The abiogenic theory is illustrated in Figure 5-13. If the abiogenic theory is correct,

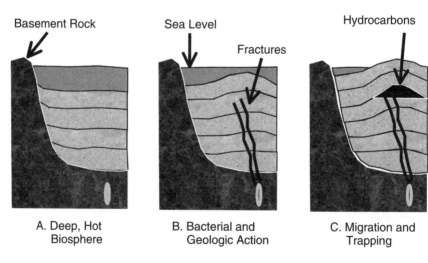

Figure 5-13. Abiogenic origin of oil and gas.

existing estimates of the volume of petroleum, which assume petroleum is a finite resource located in the crust of the earth, could be significantly understated.

ENDNOTES

1. A discussion of basic biology can be found in a variety of sources, such as Kimball [1968], Attenborough [1979], Arms and Camp [1982], Wallace [1990], and Purves, et al. [2001].
2. Many authors have proposed a definition for life. We consider the definitions presented in Orgel [1973, pg. 193], Horowitz [1986, pg. 13], Lahav [1999, pg. 113] and Smith and Szathmáry [1999, pg. 3] as representative examples. Lahav [1999] presents several other definitions of life in his Appendix A (pp. 117–121).
3. Moore [1971], Folsom [1979], Gould [1993, 2002], Ridley [1996], and Purves, et al. [2001] are basic references for much of the material on evolution. Darwin's views can be obtained directly from his own writings [for reprints, see Darwin, 1952 and 1959].
4. Brackman [1980] gives a fascinating account of "The Strange Case of Charles Darwin and Alfred Russel Wallace." Brackman argues that Wallace received less credit than he deserved because of Darwin's connections with the scientific establishment of his time.
5. See Winchester [2001] for a recent account.
6. See Gould [1989].
7. For a more detailed discussion of the formation of fossil fuels, see Selley [1998, Chapter 5] and Gold [1999, especially Chapter 3]. Geologic terms are explained further in Selley [1998], and Press and Siever [2001]. For accounts of the history of the age of the earth, see Hellman [1998, Chapter 6], and Press and Siever [2001].

EXERCISES

5-1. A conservative force $\vec{F}$ can be calculated from potential energy V using the relationship $\vec{F} = -\nabla V$. Calculate the Coulomb force from the Coulomb potential in Equation (5.1.1).

5-2. A conservative force $\vec{F}$ can be calculated from potential energy V using the relationship $\vec{F} = -\nabla V$. Calculate the nuclear force from the Yukawa potential in Equation (5.1.2).

5-3. Suppose the range of the nuclear force equals the diameter of the nucleus, and a nucleus has a diameter of 1.2×10^{-15} m. Use the value of range and Equation (5.1.4) to estimate the mass of the exchanged particle.

5-4. Consider a nucleus with a charge Zq and one electron. Examples are the helium ion He^+ with $Z = 2$ or lithium ion Li^{2+} with $Z = 3$. What is the energy of a photon emitted when an electron makes a transition from state i to state j? Hint: Repeat the Bohr model calculation with the Coulomb potential modified by the charge Zq of the nucleus.

5-5. A. Abell, et al. [1991, Section 28.4] made the following "optimistic estimates" for the variables in the Drake equation: $R = 10 \text{ year}^{-1}$, $f_P = 1, n_E = 1, f_L = 1, f_i = 0.01, f_C = 0.1$. Calculate the value of N for $L = 100$ years and $L = 10^6$ years.
B. Ward and Brownlee [2000, page 268] say that $n_E \approx 0.06$. Calculate N for $L = 100$ years and $L = 10^6$ years.

5-6. How much energy is absorbed during bacterial photosynthesis from an infrared photon with wavelength $\lambda = 1$ micron $= 1 \times 10^{-6}$ m? Assume 100% of the photon energy is absorbed.

5-7. Assume the efficiency of photosynthesis is approximately 14%. How many infrared photons with wavelength $\lambda = 0.8$ micron $= 0.8 \times 10^{-6}$ m must be absorbed to provide 1 J of energy?

5-8. A. Suppose a farm animal with a mass of 750 kg eats 1 kg of feed per day for each 50 kg of body mass. How much feed (in kg/day) does the farm animal need each day?
B. Calculate the food energy used per day by the farm animal in Part A if the energy content of the feed is 16 MJ/kg.
C. If the work done by the farm animal in Part A could help feed 10 people, how much energy per person per day was provided by the animal?

5-9. What is the energy of blue light with a wavelength equal to 475 nm? Express your answer in Joules.

5-10. Calculate the wavelength of an electron with a velocity of $0.01c$ where c is the speed of light.

CHAPTER SIX

Fossil Energy

The origin of fossil fuels was discussed in Chapter 5. Fossil energy is the energy obtained from the combustion of fossil fuels. Our goal here is to review the history of fossil fuels[1] before discussing some of the most important fossil fuels: coal, oil, and natural gas. We complete the chapter with an introduction to nonconventional fossil fuels.

6.1 THE HISTORY OF FOSSIL FUELS

Fossil fuels have been used by civilizations for millennia. Coal was the first fossil fuel to be used on a large scale. Nef [1977] describes sixteenth century Britain as the first major economy in the world that relied on coal. Britain relied on wood before it switched to coal. The transition in the period between about 1550 C.E. and 1700 C.E. was necessitated by the consumption and eventual deforestation of Britain. Coal was an alternative for wood and was the fuel of choice during the Industrial Revolution. It was used to boil steam for steam turbines and steam engines. Coal was used in transportation to provide a combustible fuel for steam engines on trains and ships. The introduction of the internal combustion engine has allowed oil to replace coal as a fuel for transportation. Coal is used today to provide fuel for many coal-fired power plants.

People have used oil since the earliest civilizations in the Middle East, such as Egypt and Mesopotamia, about 3000–2000 B.C.E. During that period, oil was collected in small amounts from surface seepages. It was used in constructing buildings, waterproofing boats and other structures, setting jewels, and mummification. Arabs began using oil to create incendiary weapons as early as 600 C.E. By the 1700s, oil was being used in Europe, lighting streets in Modena, Italy, and helping candle makers in Scotland with their trade [Shepherd and Shepherd, 1998].

Daniel Yergin [1992, page 20] selected George Bissell of the United States as the person most responsible for creating the modern oil industry.

Bissell realized in 1854 that rock oil—as oil was called in the nineteenth century to differentiate it from vegetable oil and animal fat—could be used as an illuminant. He gathered a group of investors together in the mid-1850s. The group formed the Pennsylvania Rock Oil Company of Connecticut and selected James M. Townsend to be its president.

Bissell and Townsend knew that oil was sometimes produced along with water from water wells. They believed that rock oil could be found below the surface of the earth by drilling for oil in the same way that water wells were drilled. Townsend commissioned Edwin L. Drake to drill a well in Oil Creek, near Titusville, Pennsylvania. The location had many oil seepages. The project began in 1857 and encountered many problems. By the time Drake struck oil on Aug. 27, 1859, a letter from Townsend was en route to Drake to inform him that funds were to be cut off [van Dyke, 1997].

Drake's well caused the value of oil to increase dramatically. Oil could be refined for use in lighting and cooking. The substitution of rock oil for whale oil, which was growing scarce and expensive, reduced the need to hunt whales for fuel to burn in lamps. Within fifteen months of Drake's strike, Pennsylvania was producing 450,000 barrels a year from 75 wells. By 1862, three million barrels were being produced and the price of oil dropped to ten cents a barrel [Kraushaar and Ristinen, 1993].

Industrialist John D. Rockefeller began Standard Oil in 1870 and by 1900 the company held a virtual monopoly over oil production in the United States. It took an act of Congress, the Sherman Anti-Trust Act, to break Rockefeller's grip on the oil industry, but not before oil made him famous and rich.

In 1882, the invention of the electric light bulb caused a drop in the demand for kerosene. The drop in demand for rock oil was short-lived, however. The quickly expanding automobile industry needed oil for fuel and lubrication.

The Pennsylvania oil fields provided a relatively small amount of oil to meet demand. New sources of oil were discovered in the early twentieth century. Oil was found in Ohio and Indiana, and later in the San Fernando Valley in California and near Beaumont, Texas. By 1909, the United States produced more oil than all other countries combined, producing half a million barrels per day. Up until 1950, the United States produced more than half of the world's oil supply. Discoveries of large oil deposits in Central and South America and the Middle East led to decreased United States production. Production in the United States peaked in 1970 and has since been declining. However, oil demand in the United States and elsewhere in the world has continued to grow. Since 1948, the United States has imported more oil than it exports. Today, the United States imports about half of its oil needs.

Until 1973, oil prices were influenced by market demand and the supply of oil that was provided in large part by a group of oil companies called the "Seven Sisters." This group included Exxon, Royal Dutch/Shell, British Petroleum (BP), Texaco, Mobil, Standard Oil of California (which became Chevron), and Gulf Oil. In 1960, Saudi Arabia led the formation of the Organization of Petroleum Exporting Countries, commonly known as OPEC. It was in 1973 that OPEC became a major player in the oil business by raising prices on oil exported by its members. This rise in price became known as the "first oil crisis" as prices for consumers in many countries jumped.

Today, fossil fuels are still the primary fuels for generating electrical power, but society is becoming increasingly concerned about the global dependence on finite resources and the environmental impact of fossil fuel combustion. Measurements of ambient air temperature show a rise in the average temperature of the earth's atmosphere. The rising temperature is called global warming and is attributed in large part to the emission of fossil fuel combustion byproducts into the atmosphere. Global warming is discussed in more detail in Section 7.3.

6.2 COAL

Coal is formed from organic debris by a process known as coalification.[2] When some types of organic materials are heated and compressed over time, they can form volatile products (water and gas) and a residual product (coal). In some cases, a high-molecular-weight, waxy oil is also formed. For example, swamp vegetation may be buried under anaerobic conditions and become peat. Peat is an unconsolidated deposit of partially carbonized vegetable matter in a water-saturated environment, such as a bog. If peat is overlaid by rock and subjected to increasing temperature and pressure, it can form coal.

Organisms that form coal when subjected to coalification include algae, phytoplankton and zooplankton, and the bacterial decay of plants and, to a lesser extent, animals. Organic debris is composed primarily of carbon, hydrogen, and oxygen. It may also contain minor amounts of other elements such as nitrogen and sulfur. The organic origin of coal provides an explanation for the elemental composition of coal, which ranges from $C_xH_yO_z$...S... to pure carbon where the ellipses denote other elements.

The organic material is used to create macerals, which are the basic building blocks of kerogen. Kerogen is an important constituent of coal. *Kerogen* is organic matter that is disseminated in sediments and is insoluble in a petroleum solvent such as carbon bisulfide. There are three types of macerals: vitrinite, liptinite, and inertinite. Vitrinite is derived from woody tissue. Liptinite is derived from the waxy, resinous part of plants. Inertinite is derived from degraded or burned plants. The type of maceral can be identified by its fluorescence. Vitrinite has a weak, brownish fluorescence; liptinite fluoresces under ultraviolet light; and inertinite does not fluoresce.

Coals are classified by rank. Rank is a measure of the degree of coalification or maturation in the series ranging from lignite to graphite. The lowest-rank coal is lignite, followed in order by sub-bituminous coal, bituminous coal, anthracite, and graphite. Coal rank is correlated to the maturity, or age, of the coal. As a coal matures, the ratio of hydrogen to carbon atoms and the ratio of oxygen to carbon atoms decrease. The highest-rank coal, graphite, approaches 100% carbon. Coal becomes darker and denser with increasing rank.

Coals burn better if they are relatively rich in hydrogen; this includes lower-rank coals with higher hydrogen to carbon ratios. The percentage of volatile materials in the coal decreases as coal matures. Volatile materials include water, carbon dioxide, and methane. Coal gas is gas absorbed in the coal. It is primarily methane with lesser amounts of carbon dioxide. The amount of gas that can be absorbed by the coal depends on the rank. As rank increases, the amount of methane in the coal increases because the molecular structure of higher-rank coals has a greater capacity to absorb gas and therefore can contain more gas. The Langmuir isotherm provides a measure of gas content G_C in coal. It has the form

$$G_C = V_L \frac{P}{P_L + P} \tag{6.2.1}$$

where P is pore pressure, V_L is the Langmuir volume, and P_L is Langmuir pressure. Gas content is usually expressed in standard cubic volume of gas per mass of coal, such as standard cubic meters of gas per tonne of coal.

Coals are organic sedimentary rocks. Figure 6-1 shows an idealized representation of the physical structure of a coal seam. A coal seam is the stratum or bed of coal. It is a collection of coal matrix blocks bounded by natural fractures. The fracture network in coalbeds consists of microfractures called cleats. An interconnected network of cleats allows coal gas to

Cleats

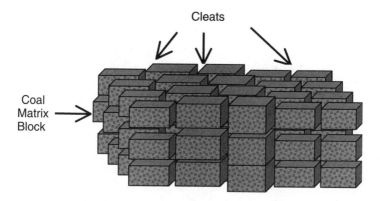

Coal
Matrix
Block

Figure 6-1. Schematic of a typical coal seam.

flow from the coal matrix blocks when the pressure in the fracture declines. This is an important mechanism for coalbed methane production.

Coalbeds are an abundant source of methane. Coalbed methane exists as a monomolecular layer on the internal surface of the coal matrix. Its composition is predominately methane, but can also include other constituents, such as ethane, carbon dioxide, nitrogen, and hydrogen. The outgassing of gas from coal is well known to coal miners as a safety hazard, and occurs when the pressure in the cleat system declines. The methane in the micropore structure of the coalbed is now considered a source of natural gas. Coal gas is able to diffuse into the natural fracture network when a pressure gradient exists between the matrix and the fracture network.

Gas recovery from coalbeds depends on three processes. Gas recovery begins with desorption of gas from the internal surface to the coal matrix and micropores. The gas then diffuses through the coal matrix and micropores into the cleats. Finally, gas flows through the cleats to the production well. The flow rate depends, in part, on the pressure gradient in the cleats and the density and distribution of cleats. The controlling mechanisms for gas production from coalbeds are the rate of desorption from the coal surface to the coal matrix, the rate of diffusion from the coal matrix to the cleats, and the rate of flow of gas through the cleats.

COAL EXTRACTION AND TRANSPORT

Coal is usually produced by extraction from coalbeds. Mining is the most common extraction method. There are several types of mining techniques. Some of the more important coal mining techniques are strip mining, drift mining, deep mining, and longwall mining. Strip mining is also known as surface mining, in which coal on the surface of the earth is extracted by scraping. Drift mines are used to extract coal from coal seams that are exposed by the slope of a mountain. Drift mines typically have a horizontal tunnel entrance into the coal seam. Deep mining extracts coal from beneath the surface of the earth. In deep mining, coal is extracted by mining the coal seam and leaving the bounding overburden layers and underburden layers undisturbed. One of the most common deep mining techniques is the room-and-pillar mining technique.

In the room-and-pillar mining technique, channels are cut into the coal seam. The roofs of the channels will collapse unless they are supported. Pillars of coal are left in place to function as supports for the channel roofs. Wooden beams can be used to provide additional support to prevent cave-ins. Several parallel channels are cut into a relatively wide seam to maximize the excavation of coal from the seam. The channels are connected by cross-channels. Air and ventilation shafts must be dug to allow coal gas, typically methane, to be removed from the rooms formed by excavation. If the coal seam is too deep for room-and-pillar mining, a technique called longwall mining can be used. In longwall mining, a cutting machine breaks coal from a panel of coal as it is moved back and forth across the exposed surface of the coal. A conveyor is used to bring the broken coal to the surface.

Coal is transported to consumers by ground transportation, especially by trains and, to a lesser extent, ships. A relatively inexpensive means of transporting coal is the coal slurry pipeline. Coal slurry is a mixture of water and finely crushed coal. Coal slurry pipelines are not widely used because it is often difficult to obtain rights of way for coal slurry pipelines that extend over long distances, particularly in areas where a coal slurry pipeline would compete with an existing railroad right of way.

6.3   PETROLEUM FLUIDS

The types of molecules that comprise a fluid define the composition of the fluid.[3] The state of a fluid can be either gas or liquid. A pure fluid consists of a single type of molecule, such as water or methane. If a fluid contains several types of molecules, it is a fluid mixture. Petroleum is a mixture of hydrocarbon molecules. In situ water, or water found in reservoir rock, usually contains dissolved solids in ionized form and may contain dissolved gases. The composition of a fluid may be specified as a

Table 6-1
Elemental composition of petroleum fluids

Element	Composition (% by mass)
Carbon	84%–87%
Hydrogen	11%–14%
Sulfur	0.6%–8%
Nitrogen	0.02%–1.7%
Oxygen	0.08%–1.8%
Metals	0%–0.14%

list of the molecular components contained in the fluid and their relative amounts.

The relative amount of each component in a mixture is defined as the concentration of the component. Concentration is expressed as a fraction, such as volume fraction, weight fraction, or mole fraction. *Volume fraction* is the volume of a component divided by the total volume of fluid; *weight fraction* is the weight of a component divided by the total weight of the fluid; and *mole fraction* is the number of moles of the component divided by the total number of moles of the fluid. The unit of concentration should be clearly expressed to avoid errors. The symbols $\{x_i, y_i, z_i\}$ are often used to denote the mole fraction of component i in the liquid phase, gas phase, and wellstream, respectively. The *wellstream* is the fluid mixture that is passing through the wellbore that connects the reservoir to the surface facilities.

Table 6-1 summarizes the mass content of petroleum fluids for the most common elements. Petroleum is predominantly hydrocarbon, which is expected given the biogenic and abiogenic theories of its origin discussed in Chapter 5. The actual elemental composition of a petroleum fluid depends on such factors as the composition of its source, reservoir temperature, and reservoir pressure.

Hydrocarbon molecules in petroleum fluids are organic molecules. We expect the molecules in petroleum fluids to be relatively unreactive and stable because they have been present in the fluid mixture for millions of years. If they were reactive or unstable, it is likely that they would have reacted or decomposed at some point in time and their products would be present in the petroleum fluid.

Paraffins, naphthenes, and aromatics are some of the most common molecules found in petroleum fluids. These molecules are relatively stable at typical reservoir temperatures and pressures. Paraffin molecules are represented by the general chemical formula C_nH_{2n+2} and are saturated hydrocarbons. A saturated hydrocarbon has a single covalent bond between each of its carbon atoms. Examples of paraffin molecules include methane (CH_4) and ethane (C_2H_6). Napthene molecules have the general chemical formula C_nH_{2n}. They are saturated hydrocarbons with a ring structure. An example of a napthene molecule is cyclopentane C_5H_{10}. The ring structure makes it possible to retain a single bond between carbon atoms with two fewer hydrogen atoms than in paraffin molecules. Each of the carbon atoms is bonded to two other carbon atoms and two hydrogen atoms. Aromatic molecules have one or more carbon rings and are unsaturated hydrocarbons—that is, they have multiple bonds between some of the carbon atoms. Benzene is an example of an aromatic molecule. The presence of two or more bonds between two carbon atoms can make an organic molecule reactive and unstable. Aromatic molecules have a ring structure that makes them relatively stable and unreactive.

FLUID TYPE

Separator gas–oil ratio (GOR) is a useful indicator of fluid type. Gas–oil ratio is the ratio of a volume of gas divided by a volume of oil at the same temperature and pressure. The volume of a petroleum mixture depends on changes in composition as well as changes in temperature and pressure. Separator GOR is the GOR at the temperature and pressure of a separator. A *separator* is a container that allows a fluid mixture to separate into distinct fluid phases. The unit SCF/STB in Table 6-2 refers to one standard cubic foot of gas divided by one stock tank barrel of oil. A *standard volume* is a volume at standard conditions of temperature and pressure. A *stock tank volume* is a volume at the temperature and pressure of the stock tank, which is the container used for storing produced oil. The third column in Table 6-2

Table 6-2
Rules of thumb for classifying fluid types

Fluid type	Separator GOR (SCF/STB)	Pressure depletion behavior in reservoir
Dry gas	No surface liquids	Remains gas
Wet gas	> 100,000	Remains gas
Condensate	3,000–100,000	Gas with liquid drop-out
Volatile oil	1,500–3,000	Liquid with significant gas
Black oil	100–1,500	Liquid with some gas
Heavy oil	0	Negligible gas formation

describes what happens to the fluid as the pressure in the reservoir declines during production. The decline in pressure is called *pressure depletion*.

Oil is often characterized in terms of its API gravity. API gravity is calculated from oil specific gravity γ_0 at standard temperature and pressure by the equation

$$\text{API} = \frac{141.5}{\gamma_0} - 131.5 \tag{6.3.1}$$

Specific gravity is the ratio of the density of a fluid divided by a reference density. Oil specific gravity is calculated at standard conditions using the density of fresh water as the reference density.

Heavy oils do not contain much gas in solution and have a relatively large molecular weight and specific gravity γ_0. By contrast, light oils typically contain a large amount of gas in solution and have a relatively small molecular weight and specific gravity γ_0. The equation for API gravity shows that heavy oil has a larger API gravity than light oil because heavy oil has a larger oil specific gravity γ_0 than light oil.

Gas specific gravity is calculated at standard conditions using air density as the reference density. The specific gravity of gas is

$$\gamma_g = \frac{M_a\,(\text{gas})}{M_a\,(\text{air})} \approx \frac{M_a\,(\text{gas})}{29} \tag{6.3.2}$$

where M_a is apparent molecular weight. Apparent molecular weight is calculated as

$$M_a = \sum_{i=1}^{N_c} y_i M_i \tag{6.3.3}$$

where N_c is the number of components, y_i is the mole fraction of component i, and M_i is the molecular weight of component i.

FLUID VOLUME

The volume of a fluid phase can have a sensitive dependence on changes in pressure and temperature. For example, gas formation volume factor is often determined with reasonable accuracy using the real gas equation of state $PV = ZnRT$ where n is the number of moles of gas in volume V at pressure P and temperature T. The gas is an ideal gas if gas compressibility factor $Z = 1$. The gas is a real gas if $Z \neq 1$. A quantity that accounts for the change of volume of a fluid phase is the formation volume factor. *Formation volume factor* is the ratio of the volume occupied by a fluid phase at reservoir conditions divided by the volume occupied by the fluid phase at surface conditions. Surface conditions are typically stock tank or standard conditions. Formation volume factors for oil and water usually range from 1 to 2 units of reservoir volume for each unit of volume at surface conditions. Gas formation volume factor varies over a wider range because gas volume is more sensitive to changes in pressure from reservoir to surface conditions. The real gas law shows that gas volume is inversely proportional to pressure.

GAS HEATING VALUE

The heating value of a gas can be estimated from the composition of the gas and heating values associated with each component of the gas. Heating value of the gas mixture H_m is defined as

$$H_m = \sum_{i=1}^{N_c} y_i H_i \tag{6.3.4}$$

where N_c is the number of components, y_i is the mole fraction of component i, and H_i is the heating value of component i. The heating value of a typical natural gas is often between 1000 BTU/SCF (3.72×10^7 J/m^3) to 1200 BTU/SCF (4.47×10^7 J/m^3). Heating values of molecular components in a mixture are tabulated in reference handbooks.

FLUID PHASE BEHAVIOR

The phase behavior of a fluid is generally presented as a function of the three variables pressure (P), volume (V), and temperature (T). The resulting PVT diagram is often simplified for petroleum fluids by preparing a pressure-temperature (P-T) projection of the PVT diagram. An example of a pressure-temperature (P-T) diagram is shown in Figure 6-2.

The P-T diagram in Figure 6-2 includes both single-phase and two-phase regions. The curve separating the single-phase region from the two-phase region is called the *phase envelope*. The pressures associated with the phase envelope are called *saturation pressures*. A petroleum fluid at a temperature below the critical point temperature T_c and pressures above the saturation pressure exists as a single-phase liquid. Saturation pressures

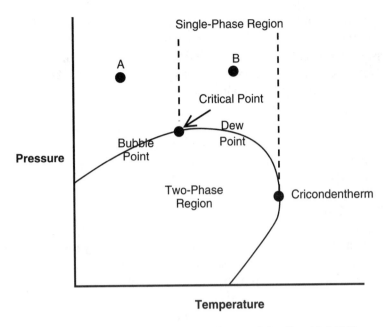

Figure 6-2. Pressure-temperature diagram [after Fanchi, 2001].

at temperatures below T_c are called *bubble point pressures*. If the pressure drops below the bubble point pressure, the single-phase liquid will make a transition to two phases: gas and liquid. At temperatures below T_c and pressures above the bubble point pressure, the single-phase liquid is called *black oil* (point A in Figure 6-2). If we consider pressures in the single-phase region and move to the right of the diagram by letting temperature increase towards the critical point, we encounter volatile oils.

The behavior of the petroleum fluid at temperatures above the critical point depends on the location of the cricondentherm. The *cricondentherm* is the maximum temperature at which a fluid can exist in both the gas and liquid phases. Reservoir fluids are called condensates if the temperature is less than the cricondentherm but greater than T_c (point B in Figure 6-2). A *condensate* is a gas at reservoir conditions, but it contains enough high-molecular-weight molecules that a liquid phase can form when the pressure in the reservoir drops below the saturation pressure. The saturation pressure of a condensate is called the *dew point pressure*. Gas reservoirs are encountered when reservoir temperature is greater than the cricondentherm.

Changes in phase behavior as a result of changes in pressure can be anticipated using the P-T diagram. Suppose a reservoir contains hydrocarbons at a pressure and temperature corresponding to the single-phase black oil region. If reservoir pressure declines at constant temperature, the reservoir pressure will eventually cross the bubble point pressure curve and enter the two-phase gas–oil region. A free gas phase will form in the two-phase region. Similarly, if we start with a single-phase gas condensate and allow reservoir pressure to decline at constant temperature, the reservoir pressure will eventually cross the dew point pressure curve to enter the two-phase region. In this case, a free phase liquid drops out of the condensate gas. Once liquid drops out, it is very difficult to recover. If the pressure declines further, some hydrocarbon mixtures will undergo retrograde condensation, that is, the condensate can reenter the gas phase.

The P-T diagram may also be applied to temperature and pressure changes in a wellbore. Reservoir fluid moves from relatively high temperature and pressure at reservoir conditions to relatively low temperature and pressure at surface conditions. As a result, it is common to see single-phase reservoir fluids become two-phase fluids by the time they reach the surface. If the change from single-phase to two-phase occurs quickly in the wellbore, which is common, then the fluid is said to have undergone a flash from one to two phases.

6.4   PETROLEUM EXPLORATION

Familiarity with theories of the earth's formation and concepts from petroleum geology are important prerequisites to finding hydrocarbon-bearing reservoirs. The formation of the earth was discussed in Chapter 4. Our focus here is on petroleum geology and geophysics, the underlying sciences of petroleum exploration.

FORMATION AND FACIES

The environment under which a rock forms is called the *environment of deposition*. If the depositional environment moves from one location to another during geologic time, it leaves a laterally continuous progression of rock that is distinctive in character. For example, a moving shoreline might leave a layer of quartz sandstone. These progressions of rocks can extend for hundreds of miles. If the progression is large enough to be mapped, it can be called a *formation*.

Formations are the basic descriptive units for a sequence of sediments. The formation represents a rock unit that was deposited under a uniform set of conditions at one time. The rock unit should be recognizable as a unit and it should be possible to map it, which implies a degree of continuity within the unit. A formation should represent a dominant set of depositional conditions even though the rocks in a formation may consist of more than one rock type and represent more than one type of depositional environment. If the different rock types within the formation can be mapped, they are referred to as members.

Formations can be a few feet thick, or they can be hundreds of feet thick. The thickness of a formation is related to the length of time an environment was in a particular location, and the amount of subsidence that occurred during that period.

A distinct rock sequence characterizes each depositional environment. A fluvial (river) environment may deposit sandstone. A deltaic environment may also deposit sandstone. However, the sandstone that is deposited in each environment has a very different character. For example, a fluvial system can deposit rocks that follow a meandering path from one location to another, while a delta tends to be fixed in place and deposits sediment at the mouth of the delta. In a fluvial system, the sandstones have certain characteristics. The grain size of sand deposited by a fluvial system becomes finer at shallower depths in a process called *fining upwards*. In addition, there is a coarser deposit at the base of the formation that distinguishes the fluvial system from a delta deposit.

The characteristics of rocks that can be used to identify their depositional environment are called *facies*. In the fluvial environment of deposition example, the characteristics of sandstone that indicate its environment of deposition also define it as a facies. A facies is distinguished by characteristics that can be observed in the field. If pebbly sandstone is seen independently of surrounding rocks, it could easily be classified into a fluvial environment. Adding information that it is surrounded by hemipelagic (deep sea) mud, a geologist may revise the interpretation to be a turbidite depositional environment in which sediments are suspended in a fast-flowing current. The turbidite depositional environment is associated with high-energy turbidity currents. Integration of all available information will lead to better characterization of the rock.

STRUCTURES AND TRAPS

Hydrocarbons can migrate along permeable pathways. Petroleum fluids are usually less dense than water and will migrate towards the surface in a process called *gravity segregation*. The migration is stopped when fluids

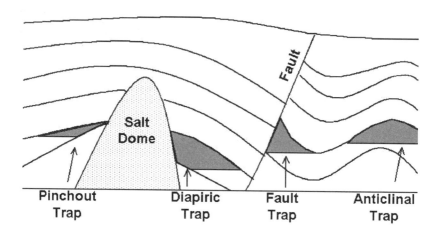

Figure 6-3. Traps.

encounter a barrier to flow. These barriers to flow are called *traps* and some of them are illustrated in Figure 6-3. Petroleum fluids will accumulate in traps. Traps are locations where the movement of petroleum fluid is stopped. There are two primary types of traps: structural traps, and stratigraphic traps. Structural traps occur where the reservoir beds are folded and perhaps faulted into shapes that can contain commercially valuable fluids such as oil and gas. Anticlines are a common type of structural trap.

Stratigraphic traps are the other principal kind of trap. Stratigraphic traps occur where the fluid flow path is blocked by changes in the character of the formation. The change in formation character must create a barrier to fluid flow that prevents hydrocarbon migration. Types of stratigraphic traps include a sand thinning out, or porosity reduction because of diagenetic changes. *Diagenesis* refers to processes in which the lithology of a formation is altered at relatively low pressures and temperatures when compared with the metamorphic formation of rock. Diagenesis includes processes such as compaction, cementation, and dolomitization. *Dolomitization* is the process of replacing a calcium atom in calcite (calcium carbonate) with a magnesium atom to form dolomite. The resulting dolomite is smaller than the original calcite and results in the formation of secondary porosity in carbonate reservoirs.

In addition to structural and stratigraphic traps, there are many examples of traps formed by a combination of structural and stratigraphic features. These traps are called combination traps. Another common trap in the Gulf of Mexico is a diapiric trap. The diapiric trap shown in Figure 6-3 was formed when relatively low-density salt moved upward and displaced higher-density sediments.

RESERVOIR FACTORS

The rock in a petroleum reservoir must have two important characteristics to qualify as an economically viable reservoir: porosity and permeability. Porosity and permeability were introduced in Chapter 4. Porosity is the ratio of the volume of void space to the total volume of rock plus pore space. Porosity is a factor that defines the capacity of the porous medium to store fluid. Permeability is a measure of interconnected pore space and has the unit of cross-sectional area. It represents the flow capacity of the rock. Sedimentary rocks are usually permeable, and igneous and metamorphic rocks are usually impermeable. Petroleum may sometimes be found in the fractures of fractured igneous rocks. If the igneous rocks are not fractured, they usually do not have enough interconnected pore space to form a pathway for petroleum to flow to a wellbore. Similar comments apply to metamorphic rocks.

Petroleum fluids are usually found in the pore space of sedimentary rocks. Igneous and metamorphic rocks originate in high pressure and temperature conditions that do not favor the formation or retention of petroleum fluids. Any petroleum fluid that might have occupied the pores of a metamorphic rock is usually cooked away by heat and pressure.

Several key factors must be present to allow the development of a hydrocarbon reservoir:

1. A source for the hydrocarbon must be present. For example, one source of oil and gas is thought to be the decay of single-celled aquatic life. Shales formed by the heating and compression of silts and clays are often good source rocks. Oil and gas can form when the remains of an organism are subjected to increasing pressure and temperature.
2. A flow path must exist between the source rock and reservoir rock.
3. Once hydrocarbon fluid has migrated to a suitable reservoir rock, a trapping mechanism becomes important. If the hydrocarbon fluid is not stopped from migrating, buoyancy and other forces will cause it to move toward the surface.
4. Overriding all of these factors is timing. A source rock can provide large volumes of oil or gas to a reservoir, but the trap must exist at the time oil or gas enters the reservoir.

EXPLORATION AND DEVELOPMENT GEOPHYSICS

Historically, geophysical techniques have been used to develop a picture of the large-scale structure of the subsurface prior to drilling. An image of the subsurface, including the structure of the reservoir, is obtained by initiating a disturbance that propagates through the earth's crust. The disturbance is reflected at subsurface boundary interfaces. The reflected signal is acquired, processed, and interpreted. The most common disturbance used in exploration geophysics is the seismic wave.

Seismic waves are vibrations, or oscillating displacements from an undisturbed position, that propagate from a source, such as an explosion or mechanical vibrator, through the earth. Energy sources such as dynamite or weight-dropping equipment are used to initiate vibrations in the earth. The energy released by the disturbance propagates away from the source of the disturbance as seismic waves. The two most common types of vibrations, or seismic waves, are compressional (P-) waves and shear (S-) waves. P-waves are longitudinal waves and S-waves are transverse waves. *Longitudinal waves* are waves in which the particles of the disturbed medium are displaced in a direction that is parallel to the direction of propagation of the wave. *Transverse waves* are waves in which the particles of the disturbed medium are displaced in a direction that is perpendicular to the direction of propagation of the wave. S-waves do not travel as quickly as P-waves and S-waves do not propagate through fluids. Compressional or P-waves are sometimes called primary or pressure waves, and shear or S-waves are sometimes called secondary waves.

Seismic waves are vibrations that propagate through the earth from a mechanical source. When the seismic wave encounters a reflecting surface, it is partially transmitted and partially reflected. A seismic reflection occurs at the interface between two regions with different acoustic impedances. Acoustic impedance Z is defined as the product of bulk density ρ_B and compressional velocity V_P, thus

$$Z = \rho_B V_P \qquad (6.4.1)$$

The acoustic impedance Z is also called compressional impedance because of its dependence on compressional velocity. A similar definition can be made using shear velocity to define shear impedance as

$$Z_S = \rho_B V_S \qquad (6.4.2)$$

where V_S is shear velocity. The reflection coefficient RC at the interface between two layers with acoustic impedances Z_1 and Z_2 is given by

$$RC = \frac{Z_2 - Z_1}{Z_2 + Z_1} \qquad (6.4.3)$$

Seismic images display contrasts in acoustic properties because the reflection coefficient depends on the difference between acoustic impedances in two adjacent layers. If the change in acoustic impedance is large enough, the reflected wave can be detected at the surface.

The ability to distinguish two features that are very close together is the essence of resolution. This becomes important in seismic measurements that are designed to obtain an image of two reflecting surfaces. Reflecting surfaces can be oriented vertically, horizontally, or at some angle in between. They can be within a few meters of each other or separated by much greater distances. These reflecting surfaces can represent facies changes, fluid contacts, or any change in acoustic impedance that is relevant to reservoir characterization. The quality of the seismic resolution will determine the usefulness of seismic surveys.

Receivers called seismometers are used to detect seismic waves. A *geophone* is a seismometer used on land and a *hydrophone* is a seismometer used in a marine environment. Surface receivers, receivers in wellbores, or receivers on seabeds are used to detect the vibrations generated by controlled sources. Figure 6-4 illustrates a 3-D seismic survey using a 2-D array of receivers on land. Travel time from source to receiver is the primary information recorded by receivers during the data acquisition phase of a seismic survey. The output from the receiver is transmitted to a recording station where the signals are recorded by a seismograph. The graph representing the motion of a single receiver is called the *trace* of the seismometer. The display of the output is called the *seismogram* or *seismic section*. Other information such as amplitude and signal attenuation can also be acquired and used in the next step: data processing.

Data processing is used to transform travel time images to depth images. This requires the conversion of time measurements to depths, which in turn depend on the velocity of propagation of the acoustic signal through the earth. The relationship between travel time and depth requires the preparation of a velocity model. The velocity model defines the dependence of seismic wave velocity on depth. The velocity distribution depends on the

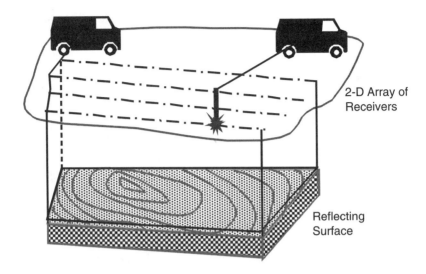

Figure 6-4. Seismic survey.

stratigraphic description of the subsurface geology. Once the seismic image has been transformed from the time to a depth representation, it is ready for the third step in the seismic survey process: interpretation.

The subsurface structure obtained from seismic measurements is part of the observational information that is used to develop a geologic model. The geologic model is an interpretation of the seismic image as a function of depth. Computers are needed to manipulate the large volume of information obtained from modern seismic surveys, and computer visualization is the most effective way to view the data. Optimum reservoir characterization is achieved by integrating seismic information with all other available information, such as regional geology and well information.

Geophysical techniques are being used today in exploration and development. Development geophysical surveys are conducted in fields where wells have penetrated the target horizon. The amount of seismic data is as great or greater than exploration geophysics, and it is possible to tie seismic lines to well information, such as well logs. The result is a set of seismic data that has been calibrated to "hard data" from the target horizons. The availability of well control data makes it possible to extract more detailed information from seismic data. Well information is obtained from measurements of fluid production or injection volumes, cores, and well logs. Well logs are obtained by running a device called a tool into the wellbore. The tool can detect physical properties such as temperature, electrical current, radioactivity, or sonic reflections.

6.5   PETROLEUM PRODUCTION

The stages in the life of a reservoir begin when the first discovery well is drilled. Prior to the discovery well, the reservoir is an exploration target. After the discovery well, the reservoir is a resource that may or may not be economical. The production life of the reservoir begins when fluid is withdrawn from the reservoir. Reservoir boundaries are established by seismic surveys and delineation wells. *Delineation wells* are wells that are originally drilled to define the size of the reservoir, but can also be used for production or injection later in the life of the reservoir. Production can begin immediately after the discovery well is drilled, or years later after several delineation wells have been drilled. The number of wells used to develop the field, the location of the wells, and their flow characteristics are among the many issues that must be addressed by reservoir management.

HISTORY OF DRILLING METHODS[4]

The first method of drilling for oil in the modern era was introduced by Edwin Drake in the 1850s and is known as cable-tool drilling. In this method, a rope connected to a wood beam has a drill bit attached to the end. The beam is raised and lowered, lifting and dropping the bit, which digs into the ground when it is dropped. Cable-tool drilling can work quickly, but the driller must periodically pull the bit out of the hole and clean the hole. The method does not work in soft-rock formations where the sides of the hole can collapse. Although cable-tool drilling is still used occasionally for drilling shallow wells and for maintenance operations, it has been largely replaced by rotary drilling.

Developed in France in the 1860s, rotary drilling was first used in the United States in the 1880s because it could drill into the soft-rock formations of the Corsicana oilfield in Texas. Rotary drilling uses a rotating drill bit with nozzles for shooting out drilling mud to penetrate into the earth. Drilling mud is a combination of clay and water that pushes rock cuttings out of the way of the bit and carries them up the wellbore to the surface.

Rotary drilling gained great popularity after Captain Anthony F. Lucas drilled the Lucas 1 well at Spindletop, near Beaumont, Texas. Lucas was born on the Dalmatian coast of the Austro-Hungarian Empire and served in the Austrian navy before immigrating to the United States. The Lucas 1 well was a discovery well and a "gusher." Gas and oil flowed up the well and engulfed the drilling derrick. Instead of flowing at the expected 50 barrels of oil per day, the well was producing up to 75,000 barrels per day. The Lucas gusher began the Texas oil boom [Yergin, 1992, pages 83–85]. Since then, rotary drilling has become the primary means of drilling.

Once a hole has been drilled, it is necessary to complete the well. A well is completed when it is prepared for production. The first completed well of the modern era was completed in 1808 when two American brothers, David and Joseph Ruffner, used wooden casings to prevent low-concentration salt water from diluting the high-concentration salt water they were extracting from deeper in their saltwater well [Van Dyke, 1997, pages 145–146].

To extract oil from wells, it is sometimes necessary to provide energy using different types of pumps or to inject gas to increase the buoyancy of the gas–oil mixture. The earliest pumps used the same wooden beams that were used for cable-tool drilling. Oil companies developed central pumping power in the 1880s. Central pumping power used a prime mover, a power source, to pump several wells. In the 1920s, demand for the replacement of on-site rigs led to the use of a beam pumping system for pumping wells. A beam pumping system is a self-contained unit that is mounted at the surface of each well and operates a pump in the hole. More modern techniques include gas-lifting and electric submersible pumps.

MODERN DRILLING METHODS

Advances in drilling technology are extending the options available for prudently managing subsurface reservoirs and producing fossil fuels, especially oil and gas. Four areas of drilling technology are introduced in the following paragraphs: infill drilling, horizontal wells, multilateral wells, and geosteering.

Infill drilling is the process of increasing the number of wells in an area by drilling wells in spaces between existing wells. The increase in well density, or number of wells per unit area, can improve recovery efficiency by providing fluid extraction points in parts of the reservoir that have not produced. Changes to well patterns and the increase in well density can alter sweep patterns in displacement processes and enable the displacement of in situ fluids by injected fluids. Infill drilling is especially useful in heterogeneous reservoirs.

A well is a string of connected, concentric pipes. The path followed by the string of pipes is called the *trajectory* of the well.

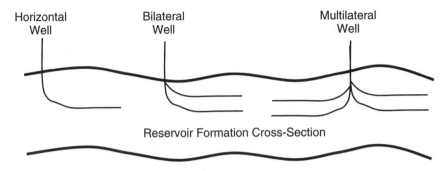

Figure 6-5. Multilateral wells.

Historically, wells were drilled vertically into the ground and the well trajectory was essentially a straight, vertical line. Today, wells are drilled so that the well trajectory is curved. This is possible because the length of each straight pipe that makes up the well is small compared to the total well length. The length of a typical section of pipe in a well is 40 ft (12.2 m). The length of a well away from the surface can be as long as 10 km. These long wells are called *extended reach wells*.

Wells can also be drilled so that the path of the well is horizontal. A horizontal well is sketched in Figure 6-5. When the well reaches a specified depth called the *kick-off point* (KOP), mechanical wedges (whip-stocks), or other downhole tools are used to deviate the drill bit and begin curving the well path. The horizontal section of the well is called the *reach*. Wells with more than one hole can be drilled. A well with more than one hole, called *branches*, is called a *multilateral well*. Figure 6-5 shows several examples of modern multilateral well trajectories. The vertical section of the well is called the *main (mother) bore* or *trunk*. The point where the main bore and a lateral meet is called a *junction*. For example, a bilateral well is a well with two branches. Multilateral well technology is revolutionizing extraction technology and reservoir management.

Multilateral wells make it possible to connect multiple well paths to a common wellbore. Multilateral wells have many applications. For example, multilateral wells are used in offshore environments where the number of well slots is limited by the amount of space available on a platform. They are used to produce fluids from reservoirs that have many compartments. A *compartment* in a reservoir is a volume that is isolated from other parts of the reservoir by barriers to fluid flow. These barriers can be sealing or partially sealing faults, or formation pinchouts.

Horizontal, extended reach, and multilateral wellbores that follow sub-surface formations are providing access to more parts of the reservoir from fewer well locations. This provides a means of minimizing environmental impact associated with drilling and production facilities, either on land or at sea. Extended reach wells make it possible to extract petroleum from beneath environmentally or commercially sensitive areas by drilling from locations outside of the environmentally sensitive areas.

Geosteering is the technology that makes it possible to accurately steer the well to its targeted location and is a prerequisite for successful extended reach drilling. Microelectronics is used in the drilling assembly to provide information to drill rig operators at the surface about the location of the drill bit as it bores a hole into the earth. Operators can modify the trajectory of the well while it is being drilled based on information from these measurement-while-drilling (MWD) systems. Extended reach drilling projects have drilled wells to locations as far away from the drilling rig as 10 km of horizontal displacement. For example, the energy company BP has used extended reach drilling in the Wytch Farm field on the southern coast of England.[5] Geosteering and extended reach drilling can reduce costs associated with the construction of expensive, new offshore platforms by expanding the volume of reservoir that is directly accessible from a given drilling location. In some cases, wells drilled from onshore drilling rigs can be used to produce coastal offshore fields that are within the range of extended reach drilling.

PRODUCTION SYSTEMS

A production system can be thought of as a collection of subsystems, as illustrated in Figure 6-6. Fluids are taken from the reservoir using wells. Wells must be drilled and completed. The performance of the well depends on the properties of the reservoir rock, the interaction between the rock and fluids in the reservoir, and properties of the fluids in the reservoir. Reservoir fluids include the fluids originally contained in the reservoir as well as fluids that may be introduced as part of the reservoir management process described in the following section. Well performance also depends on the properties of the well itself, such as its cross-section, length, trajectory, and completion. The completion of the well establishes the connection between the well and the reservoir. A completion can be as simple as an open-hole completion where fluids are allowed to drain into the wellbore from consolidated reservoir rock, to completions that require

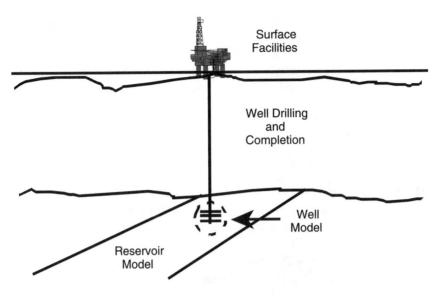

Figure 6-6. Production system.

the use of tubing with holes punched through the walls of the tubing using perforating guns to allow fluid to flow between the tubing and the reservoir.

Surface facilities are needed to drill, complete, and operate wells. Drilling rigs may be moved from one location to another on trucks, ships, or offshore platforms; or drilling rigs may be permanently installed at specified locations. The facilities may be located in desert climates in the Middle East, stormy offshore environments in the North Sea, arctic climates in Alaska and Siberia, and deepwater environments in the Gulf of Mexico and off the coast of West Africa.

Produced fluids must be recovered, processed, and transported to storage facilities and eventually to the consumer. Processing can begin at the well site where the produced wellstream is separated into oil, water, and gas phases. Further processing at refineries separates the hydrocarbon fluid into marketable products, such as gasoline and diesel fuel. Transportation of oil and gas may be by a variety of means, including pipelines, tanker trucks, double-hulled tankers, and ships capable of carrying liquefied natural gas.

DECLINE CURVE ANALYSIS

In the early 1900s, production analysts observed that production from oil wells could be predicted by fitting an exponential equation to historical decline rates [Fanchi, 2002, Chapter 13]. The exponential equation worked well for many reservoirs in production at the time. The production from other wells could be better fit using a hyperbolic decline equation. Assuming constant flowing pressure, a general equation for the empirical exponential and hyperbolic relationships used in decline curve analysis is

$$\frac{dq}{dt} = -aq^{n+1} \qquad (6.5.1)$$

where a and n are empirically determined constants, q is flow rate and t is time. The empirical constant n ranges from 0 to 1. Solutions to Equation (6.5.1) show the expected decline in flow rate q as the production time t increases. Three decline curve solutions have been identified based on the value of n [Economides, et al., 1994]. We consider only the exponential decline ($n = 0$) solution here, namely

$$q = q_i\, e^{-at} \qquad (6.5.2)$$

where q_i is initial rate and a is a factor that is determined by fitting Equation (6.5.2) to well or field data.

Decline curves are fit to observed well production data by plotting the natural logarithm of flow rate q versus production time t. A common decline curve has the shape of a straight line on a semilogarithmic plot because the natural logarithm of the exponential decline solution

$$\ln q = \ln q_i - at \qquad (6.5.3)$$

has the form $y = mx + b$ for a straight line with slope m and intercept b. The factor a in Equation (6.5.3) is the slope m of the straight line obtained by plotting the natural logarithm of q versus time t. The intercept b is the natural logarithm of the initial flow rate.

6.6   RESERVOIR MANAGEMENT

Management of subsurface resources, especially oil and gas fields, is called *reservoir management*. Reservoir management may be defined as the allocation of resources that are needed to optimize the production of commercially valuable fluids from a reservoir. The reservoir management plan should be flexible enough to accommodate technological advances, changes in economic factors, new information obtained during the life of the reservoir, and to address all relevant operating issues, including governmental regulations. One of the objectives of reservoir management is to develop a plan for maximizing recovery efficiency.

RECOVERY EFFICIENCY

Recovery efficiency is a measure of the amount of resource recovered relative to the amount of resource originally in place. It is defined by comparing initial and final in situ fluid volumes. An estimate of expected recovery efficiency can be obtained by considering the factors that contribute to the recovery of a subsurface fluid. Two factors are especially useful: displacement efficiency, and volumetric sweep efficiency.

Displacement efficiency E_D is a measure of the amount of mobile fluid in the system. For example, the displacement efficiency for oil depends on the difference between the initial volume of oil and the final volume of oil. Consider a reservoir with pore volume V_p, initial oil saturation S_{oi} at initial formation volume factor B_{oi}, and oil saturation at abandonment S_{oa} with the formation volume factor B_{oa}. An estimate of displacement efficiency is

$$E_D = \frac{\dfrac{V_p S_{oi}}{B_{oi}} - \dfrac{V_p S_{or}}{B_{oa}}}{\dfrac{V_p S_{oi}}{B_{oi}}} = \frac{\dfrac{S_{oi}}{B_{oi}} - \dfrac{S_{or}}{B_{oa}}}{\dfrac{S_{oi}}{B_{oi}}} \qquad (6.6.1)$$

In addition to displacement efficiency, recovery efficiency depends on the amount of in situ hydrocarbon contacted by injected fluids. Areal sweep efficiency E_A and vertical sweep efficiency E_V measure the degree of contact between in situ and injected fluids. Areal sweep efficiency is defined as

$$E_A = \frac{\text{swept area}}{\text{total area}} \qquad (6.6.2)$$

and vertical sweep efficiency is defined as

$$E_V = \frac{\text{swept net thickness}}{\text{total net thickness}} \qquad (6.6.3)$$

Volumetric sweep efficiency E_{Vol} expresses the efficiency of fluid recovery in terms of areal sweep efficiency and vertical sweep efficiency, thus

$$E_{Vol} = E_A \times E_V \qquad (6.6.4)$$

Recovery efficiency RE is the product of displacement efficiency, areal sweep efficiency, and vertical sweep efficiency:

$$RE = E_D \times E_{Vol} = E_D \times E_A \times E_V \qquad (6.6.5)$$

Each of the recovery efficiencies is a fraction that varies from 0 to 1. If one or more of the factors that enter into the calculation of recovery efficiency is small, recovery efficiency will be small. On the other hand, each of the factors can be relatively large, and the recovery efficiency still be small because it is a product of factors that are less than 1. In many cases, technology is available for improving recovery efficiency, but may not be implemented because it is not economical. Some of the technology is described in the following sections. The application of technology and the ultimate recovery of fossil fuels depend on the economic value of the resource. This statement can be applied to all energy sources.

PRIMARY PRODUCTION

The life of a reservoir has traditionally been divided into a set of production stages. The first stage of production is ordinarily called *primary production*. It relies entirely on natural energy sources. To remove petroleum from the pore space it occupies, the petroleum must be replaced by another fluid, such as water, natural gas, or air. Oil displacement is caused by the expansion of in situ fluids as pressure declines during primary reservoir depletion. The natural forces involved in the displacement of oil during primary production are called *reservoir drives*. The most common reservoir drives for oil reservoirs are water drive, solution or dissolved gas drive, and gas cap drive.

The most efficient drive mechanism is water drive. In this case, water displaces oil as oil flows to production wells. An effective reservoir management strategy for a water drive reservoir is to balance oil withdrawal with the rate of water influx. Water drive recovery typically ranges from 35% to 75% of the original oil in place (OOIP).

In a solution gas drive, gas dissolved in the oil phase at reservoir temperature and pressure is liberated as pressure declines. Some oil moves with the gas to the production wells as the gas expands and moves to the lower-pressure zones in the reservoir. Solution gas drive recovery ranges from 5% to 30% OOIP.

A *gas cap* is a large volume of gas at the top of a reservoir. When production wells are completed in the oil zone below the gas cap, the drop in pressure associated with pressure decline causes gas to move from the higher-pressure gas cap down toward the producing wells. The gas movement drives oil to the wells, and eventually large volumes of gas will be produced with the oil. Gas cap drive recovery ranges from 20% to 40% OOIP, although recoveries as high as 60% can occur in steeply dipping reservoirs with enough permeability to allow oil to drain to wells that are producing from lower in the structure.

Gravity drainage is the least common of the primary production mechanisms. In this case oil flows to a well that is producing from a point lower in the reservoir. Oil will flow to the production well if the pressure difference between the pressure in the reservoir and the pressure in the production well favors downstructure oil flow rather than oil movement upstructure due to gravity segregation. Gravity drainage can be effective in shallow, highly permeable, steeply dipping reservoirs.

In many cases, one or more drive mechanisms are functioning simultaneously. The behavior of the field depends on which mechanism is most important at various times during the life of the field. The best way to predict the behavior of such fields is with sophisticated reservoir flow models. Although the preceding discussion referred to oil reservoirs, similar comments apply to gas reservoirs. Water drive and gas expansion with reservoir pressure depletion are the most common drives for gas reservoirs. Gas reservoir recovery can be as high as 70% to 90% of original gas in place (OGIP) because of the relatively high mobility of gas. Mobility λ is the ratio of effective permeability k_{eff} to viscosity μ:

$$\lambda = \frac{k_{eff}}{\mu} \tag{6.6.6}$$

Mobility can be defined for any fluid phase, and is an important design parameter for the management of subsurface fluid resources.

Gas storage reservoirs have a different life cycle than gas reservoirs that are being depleted. Gas storage reservoirs are used to warehouse gas. If the gas is used as a fuel for power plants, it will need to be periodically produced and replenished. The performance attributes of a gas storage reservoir are verification of gas inventory, assurance of gas deliverability, and containment against migration away from the gas storage reservoir. Gas deliverability must be sufficient to account for swings in demand that arise from factors such as seasonal demand. For example, demand for energy is often higher in the summer to provide air conditioning and in the winter to provide heat.

SECONDARY PRODUCTION

Primary depletion is usually not sufficient to optimize recovery from an oil reservoir. Oil recovery can be doubled or tripled by supplementing natural reservoir energy. The supplemental energy is provided through an external energy source, such as water injection or gas injection. The injection of water or natural gas may be referred to as pressure maintenance or secondary production. The latter term arose because injection usually followed a period of primary pressure depletion, and was therefore the second production method used in a field. Many modern reservoirs incorporate pressure maintenance early in the production life of the field, sometimes from the beginning of production. In this case the reservoir is not subjected to a conventional primary production phase.

ALTERNATIVE CLASSIFICATIONS

Both primary and secondary recovery processes are designed to produce oil using immiscible methods. An *immiscible displacement process* is a process in which there is a distinct interface between the injected displacing fluid and the in situ displaced fluid. Additional methods may be used to improve oil recovery efficiency by reducing residual oil saturation. *Residual oil saturation* is the fraction of oil that remains in the rock after a displacing fluid is used to flood the rock. The reduction of residual oil saturation requires a change in factors that govern the interaction between the fluid and the rock.

Methods designed to reduce residual oil saturation have been referred to in the literature as tertiary production, enhanced oil recovery, and improved oil recovery. The term *tertiary production* was originally used to identify the third stage of the production life of the field. Typically, the third stage occurred after waterflooding. The third stage of oil production involved a process that was designed to mobilize waterflood residual oil. An example of a tertiary production process is a miscible flood process, such as carbon dioxide flooding. Tertiary production processes were designed to improve displacement efficiency by injecting fluids or heat. They were referred to as enhanced recovery processes. It was soon learned, however, that some fields performed better if enhanced recovery processes were implemented before the third stage in the production life of the field. In addition, it was found that enhanced recovery processes were often more expensive than just drilling more wells in a denser pattern.

The drilling of wells to reduce well spacing and increase well density is called *infill drilling*. The origin of the term *infill drilling* coincide with the birth of another term, *improved recovery*. Improved recovery includes enhanced oil recovery and infill drilling. Some major improved recovery processes are waterflooding, gasflooding, chemical flooding, and thermal recovery [van Dyke, 1997].

Improved recovery technology includes traditional secondary recovery processes such as waterflooding and immiscible gas injection, as well as enhanced oil recovery (EOR) processes.[6] EOR processes are usually classified as one of the following processes: chemical, miscible, thermal, and microbial. Chemical flooding methods use injected chemicals such as long-chain, high-molecular-weight polymers and detergent-like surfactants to help mobilize and displace oil. Miscible flooding methods include carbon dioxide injection, natural gas injection, and nitrogen injection. Miscible gas injection must be performed at a high enough pressure to ensure miscibility between the injected gas and in situ oil. Miscibility is achieved when interfacial tension (IFT) between the aqueous (water-based) and oleic (oil-based) phases is significantly reduced. Any reduction in IFT can improve displacement efficiency. Thermal flooding methods include hot water injection, steam drive, steam soak, and in situ combustion. The injection or generation of heat in a reservoir is designed to reduce the viscosity of in situ oil and improve the mobility of the displaced fluid. Electrical methods can also be used to heat fluids in relatively shallow reservoirs containing high-viscosity oil, but electrical methods are not as common as hot-fluid injection methods. Microbial EOR uses the injection of microorganisms and nutrients in a carrier medium to increase oil recovery and/or reduce water production in petroleum reservoirs.

6.7   NONCONVENTIONAL FOSSIL FUELS

Clean energy refers to energy that is generated with little environmental pollution. Natural gas is a source of clean energy. Oil and gas fields are considered conventional sources of natural gas. Nonconventional sources of natural gas include coalbed methane discussed in Section 6.2, methane hydrates, tight gas sand, and shale gas. Other nonconventional fossil fuels include shale oil and tar sands. In this section, we provide an introduction to nonconventional fossil fuels.[7]

METHANE HYDRATES

The entrapment of gas molecules in ice at very low temperatures forms an icelike solid called a gas hydrate. Gas hydrates are *clathrates*: a chemical complex that is formed when one type of molecule completely encloses another type of molecule in a lattice. In the case of gas hydrates, hydrogen-bonded water molecules form a cagelike structure in which mobile molecules of gas are absorbed or bound. Gas hydrates with bound methane are called methane hydrates.

Gas hydrates are generally considered a problem for oil and gas field operations. For example, the existence of hydrates on the ocean floor can affect drilling operations in deep water. The simultaneous flow of natural gas and water in tubing and pipelines can result in the formation of gas hydrates that impede or completely block the flow of fluids through pipeline networks. Heating the gas or treating the gas–water system with chemical inhibitors can prevent the formation of hydrates, but increases operating costs.

The commercial potential of methane hydrates as a clean energy resource is changing the industry perception of gas hydrates. Methane hydrates contain a relatively large volume of methane in the hydrate complex. The hydrate complex contains about 85 mole percent water and approximately 15 mole percent guests, where a guest is methane or some other relatively low-molecular-weight hydrocarbon.

Methane hydrates can be found throughout the world. They exist on land in sub-Arctic sediments and on seabeds where the water is near freezing. Difficulties in cost-effective production of methane hydrates have hampered the production of methane from hydrates.

TIGHT GAS SANDS AND SHALE GAS

Nonconventional gas resources include coalbed methane, tight gas sands, and fractured gas shales. Coalbed methane was discussed in Section 6.2. Both tight gas sands and gas shales are characterized by low permeabilities, that is, permeabilities that are a fraction of a millidarcy (less than 10^{-15} m^2). The low permeability associated with nonconventional gas resources makes it more difficult to produce the gas at economical rates.

Economic production of gas from a gas shale or tight gas sand often requires the creation of fractures by a process known as hydraulic fracturing. In this process, a fluid is injected into the formation at a pressure that exceeds the fracture pressure of the formation. Once fractures have been created in the formation, a proppant such as coarse grain sand or manmade pellets are injected into the fracture to prevent the fracture from closing, or healing, when the injection pressure is removed. The proppant provides a higher-permeability flow path for gas to flow to the production well. Nonconventional low-permeability gas sands and shales often require more wells per unit area than do conventional higher permeability gas reservoirs. The key to managing a nonconventional gas resource is to develop the resource with enough wells to maximize gas recovery without drilling unnecessary wells.

SHALE OIL AND TAR SANDS

Shale oil is high-API-gravity oil contained in porous, low-permeability shale. Sand grains that are cemented together by tar or asphalt are called *tar sands*. Tar and asphalt are highly viscous, plastic or solid hydrocarbons. Extensive shale oil and tar sand deposits are found throughout the Rocky Mountain region of North America, and in other parts of the world. Although difficult to produce, the volume of hydrocarbon in tar sands has stimulated efforts to develop production techniques.

The hydrocarbon in shale oil and tar sands can be extracted by mining when oil shales and tar sands are close enough to the surface. Tar pits have been found around the world and have been the source of many fossilized dinosaur bones. In locations where oil shales and tar sands are too deep to mine, it is necessary to increase the mobility of the hydrocarbon.

An increase in permeability or a decrease in viscosity can increase mobility. Increasing the temperature of high-API-gravity oil, tar, or asphalt can significantly reduce viscosity. If there is enough permeability to allow injection, steam or hot water can be used to increase formation temperature and reduce hydrocarbon viscosity. In many cases, however, permeability is too low to allow significant injection of a heated fluid. An alternative to fluid injection is electromagnetic heating. Radio frequency heating has been used in Canada, and electromagnetic heating techniques are being developed for other parts of the world.

ENDNOTES

1. References for the history of fossil fuels include Nef [1977], Yergin [1992, especially Chapter 1], Kraushaar and Ristinen [1993, Chapter 2], van Dyke [1997, Chapters 4–6], and Shepherd and Shepherd [1998, Chapters 4 and 5].
2. Coal references include Shepherd and Shepherd [1998, Chapter 4], Selley [1998, especially Chapter 5], Wiser [2000, Chapters 4 and 5], and Press and Siever [2001].
3. Much of the discussion of petroleum fluids is taken from Fanchi [2001, 2002].
4. References for the history of drilling include Yergin [1992, especially Part 1], and van Dyke [1997, Chapters 4–6].
5. For a discussion of physics in oil exploration and development, see Clark and Kleinberg [2002].
6. More detailed discussions of EOR processes are presented in such references as Lake [1989], and Green and Willhite [1998].
7. For more discussion of nonconventional fossil fuels, see Cassedy and Grossman [1998, Appendix C], Selley [1998, Chapter 9], and Wiser [2000, Chapter 6].

EXERCISES

6-1. A. A coal seam is 800 feet wide, 1 mile long, and 10 feet thick. The volume occupied by the fracture network is 1%. What is the volume of coal in the coal seam? Express your answer in m^3.
B. If the density of coal is 1.7 lbm/ft^3, how many tonnes of coal are in the coal seam?

6-2. A. Assume the Langmuir isotherm for the coal seam in Exercise 6-1 has a Langmuir volume of 600 standard cubic feet per ton of coal (SCF/ton) and a Langmuir pressure of 450 psia. Calculate the volume of gas per ton of coal at a pressure of 1000 psia. Express your answer in SCF/ton where 1 ton = 2000 lbm.
B. How much gas is contained in the coal? Express your answer in cubic meters.

6-3. A barrier island is a large sand body. Consider a barrier island that averages 3 miles wide, 10 miles long, and is 30 feet thick. The porosity of the sand averages almost 25%. What is the pore volume of the barrier island? Express your answer in bbl and m^3.

6-4. Plot API gravity as a function of oil specific gravity for the range $0.1 \leq \gamma_o \leq 1.0$.

6-5. A. Use the real gas law $PV = ZnRT$ to find a general expression for gas formation volume factor B_g. Use subscripts "s" and "r" to denote surface conditions and reservoir conditions, respectively.
B. Calculate B_g using $\{P_s = 14.7 \text{ psia}, T_s = 60° \text{ F}, Z_s = 1\}$ and $\{P_r = 2175 \text{ psia}, T_r = 140° \text{ F}, Z_r = 0.9\}$. Express B_g as reservoir cubic feet per standard cubic feet (RCF/SCF).
C. Calculate B_g using $\{P_s = 1 \text{ atm}, T_s = 20° \text{ C}, Z_s = 1\}$ and $\{P_r = 15 \text{ MPa}, T_r = 60° \text{ C}, Z_r = 0.9\}$. Express B_g as reservoir cubic meters per standard cubic meter (Rm^3/Sm^3).
D. What is the difference between the calculation in Part B and the calculation in Part C?

6-6. We want to drill a 5,000 ft deep vertical well. We know from previous experience in the area that the drill bit will be effective for 36 hrs before it has to be replaced. The average drill bit will penetrate 20 ft of rock in the area for each hour of drilling. Again based on previous experience, we expect the average trip to replace the drill bit to take about 8 hrs. A *trip* is the act of withdrawing the drill pipe, replacing the drill bit, and then returning the new drill bit to the bottom of the hole. Given this information, estimate how long it will take to drill the 5,000 ft deep vertical well. Hint: Prepare a table as follows:

Incremental time (hrs)	Incremental depth (ft)	Cumulative time (hrs)	Cumulative depth (ft)

6-7. A. Show that Equation (6.5.2) is a solution of Equation (6.5.1).
B. Plot oil flow rate as a function of time for a well that produces 10,000 barrels per day with a decline factor $a = 0.06$ per year. Time should be expressed in years, and should range from 0 to 50 years.
C. When does the flow rate drop below 1000 barrels per day?

6-8. Show that $q^{-1} = at + q_i^{-1}$ is a solution of Equation (6.5.1) with $n = 1$. This solution is known as the *harmonic decline solution*.

6-9. A. Calculate volumetric sweep efficiency E_{Vol} and recovery efficiency RE from the following table:

S_{oi}	0.75
S_{oa}	0.30
Area swept	750 hectares
Total area	1000 hectares
Thickness swept	10 meters
Total thickness	15 meters
Neglect swelling effects: $B_{oi} \approx B_{oa}$	

B. Discuss how recovery efficiency could be improved.

6-10. Calculate the reflection coefficient at the interface between two layers with the following properties:

Layer 1: bulk density = 2.3 g/cc and compressional velocity = 5000 m/s

Layer 2: bulk density = 2.4 g/cc and compressional velocity = 5500 m/s

6-11. A wildcat well has been drilled to a total vertical depth of 9500 feet. An oil reservoir is discovered at this depth. The initial reservoir pressure is estimated to be 6000 psia. What is the pressure gradient in the wellbore if the pressure at the surface of the well is 14.7 psia? Express your answer in psia per foot. The pressure gradient is the change in pressure divided by the change in depth.

6-12. A. Suppose a reservoir covers an area of 1100 acres and is 30 feet thick. The average porosity of the reservoir is 18% and the average water saturation is 35%. The remaining pore volume is occupied by oil with a formation volume factor of 1.2 RB/STB where RB denotes reservoir barrel and STB denotes stock tank barrel. Estimate the volume of oil in the reservoir and express your answer in STB.
B. If the expected recovery factor is 30%, how much oil will be recovered? Express your answer in MMSTB and m^3. Note: 1 MMSTB = 1 million STB.

CHAPTER SEVEN

Solar Energy

Fossil energy and nuclear energy are considered nonrenewable energy types. Nonrenewable energy is obtained from sources at a rate that exceeds the rate at which the sources are replenished. For example, if the biogenic origin of fossil fuels is correct, we could consider fossil fuels renewable over a period of millions of years, but the existing store of fossil fuels is being consumed over a period of centuries. Because we are consuming fossil fuels at a rate that exceeds the rate of replenishment, we consider fossil fuels nonrenewable. Similar comments apply to nuclear fuels such as uranium, as we observe in later chapters. Solar energy is considered a renewable energy for the following reasons.[1]

Renewable energy is energy obtained from sources at a rate that is less than or equal to the rate at which the source is replenished. In the case of solar energy, we can use only the amount of energy provided by the sun. Because the remaining lifetime of the sun is measured in millions of years, many people consider solar energy an inexhaustible supply of energy. In fact, solar energy from the sun is finite, but should be available for use by many generations of people. Solar energy is therefore considered renewable. Energy sources that are associated with solar energy, such as wind and biomass, are also considered renewable.

Solar radiation may be converted to other forms of energy by several conversion processes. Thermal conversion relies on the absorption of solar energy to heat a cool surface. Biological conversion of solar energy relies on photosynthesis. Photovoltaic conversion generates electrical power by the generation of an electrical current as a result of a quantum mechanical process. Wind power and ocean energy conversion rely on atmospheric pressure gradients and oceanic temperature gradients to generate electrical power. In this chapter we focus on thermal conversion. We first discuss the source of available solar energy, and then consider solar energy technology in two of its three forms: passive solar, and active solar. The third form of solar energy, solar electric, is discussed in the next chapter. We end this chapter with a discussion of solar power plants.

7.1 NUCLEAR FUSION: THE SOURCE OF SOLAR ENERGY

The energy emitted by a star like the sun is generated by a fusion reaction [Bernstein, et al., 2000, Section 18-5]. If we want to understand the source of solar energy, we need to know something about nuclear fusion.[2] In this section, we briefly describe the discovery of the nucleus and introduce the concepts of nuclear mass and nuclear binding energy. We then describe the fusion process in the sun.

DISCOVERY OF THE NUCLEUS

German physicist Wilhelm Roentgen observed a new kind of radiation called X-rays in 1895. Roentgen's X-rays could pass through the body and create a photograph of the interior anatomy. Frenchman Henri Becquerel discovered radioactivity in 1896 while looking for Roentgen's X-rays in the fluorescence of a uranium salt. French physicist and physician Marie Curie (formerly Maria Sklodowska from Warsaw, Poland) and her husband Pierre Curie were the first to report the discovery of a new radioactive element in 1898. They named the element polonium after Marie's homeland. Ernest Rutherford identified the "rays" emitted by radioactive elements and called them α, β, and γ rays. Today we know that the α ray is the helium nucleus, the β ray is an electron, and the γ ray is an energetic photon. By 1913, Rutherford and his colleagues at the Cavendish Laboratory in Cambridge England had used α particles to bombard thin metallic foils and discovered the nucleus. The constituents of the nucleus were later identified as the proton and a new, electrically neutral particle, the neutron. James Chadwick discovered the neutron in 1932 while working in Rutherford's laboratory.

The proton and neutron are classified as nucleons, or nuclear constituents. The notation for the number of constituents in the nucleus is the atomic symbol $^A_Z X_N$ where X is the symbol of the chemical element, Z is the atomic number (number of protons), N is the number of neutrons, and the mass number A satisfies the relation

$$A = N + Z \tag{7.1.1}$$

In an electrically neutral atom, the number of negatively charged electrons is equal to the number of positively charged protons Z. The mass of the atom or nucleus is expressed in terms of the atomic mass unit, or amu.

One amu is defined so that the mass of a neutral atom of carbon-12 with atomic symbol $^{12}_{6}C_6$ is exactly 12. The current value of one atomic mass unit is 1.6605×10^{-27} kg.

The possible combinations of nuclear constituents have led to a classification scheme based on the terms *nuclide, isotope, isotone,* and *isobar.* A species of atoms with the same atomic number Z and mass number A is called a nuclide. Isotopes are nuclei with the same atomic number Z but different numbers of neutrons N. For example, the naturally occurring isotopes of carbon are carbon-12 $\left(^{12}_{6}C_6\right)$, carbon-13 $\left(^{13}_{6}C_7\right)$, and carbon-14 $\left(^{14}_{6}C_8\right)$. Carbon-13 and carbon-14 are radioactive and therefore unstable. A trio of isotopes that is important in nuclear fusion is the set of isotopes of hydrogen: hydrogen with a proton nucleus $\left(^{1}_{1}H_0\right)$, deuterium with a deuteron nucleus $\left(^{2}_{1}H_1\right)$, and tritium with a triton nucleus $\left(^{3}_{1}H_2\right)$. Isotones are nuclei with the same number of neutrons N but different numbers of protons Z. Isobars are nuclei with the same mass number A. The term *isobar* used to describe nuclei should not be confused with the term *isobar* when used to denote constant pressure in thermodynamics. It should be clear from the context which meaning is implied.

NUCLEAR MASS AND BINDING ENERGY

The mass of a nucleus with mass number A and atomic number Z is given by the formula

$$m(N, Z) = Nm_n + Zm_p - \frac{B(N, Z)}{c^2} \qquad (7.1.2)$$

where m_n is the mass of the neutron, m_p is the mass of the proton, c is the speed of light in vacuum, and $B(N, Z)$ is the binding energy of the nucleus. Binding energy is the energy needed to separate the nucleus into its constituent nucleons.

One way to estimate binding energy is to view the nucleus as a drop of liquid. The liquid drop model of the nucleus was first developed by C.F. von Weizsächer in 1935 and treats the nucleons as if they were atoms in a drop of liquid. The liquid drop model has been used to develop the semi-empirical binding energy formula [Serway, et al., 1997, page 542]

$$B(N, Z) = c_1 A + c_2 A^{2/3} + c_3 \frac{Z(Z-1)}{A^{1/3}} + c_4 \frac{(N-Z)^2}{A} \qquad (7.1.3)$$

Mass number A is the function of N and Z given by Equation (7.1.1). Equation (7.1.3) provides a reasonable approximation of the binding energy. The equation is referred to as "semi-empirical" because each of the terms is associated with a physical concept, but the coefficients $\{c_1, c_2, c_3, c_4\}$ multiplying each term are determined by fitting the equation to experimental data. The first term $(c_1 A)$ on the right-hand side of Equation (7.1.3) is called the volume term and represents the proportionality between binding energy and atomic number, which is proportional to the volume of the nucleus. The second term $(c_2 A^{2/3})$ is called the surface term and accounts for the fact that nucleons on the surface of the nucleus in the liquid drop model are not completely surrounded by other nucleons. The third term $(c_3 Z(Z-1)/A^{1/3})$ accounts for Coulomb repulsion of the positively charged protons in the nucleus. The fourth term $[c_4(N-Z)^2/A]$ accounts for a decrease in binding energy observed in heavy nuclei with an excess of neutrons. The fourth term is zero when $N = Z$, and becomes increasingly negative when $N \neq Z$. The coefficients in Equation (7.1.3) are

$$c_1 = 2.5 \times 10^{-12} \text{ J}, \quad c_2 = -2.6 \times 10^{-12} \text{ J}, \quad c_3 = -0.114 \times 10^{-12} \text{ J},$$

$$c_4 = -3.8 \times 10^{-12} \text{ J} \tag{7.1.4}$$

Other terms can be added to the semi-empirical binding energy formula to refine the calculation of binding energy, but are beyond the scope of this book.

Figure 7-1 shows the binding energy per nucleon obtained from the semi-empirical mass formula for nuclei with $N = Z$. The curve in Figure 7-1 is most accurate for $A > 24$. Below this value, the observed binding energy per nucleon has several peaks corresponding to the nuclides $\{^4\text{He}, {}^8\text{Be},$ $^{12}\text{C}, {}^{16}\text{O}, {}^{20}\text{Ne}, {}^{24}\text{Mg}\}$. These nuclides have mass numbers that are integer multiples of the mass number $A = 4$ of the helium nucleus. The large binding energy per nucleon for each of these nuclides shows that they are particularly stable; a great deal of energy must be provided to split the nucleus into its constituent parts.

The largest binding energy per nucleon occurs near the maximum at $A \approx 45$. The nuclear force binding the nucleons is represented by the volume term and is responsible for the positive slope of the binding energy per nucleon for $A < 45$. Coulomb repulsion between protons is the principal cause of the decline in binding energy per nucleon when $A > 45$. Energy can be released by splitting the nucleus into two comparable fragments when A is sufficiently large, which is experimentally observed to occur at mass

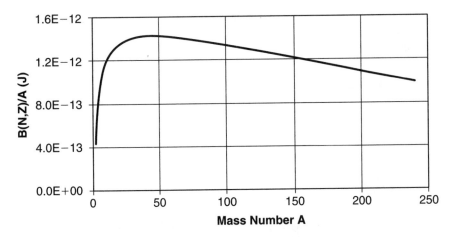

Figure 7-1. Binding energy per nucleon for $N = Z$.

numbers as low as $A \approx 120$ or $Z \approx 60$. The separation of a large nucleus into two comparable fragments is an example of spontaneous fission. Fission can also occur if very massive nuclides such as uranium isotopes ^{235}U or ^{238}U are excited to a higher energy state. By contrast, energy is released in the fusion process when two light nuclei with very small A are combined.

SOLAR FUSION

The temperature and density of matter in the interior of a star are very high. Two neutrons and two protons follow a sequence of reactions to form the nucleus of the most common isotope of helium, ^{4}He. In this fusion process, there is a net loss in mass that is converted into radiant energy.

The first step in the fusion reaction sequence is to combine two protons to form the deuterium nucleus in the reaction

$$p + p \rightarrow d + e^{-} + \nu_{e} \tag{7.1.5}$$

where ν_{e} is an electron neutrino. The deuterium nucleus consists of a bound proton and neutron. Once formed, the deuterium nucleus can interact with a proton in the reaction

$$d + p \rightarrow {}^{3}He + \gamma \tag{7.1.6}$$

to form the nucleus of the light helium isotope ^{3}He and a gamma ray photon γ. Two light helium isotopes combine in the last reaction of the fusion sequence

$$^3\text{He} + {}^3\text{He} \rightarrow {}^4\text{He} + 2p \qquad (7.1.7)$$

to form ^{4}He and two protons. The overall fusion reaction may be written as

$$4p \rightarrow {}^4\text{He} + 2e^- + 2\nu_e \qquad (7.1.8)$$

The fusion reaction releases 26.7 MeV (4.27×10^{-12} J) in kinetic energy and electromagnetic radiation. The radiant energy begins as gamma rays and is eventually transformed into the solar spectrum that irradiates the earth.

Another fusion process called the carbon cycle occurs in some stars [Bernstein, et al., 2000, Section 15-6]. The carbon cycle uses carbon-12, ^{12}C, to catalyze the formation of ^{4}He in a multistep process that includes the temporary formation of isotopes of nitrogen and oxygen. The carbon cycle can occur only in stars that contain enough carbon-12 to act as a catalyst in the carbon cycle. Because carbon is produced in another set of nuclear reactions, stars that generate fusion in the carbon cycle did not appear until relatively late in the evolution of the universe.

7.2   THE LIFE OF A STAR

The length of time that solar energy will be available as a source of energy depends on the lifetime of a star. Stars pass through different stages during their lifetime. The behavior of a star as a function of the age of the star is called *stellar evolution*. Astronomers have learned that stellar evolution depends on such factors as the initial size and mass of the star. Figure 7-2 is a plot of star brightness versus star color. Plots like Figure 7-2 were first made by the Danish astronomer Ejnar Hertzsprung, and independently by the American astronomer Henry Norris Russell. They are called Hertzsprung–Russell diagrams and have proven to be valuable tools for understanding stellar evolution. Using such diagrams, we can project an evolutionary sequence for our sun.

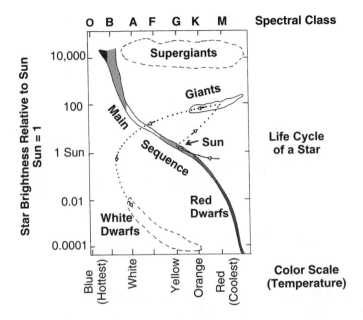

Figure 7-2. Hertzsprung–Russell (H–R) diagram.

Figure 7-2 shows that our sun is presently in the Main Sequence of stars. The future evolutionary course of the sun is depicted by the dotted line in the figure. A few billion years from now, perhaps five, our sun will exhaust its hydrogen. As this nuclear fuel depletes, the solar core will shrink and the solar surface will expand. Temperatures will rise in the shrinking core; the surface of the sun will expand and brighten. Eventually the expanding stellar surface will engulf the inner planets, including Earth, as the sun becomes a giant star. In time, the sun will exhaust its remaining nuclear fuel, helium, and undergo gravitational collapse. The radiance of energy from the sun will diminish as the collapse transforms the sun into a white dwarf. The collapse will continue until both the collapse and the radiance cease. The dead sun will be a black dwarf, and much of its mass-energy will be traversing the universe in search of a new gravitational home.

The H–R diagram is an empirical result that is qualitatively explained by considering the luminosity of a star, or total energy radiated per second. The luminosity of the sun is approximately 3.8×10^{26} W. The amount of radiation from the sun that reaches the earth's atmosphere is approximately 1370 watts per square meter. The value 1370 W/m^2 is called the *solar constant*.

Stellar luminosity L_{star} is roughly proportional to the cube of the star's mass M_{star} for stars on the Main Sequence of the H–R diagram. We can write

$$L_{star} = L_S \left(\frac{M_{star}}{M_S} \right)^3 \qquad (7.2.1)$$

where L_S is the luminosity of our sun and M_S is its mass. More-massive stars have a greater luminosity and require the consumption of more nuclear fuel than less-massive stars. As a consequence, more-massive stars tend to have shorter lifetimes than less-massive stars. Silk [2001, page 439] estimated the lifetime of a hydrogen-burning star on the Main Sequence of the H–R diagram to be

$$t_{star} = 10^{10} \left(\frac{M_{star}}{M_S} \right)^{-2} \text{ yrs} \qquad (7.2.2)$$

When stars approach the end of their lifetime, which corresponds roughly to the time when their nuclear fuel is exhausted, they are subjected to one of several fates.

Some stars explode. By-products of the explosion—which is called a nova or supernova—include many heavy elements as debris. The debris of a supernova can form a nebula from which other stars may evolve. A supernova explosion is believed to be the source of material that formed our sun, its planets, including Earth, and life. Other stars simply contract as they cool and become white dwarfs. Still other stars undergo gravitational collapse under their own weight until they enter a very dense state called the neutron-star state. In a few instances, black holes are the end result of the gravitational collapse of a star.

Some of the bodies that are formed by the coalescence of matter in nebular clouds are not dense enough to ignite nuclear reactions. They may be captured by stars and become satellites, or planets, of the star. Still smaller disturbances of the matter distribution in a nebula may be captured by the planets and become moons. Yet other disturbances become asteroids, comets, and meteors, a process that continues in parts of the universe today.

7.3   SOLAR ENERGY

Solar energy is energy emitted by a star. Figure 7-3 shows the anatomy of a star. We saw in the preceding section that the energy emitted by a star

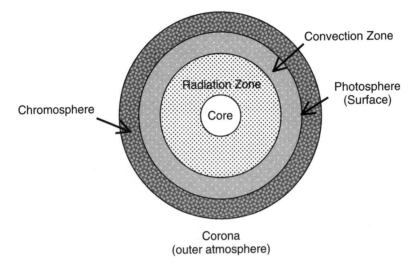

Figure 7-3. Anatomy of a star.

is generated by nuclear fusion. The fusion process occurs in the core, or center, of the star. Energy released by the fusion process propagates away from the core by radiating from one atom to another in the radiation zone of the star. As the energy moves away from the core and passes through the radiation zone, it reaches the part of the star where energy continues its journey toward the surface of the star as heat associated with thermal gradients. This part of the star is called the *convection zone*. The surface of the star, called the *photosphere*, emits light in the visible part of the electromagnetic spectrum. The star is engulfed in a stellar atmosphere called the *chromosphere*. The chromosphere is a layer of hot gases surrounding the photosphere.

The luminosity of a star is the total energy radiated per second by the star. The luminosity of the sun is approximately 3.8×10^{26} W. Radiation from the sun is comparable to the radiation emitted by a black body at $6000°$ K [Sørenson, 2000, page 26]. The amount of radiation from the sun that reaches the earth's atmosphere is called the *solar constant* and is approximately equal to 1370 watts per square meter. The solar constant varies with time because the earth's axis is inclined and the earth follows an elliptical orbit around the sun. The distance between a point on the surface of the earth and the sun varies throughout the year. To account for the time dependence, we write the solar constant as a function of time $S(t)$.

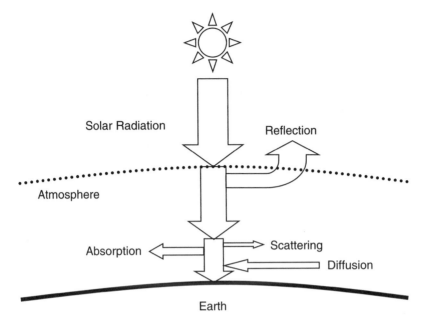

Figure 7-4. Solar radiation and the earth-atmosphere system.

The amount of solar radiation that reaches the surface of the earth depends on the factors illustrated in Figure 7-4. The flux of solar radiation incident on a surface placed at the edge of the earth's atmosphere depends on the time of day and year, and the geographical location of the surface. The geographical location of the surface can be identified by its latitude θ_{lat} and longitude θ_{long}. The incident flux of solar radiation at the edge of the atmosphere is

$$E_{inc}(t, \theta_{lat}, \theta_{long}) = S(t)\cos\theta(t, \theta_{lat}, \theta_{long}) \qquad (7.3.1)$$

where the angle $\theta(t, \theta_{lat}, \theta_{long})$ is the angle between the incident solar flux at time t and the normal to the surface at latitude θ_{lat} and longitude θ_{long}.

Some incident solar radiation is reflected by the earth's atmosphere. The fraction of solar radiation that is reflected back into space by the earth-atmosphere system is called the *albedo*. The albedo is approximately 0.35, which is due to clouds (0.2), atmospheric particles (0.1), and reflection by the earth's surface (0.05). The flux that enters the atmosphere is reduced by the albedo, thus

$$E_{atm}(t, \theta_{lat}, \theta_{long}) = E_{inc}(t, \theta_{lat}, \theta_{long})(1 - a) \qquad (7.3.2)$$

Once in the atmosphere, solar radiation can be absorbed in the atmosphere or scattered away from the earth's surface by atmospheric particulates such as air, water vapor, dust particles, and aerosols. Some of the scattered light eventually reaches the surface of the earth as diffused light. Solar radiation that reaches the earth's surface from the disk of the sun is called *direct solar radiation* if it has experienced negligible change in its original direction of propagation.

GLOBAL WARMING

Solar radiation heats the earth. The average temperature at the earth's surface is approximately 287° K, and typically varies from 220° K to 320° K [Sørenson, 2000, page 26]. Daily variations in temperature are due to the rotation of the earth around its axis. Seasonal variations in temperature are due to the rotation of the earth around the sun and the angle of inclination of the earth's axis relative to the ecliptic plane.

We noted in the previous chapter that measurements of ambient air temperature show a global warming effect that corresponds to an increase in the average temperature of the earth's atmosphere.[3] The increase in atmospheric temperature can be traced to the beginning of the twentieth century [Lide, 2002, page 14–32] and is associated with the combustion of fossil fuels. When a carbon-based fuel burns, carbon can react with oxygen and nitrogen in the atmosphere to produce carbon dioxide, carbon monoxide, and nitrogen oxides (often abbreviated as NO_x). The combustion by-products, including water vapor, are emitted into the atmosphere in gaseous form. Some of the gaseous by-products are called greenhouse gases because they capture the energy in sunlight that is reflected by the earth's surface and reradiate the energy in the form of infrared radiation. Greenhouse gases include carbon dioxide, methane, and nitrous oxide, as well as other gases such as volatile organic compounds and hydrofluorocarbons.

Global warming due to the absorption of reflected sunlight and subsequent emission of infrared radiation is called the *greenhouse effect* because greenhouse walls allow sunlight to enter the greenhouse and then trap reradiated infrared radiation. The greenhouse effect is illustrated in Figure 7-5. Some of the incident solar radiation from the sun is absorbed by the earth, some is reflected into space, and some is captured by chemicals in the atmosphere and reradiated as infrared radiation (heat). The reradiated energy would have escaped the earth as reflected sunlight if greenhouse gases were not present in the atmosphere.

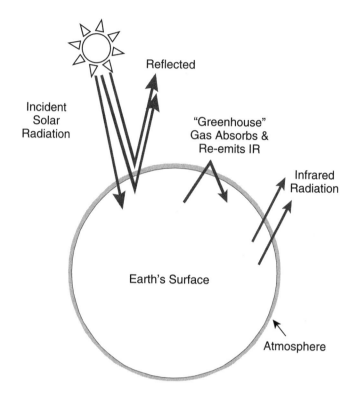

Figure 7-5. The greenhouse effect.

Carbon dioxide (CO_2) is approximately 83% of the greenhouse gases emitted by the United States as a percent of tonnes of carbon or carbon equivalent. Wigley, Richels and Edmonds [1996] projected ambient CO_2 concentration through the twenty-first century. Preindustrial atmospheric CO_2 concentration was on the order of 288 parts per million. Atmospheric CO_2 concentration is currently at 340 parts per million. The concentration of CO_2 that would establish an acceptable energy balance is considered to be 550 parts per million. To achieve the acceptable concentration of CO_2 through the next century, society would have to reduce the volume of greenhouse gases entering the atmosphere. The Kyoto Protocol is an international treaty that was negotiated in Kyoto, Japan in 1997 to establish limits on the amount of greenhouse gases a country can emit into the atmosphere. The Kyoto Protocol has not been accepted worldwide. Some countries believe the greenhouse gas emission limits are too low, and would adversely affect national and world economies without solving the problem of global warming. Research is underway to develop the technology needed to capture and store greenhouse gases in geologic formations as an economically viable means of mitigating the increase in greenhouse gas concentration in the atmosphere. The storage of greenhouse gases in an environmentally acceptable manner is called sequestration.

7.4   PASSIVE SOLAR

Passive solar energy technology integrates building design with environmental factors that enable the capture or exclusion of solar energy. Mechanical devices are not used in passive solar energy applications. We illustrate passive solar energy technology by considering two simple but important examples: the roof overhang and thermal insulation.

ROOF OVERHANG

Sunlight that strikes the surface of an object and causes an increase in temperature of the object is an example of direct solar heat. Direct solar heating can cause an increase in temperature of the interior of buildings with windows. The windows that allow in the most sunlight are facing south in the northern hemisphere and facing north in the southern hemisphere. Figure 7-6 illustrates two seasonal cases. The figure shows that the

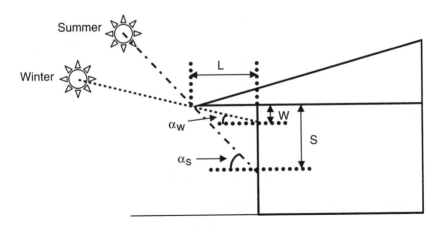

Figure 7-6. Roof overhang.

maximum height of the sun in the sky varies from season to season because of the angle of inclination of the earth's axis of rotation relative to the ecliptic plane. The earth's axis of rotation is tilted 23.5° from a line that is perpendicular to the ecliptic plane.

One way to control direct solar heating of a building with windows is to build a roof overhang. The roof overhang is used to control the amount of sunlight entering the windows. Figure 7-6 illustrates a roof overhang. The bottom of the window is at a distance S from the base of the roof. The length L of the roof overhang can be calculated using the trigonometric relationship

$$L = \frac{S}{\tan \alpha_S} \quad \text{or} \quad L = \frac{W}{\tan \alpha_W} \qquad (7.4.1)$$

where the angles and distances are defined in Figure 7-6. The shortest roof overhang is obtained if the summer angle α_S is used, while the longest roof overhang is obtained if the winter angle α_W is used.

Passive solar cooling is achieved when the roof overhang casts a shadow over the windows facing the sun. In this case, the roof overhang is designed to exclude sunlight, and its associated energy, from the interior of the building. Alternatively, the windows may be tinted with a material that reduces the amount of sunlight entering the building. Another way to achieve passive solar cooling is to combine shading with natural ventilation.

Passive solar heating is the capture and conversion of solar energy into thermal energy. The technology for passive solar heating can be as simple as using an outdoor clothesline to dry laundry, or designing a building to capture sunlight during the winter. In the latter case, the building should be oriented to collect sunlight during cooler periods. Sunlight may enter the building through properly positioned windows that are not shaded by a roof overhang, or through sunroofs. The sunlight can heat the interior of the building, and it can provide natural light. The use of sunlight for lighting purposes is called *daylighting*. An open floor plan in the building interior maximizes the affect of daylighting and can substantially reduce lighting costs.

THERMAL CONDUCTIVITY AND INSULATION

Solar energy may be excluded from the interior of a structure by building walls that have good thermal insulation. Thermal insulation can reduce the consumption of energy for air conditioning in the summer and for heating

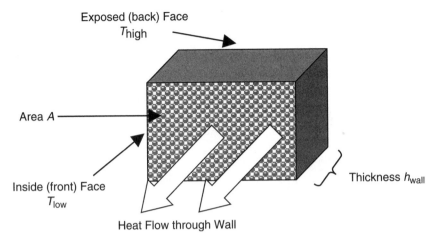

Figure 7-7. Thermal conductivity of an insulated wall.

in the winter. The quality of thermal insulation for a wall with the geometry shown in Figure 7-7 can be expressed in terms of thermal conductivity and thermal resistance.[4]

The rate of heat flow through the insulated wall shown in Figure 7-7 depends on wall thickness h_{wall}, the cross-sectional area A transverse to the direction of heat flow, and the temperature difference between the exposed and inside faces of the wall. We can write the temperature difference across the wall ΔT_{wall} as the temperature of the exposed back face T_{high} minus the temperature of the inside front face T_{low}, thus

$$\Delta T_{wall} = T_{high} - T_{low} \tag{7.4.2}$$

The rate of heat flow through the insulated wall H_{wall} is

$$H_{wall} = k_{wall}A\frac{\Delta T_{wall}}{h_{wall}} = k_{wall}A\frac{T_{high} - T_{low}}{h_{wall}} \tag{7.4.3}$$

where k_{wall} is a constant of proportionality called the *thermal conductivity*. Thermal conductivity is a measure of heat flow through a material and depends on the material. Metals have relatively high thermal conductivities. For example, the thermal conductivity of copper is 385 W/(m·°K). By contrast, the thermal conductivity of insulating brick is 0.15 W/(m·°K).

A more general form of Equation (7.4.3) is

$$H_{\text{wall}} = \frac{dQ_{\text{wall}}}{dt} = -k_{\text{wall}} A \nabla T \tag{7.4.4}$$

where Q_{wall} is the flow of heat through the wall and ∇T is the temperature gradient across the wall.

Equation (7.4.3) can be written in terms of thermal resistance R as

$$H_{\text{wall}} = A \frac{T_{\text{high}} - T_{\text{low}}}{R} \tag{7.4.5}$$

Thermal conductivity and thermal resistance obey the inverse relationship

$$R = \frac{h_{\text{wall}}}{k_{\text{wall}}} \tag{7.4.6}$$

Thermal resistance R is proportional to the thickness h_{wall} of the thermal insulator. An increase in thickness of the thermal insulator increases thermal resistance and reduces heat flow through the insulated wall. Using the preceding examples, a 1 m thick wall of copper with a thermal conductivity of 385 W/(m·°K) has a thermal resistance of 1 m / [385 W/(m·°K)] = 0.0026 m²·°K/W. For comparison, a 1 m thick wall of insulating brick with a thermal conductivity of 0.15 W/(m·°K) has a thermal resistance of 1 m / [0.15 W/(m·°K)] = 6.7 m²·°K/W.

We have considered thermal insulation here as a passive solar technology example. Thermal insulation is also an energy conservation technology. Thermal insulation in walls can keep heat out of a structure during the summer and keep heat in during the winter. Consequently, thermal insulation can reduce the demand for energy to cool a space during the summer and heat a space during the winter. This reduces the demand for energy, and makes it possible to conserve, or delay the use of, available energy.

7.5 ACTIVE SOLAR

Active solar energy refers to the design and construction of systems that collect and convert solar energy into other forms of energy such as heat and electrical energy. Active solar energy technologies are typically mechanical systems that are used to collect and concentrate solar energy. We study solar heat collectors here as an illustration of active solar energy technology. Another example, a solar power plant, is discussed later in this chapter.

SOLAR HEAT COLLECTORS

Solar heat collectors capture sunlight and transform radiant energy into heat energy. Figure 7-8 is a diagram of a solar heat collector. Sunlight enters the collector through a window made of a material like glass or plastic. The window is designed to take advantage of the observation that sunlight is electromagnetic radiation with a distribution of frequencies. The window in a solar heat collector is transparent to incident solar radiation and opaque to infrared radiation.

The heat absorber plate in the solar heat collector is a dark surface, such as a blackened copper surface, that can be heated by the absorption of solar energy. The surface of the heat absorber plate emits infrared radiation as it heats up. Sunlight enters through the window, is absorbed by the heat absorber plate, and is reradiated in the form of infrared radiation. Greenhouses work on the same principle; the walls of a greenhouse allow sunlight to enter and then trap reradiated infrared radiation. The window of the solar heat collector is not transparent to infrared radiation, so the infrared radiation is trapped in the collector.

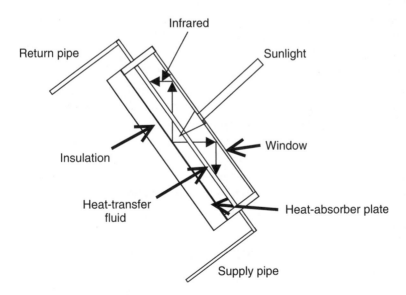

Figure 7-8. Solar heat collector.

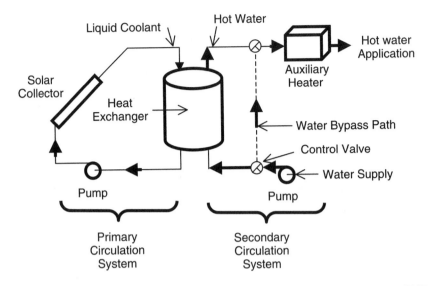

Figure 7-9. Solar heating system [after Cassedy and Grossman [1998, page 282]; and Energy Technologies and the Environment, U.S. Department of Energy Report No. DOE/EP0026 [June 1981].

The solar heat collector must have a means of transferring collected energy to useful energy. A heat transfer fluid such as water is circulated through the solar heat collector in Figure 7-8 and carries heat away from the solar heat collector for use elsewhere. Figure 7-9 illustrates a solar heating system for residential or commercial use.

The solar heating system sketched in Figure 7-9 uses solar energy to heat a liquid coolant such as water or anti-freeze. The heat exchanger uses heat from the liquid coolant in the primary circulation system to heat water in the secondary circulation system. The control valve in the lower right of the figure allows water to be added to the secondary circulation system. An auxiliary heater in the upper right of the figure is included in the system to supplement the supply of heat from the solar collector. It is a reminder that solar energy collection is not a continuous process. A supplemental energy supply or a solar energy storage system must be included in the design of the heating system to assure continuous availability of heat from the solar heating system.

ENERGY CONVERSION EFFICIENCY

The temperature of a solar heat collector does not increase indefinitely because the window and walls of the solar heat collector cannot prevent energy from escaping by conduction and radiation. The collector will emit thermal radiation according to the Stefan–Boltzmann law when its temperature is greater than ambient temperature. The Stefan–Boltzmann law says that the net energy ΔQ_{rad} radiated through a surface area A by an object at absolute temperature T with surroundings at absolute temperature T_e during a time interval Δt is

$$\frac{\Delta Q_{\text{rad}}}{\Delta t} = A e \sigma \left(T^4 - T_e^4 \right) \tag{7.5.1}$$

where σ is the Stefan–Boltzmann constant, and e is the thermal emissivity of the object at absolute temperature T. Thermal emissivity is a dimensionless quantity, and the thermal emissivity of a black body is 1. The temperature of the solar heat collector will increase until thermal equilibrium is established.

The energy balance for thermal equilibrium must include energy output as well as energy loss, thus

$$E_{\text{input}} = E_{\text{output}} + E_{\text{loss}} \tag{7.5.2}$$

The energy conversion efficiency η_{shc} of the solar heat collector is then given by

$$\eta_{\text{shc}} = \frac{E_{\text{output}}}{E_{\text{input}}} = 1 - \frac{E_{\text{loss}}}{E_{\text{input}}} \tag{7.5.3}$$

The efficiency η_{shc} depends on the increase in temperature relative to ambient temperature, the intensity of solar radiation, and the quality of thermal insulation. An example of an expression for the efficiency η_{shc} for a solar heat collector with commercial insulation is

$$\eta_{\text{shc}} = a_0 + b_0 \left(T - T_{\text{amb}} \right) \frac{I_{\text{smax}}}{I_s} \tag{7.5.4}$$

where a_0, b_0 are empirical constants, T_{amb} is ambient temperature in degrees Celsius, T is the temperature in degrees Celsius of the solar heat collector at a given time, I_s is incident solar intensity at a given time, and I_{smax} is the maximum solar intensity observed at the location of the solar heat collector.

Hayden [2001, page 140] presented an energy conversion efficiency example for a solar heat collector characterized by empirical constants $a_0 = 80\%$, $b_0 = -0.89/°$ C and at a location with a maximum solar intensity $I_{smax} = 950$ W/m^2. Efficiency η_{shc} is in percent for these constants, and the temperatures T, T_{amb} are in degrees Celsius. The negative sign in empirical constant b_0 shows that an increase in the temperature of the solar heat collector T relative to ambient temperature T_{amb} causes a decrease in efficiency, and a decrease in incident solar intensity I_s from its maximum value I_{smax} causes a decrease in efficiency. The incident solar intensity I_s can decrease by more than 50% in cloudy (or smoggy) conditions relative to I_{smax}. The efficiency of converting solar energy to heat decreases because there is less solar energy impinging on the collector. An increase in solar heat collector temperature T relative to ambient temperature T_{amb} causes a decrease in intensity because of energy losses associated with convection and thermal radiation. The loss of energy by convection and radiation causes a decrease in energy conversion efficiency.

7.6 SOLAR POWER PLANTS

Society is beginning to experiment with solar power plants, and a few are in commercial operation. Solar power plants are designed to provide electrical power on the same scale as plants that rely on nuclear or fossil fuel. They use reflective materials like mirrors to concentrate solar energy. The solar power tower and the Solar Electric Generating Station in Southern California are examples of solar power plants. They are described in the following sections.

SOLAR POWER TOWER

Figure 7-10 is a sketch of a solar power tower with a heliostat field. The heliostat field is a field of large, sun-tracking mirrors called heliostats arranged in rings around a central receiver tower. The heliostats concentrate sunlight on a receiver at the top of the tower. The solar energy heats a fluid inside the receiver.

Figure 7-11 is a sketch of the geometry of the sun-tracking mirrors relative to the central receiving station. The heliostats must be able to rotate to optimize the collection of light at the central receiving station. Computers control heliostat orientation. As a ring of heliostats gets farther away from the tower, the separation between the ring and adjacent, concentric rings must increase to avoid shading one ring of mirrors by an adjacent ring.

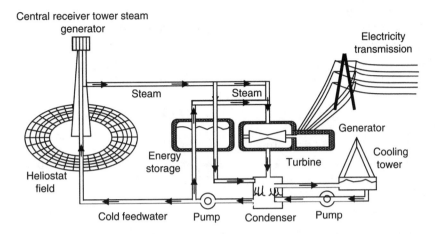

Figure 7-10. Solar power tower schematic [after Kraushaar and Ristinen [1993, page 172]; and Solar Energy Research Institute (now U.S. National Renewable Energy Laboratory), Golden, Colorado].

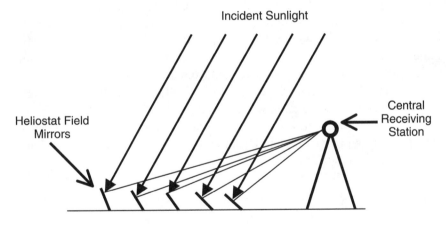

Figure 7-11. Solar tower sketch.

The first solar power plant based on the solar power tower concept was built in the Mojave Desert near Barstow, California, in the 1980s. The solar-thermal power plant at Barstow used 1900 heliostats to reflect sunlight to the receiver at the top of a 300-foot tall tower. The sunlight generates heat to create steam. The steam is used to drive a turbine or it can be stored for later use. The first solar power tower, Solar One, demonstrated the feasibility of collecting solar energy and converting it to electrical energy. Solar One was a 10-MWe plant. The heat transfer fluid in Solar One was steam.

The Solar One installation was modified to use molten nitrate salt as the heat transfer fluid. The modified installation, called Solar Two, was able to improve heat transfer efficiency and thermal storage for the 10-MWe demonstration project. The hot salt could be retrieved when needed to boil water into steam to drive a generator turbine.

SOLAR ELECTRIC GENERATING SYSTEMS

A Solar Electric Generating System (SEGS) consists of a large field of solar heat collectors and a conventional power plant. The SEGS plants in Southern California use rows of parabolic trough solar heat collectors. The collectors are sun-tracking reflector panels, or mirrors. The sunlight reflected by the panels is concentrated on tubes carrying heat transfer fluid. The fluid is heated and pumped through a series of heat exchangers to produce superheated steam. The steam turns a turbine in a generator to produce electricity.

For extended periods of poor weather, solar power plants must use auxiliary fuels in place of sunlight. A prototype SEGS plant used natural gas as an auxiliary fuel. Goswami, et al. [2000, Section 8.7] reported that, on average, 75% of the energy used by the plant was provided by sunlight, and the remaining 25% was provided by natural gas. They further reported that solar collection efficiencies ranged from 40% to 50% , electrical conversion efficiency was on the order of 40% , and the overall efficiency for solar to electrical conversion was approximately 15% .

The overall efficiency η_{SEGS} of a solar electric generating system is the product of optical efficiency η_o, thermal conversion efficiency η_c, and thermodynamic efficiency η_t:

$$\eta_{\text{SEGS}} = \eta_o \eta_c \eta_t \qquad (7.6.1)$$

The optical efficiency is a measure of how much sunlight is reflected into the system. The thermal conversion efficiency is a measure of how much sunlight entering the system is converted to heat in the system. The thermodynamic efficiency is a measure of how much heat in the system is converted to the generation of electricity.

SEGS plants are designed to supply electrical power to local utilities during peak demand periods. In Southern California, a peak demand period would be a hot summer afternoon when the demand for air conditioning is high. This is a good match for a SEGS plant because solar intensity is high. Peak demand periods also correspond to periods of high pollution.

One benefit of a SEGS plant is its ability to provide electrical power without emitting fossil fuel pollutants such as nitrous oxide (a component of smog) and carbon dioxide (a greenhouse gas).

ENDNOTES

1. The principles of nuclear physics are discussed in several sources. See Garrod [1984, Chapter 8], Williams [1991], Serway, et al. [1997, Chapter 13], Bernstein, et al. [2000, Chapter 15], and Lilley [2001, especially Chapters 1 and 10].
2. The primary references used for solar energy are Sørensen [2000], Goswami, et al. [2000], Bernstein, et al. [2000], and Serway, et al. [1997].
3. The discovery of the risk of global warming is reviewed by Weart [1997]. Global warming is discussed in several sources, including Wiser [2000, Chapter 15], Ristinen and Kraushaar [1999, Chapter 10], Cassedy and Grossman [1998, Chapter 6], and Elliott [1997, Chapter 2].
4. See Young and Freedman [2000, pages 479–480] for more details about thermal conductivity and thermal resistance.

EXERCISES

7-1. A. Calculate the binding energy per nucleon for nuclei in which the number of neutrons N equals the number of protons Z for values of Z in the range $1 \leq Z \leq 120$.
B. Plot the binding energy per nucleon as a function of mass number A.

7-2. Calculate the maximum value of binding energy per nucleon B/A with respect to mass number A. Assume the number of neutrons N equals the number of protons Z. Remember that $A = N + Z$ and calculate $\partial(B/A)/\partial A = 0$.

7-3. A. Calculate the surface area of a sphere that has the sun at its center and a radius equal to one astronomical unit, which is the mean distance from the earth to the sun.
B. Use the solar constant of 1370 W/m^2 to estimate the luminosity of the sun.

7-4. Sunlight is solar radiation that is propagating radially outward from the sun. The mean distance from the earth to the sun is about $r_s = 150 \times 10^{11}$ m, and the equatorial radius of the earth is approximately $r_e = 6.38 \times 10^6$ m. Estimate the fraction of sunlight that is intercepted by the earth. Hint: Assume the earth is a sphere with radius r_e moving in a circular orbit around the sun at distance r_s.

7-5. A. Estimate the lifetime of a hydrogen-burning star on the Main Sequence of the H–R diagram if the star mass is two times the mass of the sun. Express your answer in years.
B. Is the luminosity of the star greater than or less than the luminosity of the sun?

7-6. A. Suppose the sun is 73.5° above the horizon during the summer and 26.5° above the horizon during the winter. Calculate the roof overhang length that will completely shade a window during the summer. The base of the window is 2.0 m below the base of the roof.
B. How long must the roof overhang be if we use the average of the summer and winter angles given in Part A?

7-7. A. The thermal resistance R, or R-value, of insulation is given by

$$Q_h = \frac{A \left(T_{high} - T_{low} \right)}{R}$$

where Q_h is the flow rate of heat (BTU/hr) across a wall with cross-sectional area A (ft^2). The heat flows from the side of the wall with the higher temperature T_{high} (°F) to the side with lower temperature T_{low} (°F). Estimate the flow rate of heat across a wall that is 8 ft high and 10 ft long. The temperature difference is $T_{high} - T_{low} = 60°$ F, and the R-value of the wall is 15 (°F · ft^2/BTU/hr).
B. Express the heat flow rate calculated in Part A in SI units.

7-8. A. Use Equation (7.5.6) and the data in Section 7.5 to plot energy conversion efficiency η_{shc} for a solar heat collector with commercial insulation versus temperature difference $(T - T_{amb})$. The temperature difference should cover the range $0°$ C $\leq (T - T_{amb}) \leq 90°$ C for temperature expressed in degrees Celsius. Assume the incident solar intensity I_s is half the maximum solar intensity I_{smax}. Note: Efficiency should be in the range $0\% \leq \eta_{shc} \leq 100\%$.

B. Repeat the exercise in Part A assuming incident solar intensity I_s equals maximum solar intensity I_{smax}.

C. Compare the plots in Parts A and B.

7-9. Suppose a solar heat collector is 10° C warmer than the ambient temperature of 20° C. Use the Stefan–Boltzmann law to estimate the net energy flux F_{rad} radiated by the collector. The net energy flux is the net energy radiated per unit time per unit area. Assume thermal emissivity is 1.

7-10. A. The overall efficiency η_{SEGS} of a solar electric generating system is the product of optical efficiency η_o, thermal conversion efficiency η_c, and thermodynamic efficiency η_t. Find the minimum and maximum overall efficiencies for the efficiency range $0.70 < \eta_o < 0.80$, $0.35 < \eta_c < 0.50$, and $\eta_t \approx 0.35$.

B. If the system is exposed to 2×10^9 W of solar radiation, estimate the minimum and maximum power that can be provided by the system using the efficiencies calculated in Part A.

7-11. A. An array of solar mirrors covers an area of 50 hectares and has a peak electric power production of 10 MWe. Calculate the average power output for a capacity factor of 20%.

B. How many equivalent arrays of solar mirrors would be needed to provide 1000 MWe average power output?

C. How much area would be covered (in km^2) by those arrays?

7-12. A. The annual average solar intensity in the United States is approximately 200 W/m^2. If 100% of the incident sunlight could be converted to useful energy, how much land would be needed (in m^2) to provide the 100 exajoules of energy consumed annually by the United States at the end of the twentieth century?

B. The area of the continental United States (excluding Alaska and Hawaii) is approximately 8×10^{12} m^2. What percentage of this area would be covered by the land area calculated in Part A?

CHAPTER EIGHT

Solar Electric Technology

We discussed passive and active solar energy in the previous chapter. Here we consider solar electric technology. Solar electric technology uses solar energy to directly generate electricity. We begin our study by introducing concepts from quantum mechanics that give us the background we need to develop a relatively sophisticated understanding of solar electric technology. We then apply these ideas to two topics that are fundamental to solar electric technology: the photoelectric effect and photovoltaic devices.

8.1 HEISENBERG'S UNCERTAINTY PRINCIPLE

A revolution in physics was born at the beginning of the twentieth century that affects our understanding of energy today.[1] Prior to 1900 scientists were able to use the classical physics of Isaac Newton, James Clerk Maxwell, and many others to explain virtually all known experimental measurements. Only the observations of a few experiments seemed to withstand the concerted attack of nineteenth-century theorists. Yet an eventual understanding of these experiments required the reevaluation, if not the overthrow, of many previously unchallenged concepts. The constancy of the speed of light in vacuum, for example, was understood only when Albert Einstein rejected Newton's view of absolute and independent space and time. The resulting mathematical link between space and time has physical implications that are still being studied today. Another experimental observation, the radiation emitted by heated black bodies, was explained by the rejection of another long-established concept: the continuousness of energy. The black body experiment discussed in Chapter 3 established the idea that energy existed in discrete bundles called *quanta*. One of the most peculiar and philosophically significant consequences of the quantum theory is the uncertainty principle proposed by the German physicist Werner Heisenberg[2] in 1927.

Heisenberg suggested that a limit exists to the level of knowledge we can obtain about nature. Heisenberg pointed out that the position and momentum of a particle could not be measured with arbitrary accuracy. The uncertainties in position Δx and momentum Δp in one dimension must satisfy the inequality

$$\Delta x \Delta p \geq \frac{\hbar}{2} \qquad (8.1.1)$$

where $\hbar = h/(2\pi)$ is pronounced "h bar" and h is Planck's constant. One of the significant differences between classical physics and quantum physics is represented by Heisenberg's uncertainty principle. In particular, classical physics corresponds to the case in which $h \rightarrow 0$ and both position and momentum can be determined to infinite accuracy. The fact that Planck's constant is a non-zero, finite, albeit small, value implies that we cannot measure both position and momentum to an arbitrarily small uncertainty. The limitation pointed out by Heisenberg is not something we can reduce by improving our measuring techniques. It is a limitation imposed by nature.

Consider, as an illustration, a simple experiment. We shall attempt to measure the speed of an 8-ball rolling on a billiard table. This seems simple enough until we further require that the 8-ball motion must not be changed by our measurement. Suppose we try to measure the speed of the 8-ball by measuring the length of time it takes the 8-ball to travel a known distance between two markers A and B. This should let us calculate a speed for the 8-ball. How accurate is our calculation?

The accuracy of our calculation depends on how accurately we can measure time duration and length. If we can measure time duration and length to whatever accuracy we please, we can find the speed of the 8-ball to the same unlimited degree of accuracy. The problem is that our measurements of time duration and length cannot be made with perfect accuracy. To measure time duration and length, we must also detect the 8-ball as it passes markers A and B. How are we to know when the 8-ball passes our markers?

Suppose we fire bullets from each of the markers. The bullets are fired in a direction that is perpendicular to the path of the 8-ball and at regular time intervals. Two sand boxes are located in the lines of fire of the bullets. We fire the bullets with sufficient frequency that at least one of the bullets will hit the 8-ball as it passes a marker. The bullets are our projectiles, and the 8-ball is the target. The 8-ball is detected when the regularity of the bullets entering our sand boxes is interrupted. The problem with this technique is obvious. When the bullet strikes the 8-ball at our first marker, the 8-ball will be deflected by the collision. As a consequence, the motion of the 8-ball is significantly changed by our detection method.

Instead of bullets, let us use marbles as our projectiles. Doing this gives us a better technique for measuring the speed of the 8-ball because the 8-ball is deflected less by a marble than by a bullet. We can improve our measurement further by using an even smaller object than a marble to detect the presence of the 8-ball as it passes our markers. If we use a particle of light, a photon, we obtain a very good measurement of the speed of the 8-ball. Our measurement is still not arbitrarily accurate, however, because a submicroscopic object like a photon will still deflect the 8-ball, but to such a small extent that for most practical purposes the deflection can be ignored.

Now we repeat the experiment, but we replace our target 8-ball with an electron. In this case we cannot ignore the effect of the detection projectile, the photon, on our measurement of the speed of the electron because the dynamical properties of the electron and the photon are comparable. On the scale of submicroscopic particles, such as the electron and the photon, it is not possible to arbitrarily reduce the effect of the experimentalist on the system being observed. As Heisenberg noted:

> **Q8-1.** In classical physics it has always been assumed either that this interaction [between observer and observed object] is negligibly small, or else that its effect can be eliminated from the result by calculations based on "control" experiments. This assumption is not permissible in atomic physics; the interaction between observer and object causes uncontrollable and large changes in the system being observed, because of the discontinuous changes characteristic of atomic processes. [Sambursky, 1975, pg. 518]

Scientists now believe that Heisenberg's uncertainty principle applies not only to atomic physics, but to all phenomena describable by the quantum theory; including phenomena explained by classical physics.

The philosophical impact of Heisenberg's uncertainty principle is staggering. French physicist and Prince Louis Victor de Broglie observed that Heisenberg's uncertainty principle makes it

> **Q8-2.** "impossible to attribute simultaneously to a body a well-defined motion and a well-determined place in space and time" [de Broglie, 1966, pg. 122].

If we cannot specify with certainty both the location and the motion of an object at a given time, then the determinism of classical physics fails.

Yet we have already seen how successful classical physics is in describing a wide range of natural phenomena, particularly macroscopic phenomena. Quantum theory accounts for the success of classical physics by containing the classical theory as a special case of the more comprehensive quantum theory.

8.2   BOHR'S COMPLEMENTARITY AND WAVE–PARTICLE DUALITY

You may have noticed something unusual in our discussion of quantum theory thus far. We have used light as both a wave and a particle in our description of quantum processes. Light behaved as a wave in the black body experiment, and light was treated as a particle in our discussion of the uncertainty principle. Is light both a wave and a particle? If we perform Experiment A to determine if light is a wave, we would find light behaving like a wave. On the other hand, if we perform Experiment B to determine if light is a particle, which it should not be since light behaved like a wave in Experiment A, we would discover that light behaves like a particle in Experiment B. How can this be?

If we carefully analyze our experiments, we discover that our measuring techniques make it impossible to simultaneously measure the wave and particle properties of light. Were we to devise an experiment to measure the particle properties of light, such as locating a photon at a point in space, the information we need to simultaneously study the wave properties of light, such as extension throughout space, would be unattainable at worst, and inaccurate at best. Likewise, an experiment to measure the wave properties of light would be unable to provide accurate information about the particle properties of light. The limitations on our experiments are not due to any fault we can correct, but are natural limitations associated with Heisenberg's uncertainty principle. Consequently, we must speak of a wave–particle duality: light behaves like a wave and like a particle depending on the experimental circumstances. Furthermore, experiments have shown that wave–particle duality applies to all submicroscopic objects such as electrons, protons, and neutrons.

Wave–particle duality may be thought of as complementary descriptions of reality. Each view is correct for a limited range of applications, but both views are necessary for a complete description. This is the essence of the complementarity principle enunciated by the Danish physicist Niels Bohr.

Bohr showed that classical physics could no longer be used to adequately describe the spectrum of the atom. He used the idea of quantized energy to correctly calculate the spectrum of hydrogen, and advocated the complementarity principle.

After a series of debates with Albert Einstein, who believed until his death that quantum theory was incomplete, Bohr's view was adopted as the traditional interpretation of quantum mechanics and became known as the Copenhagen Interpretation. It represents the state of a physical system with a mathematical function known as the wavefunction. In this view, the wavefunction is a complete description of the subjective knowledge of the observer rather than the objective state of the observed system. New knowledge about the system changes the mathematical description of the relationship between physical systems. According to Heisenberg,

---

Q8-3. The act of recording . . . is not a physical, but rather, so to say, a mathematical process. With the sudden change of our knowledge also the mathematical presentation of our knowledge undergoes a course of sudden change. [Sambursky, 1975, pg. 538]

The change of our knowledge is embodied in the wavefunction and is interpreted from this perspective as a reduction of the wavefunction. Its interpretation is discussed in more detail in the following section.

8.3   BORN'S PROBABILISTIC VIEW

The German-born British physicist Max Born said that the calculational procedure used to describe observations in quantum theory should be interpreted in terms of probabilities. The probabilities of interest to us are objective probabilities that can be calculated using a repeatable, well-defined procedure. For example, the number of times an event can occur in a set of possible outcomes is the frequency of occurrence of the event. The frequency of occurrence is considered an estimate of the probability of occurrence of the event.

Probability satisfies a few basic properties. Suppose we denote the probability of observing an event A in a sample space S as $P(A)$. Probability must be non-negative; the probability of observing at least one event in the sample space is 1; and the probability of obtaining the empty set (no event in the sample space) is zero. If we use $\emptyset$ to denote the empty set, we can write the properties of probability as

$$P(A) \geq 0$$
$$P(S) = 1 \quad\quad\quad (8.3.1)$$
$$P(\emptyset) = 0$$

If events A and not A (denoted A') are complementary, then

$$P(A) + P(A') = 1 \quad\quad\quad (8.3.2)$$

Probabilities may be either discrete or continuous. A roll of a single die with six sides has six equally probable outcomes. The probability of obtaining any one of the outcomes on a single roll of the die is a discrete probability. We can define a function $f(x)$ as the probability of a discrete random variable X (such as a roll of the die) with an outcome x (the value of the side of the die that is showing after the roll). The ordered pair $\{x, f(x)\}$ is a discrete probability distribution if

$$f(x) \geq 0$$
$$\sum_x f(x) = 1 \quad\quad\quad (8.3.3)$$

where the sum is over all possible outcomes. The mean of the discrete probability distribution is the expectation value $E(X)$ of the discrete random variable X defined by

$$E(X) = \sum_x x f(x) \quad\quad\quad (8.3.4)$$

where the sum is over all possible outcomes.

A continuous probability distribution is defined in terms of a probability density $\rho(x)$ that satisfies the conditions

$$\rho(x) \geq 0$$
$$\int_x \rho(x)\, dx = 1 \quad\quad\quad (8.3.5)$$

where the integral is over all possible outcomes. Notice that the discrete probability distribution obeys a summation rule in Equations (8.3.3) and the continuous probability distribution obeys an integration rule in Equations (8.3.5). The mean of the continuous probability distribution is the expectation value $E(X)$ of the continuous random variable X defined by

$$E(X) = \int_{-\infty}^{\infty} x \rho(x) \, dx \qquad (8.3.6)$$

where the integral is over all possible outcomes.

The probability distribution associated with two or more probability distributions is called a joint probability distribution. For example, the joint probability distribution for two continuous random variables X and Y may be written as $\rho(x, y)$. It must satisfy the non-negativity requirement $\rho(x, y) \geq 0$ for all values of (x, y) and it must satisfy the normalization condition

$$\int_y \int_x \rho(x, y) \, dx \, dy = 1 \qquad (8.3.7)$$

A marginal probability distribution can be obtained from the joint probability distribution by summing over one of the discrete random variables or integrating over one of the continuous random variables. In the case of a joint probability distribution of two random variables $\rho(x, y)$, we have the two marginal probability distributions

$$\rho(y) = \int_x \rho(x, y) \, dx$$

$$\rho(x) = \int_y \rho(x, y) \, dy \qquad (8.3.8)$$

If we divide the marginal probability distribution by the joint probability distribution, we obtain conditional probability distributions $\rho(x \mid y)$ and $\rho(y \mid x)$, thus

$$\rho(y \mid x) = \frac{\rho(x, y)}{\rho(x)}$$

$$\rho(x \mid y) = \frac{\rho(x, y)}{\rho(y)} \qquad (8.3.9)$$

For a conditional probability distribution, we say that $\rho(x \,|\, y)$ is the probability of obtaining x given y.

The remarks we have made here for continuous random variables can be applied with appropriate modifications to discrete random variables. In the case of discrete random variables, the integrals over continuous random variables are replaced by sums over discrete random variables. Both discrete and continuous random variables appear in quantum mechanics. Before discussing quantum mechanics, we consider a few examples of discrete and continuous probability distributions.

Example 8.3.1: Dart Board

Consider the distribution of letters shown in the "dart board" in Table 8-1. There are 16 elements in the table. Each element contains a single letter {a, b, c} that is in either upper case {A, B, C} or lower case {a, b, c}. The table contains three different letters {a, b, c} in two different cases {U = Upper, L = Lower}. We assume that each element is equally likely to be hit by a thrown dart. The joint probability of hitting a letter denoted by the symbol λ with a case denoted by the symbol κ is

$$P(\lambda, \kappa) = \frac{N(\lambda, \kappa)}{N} \qquad (8.3.10)$$

where N is the total number of elements and $N(\lambda, \kappa)$ is the number of elements showing the letter λ with case κ. In our example, the number of elements with upper case letter "A" is $N(a, U) = 1$ and the number of elements with lower case letter "a" is $N(a, L) = 6$.

The marginal probability for case κ is

$$P(\kappa) = \sum_{\lambda} \frac{N(\lambda, \kappa)}{N} = \frac{N(\kappa)}{N} \qquad (8.3.11)$$

Table 8-1
Dart board

B	a	b	C
c	B	B	a
a	A	C	a
B	a	a	B

Table 8-2
Reduced dart board

	a	b	
c			a
a			a
	a	a	

where $N(\kappa)$ is the total number of elements with case κ. The conditional probability for letter λ to appear, given case κ, is

$$P(\lambda|\kappa) = \frac{P(\lambda,\kappa)}{P(\kappa)} = \frac{N(\lambda,\kappa)}{N(\kappa)} \qquad (8.3.12)$$

The conditional probability applies to a reduced set of possible events. As an illustration, if the case is Lower ($\kappa = L$), then we have the reduced dart board shown in Table 8-2 for the conditional probability $P(\lambda|L)$. For our example, we calculate the joint probability of obtaining the letter b and the lower case as $P(b, L) = 1/16$ and the conditional probability of obtaining the letter b given the case is Lower as $P(b | L) = 1/8$.

Example 8.3.2: Discrete, Uniform Probability Distribution

Suppose we have a discrete random variable x that can have n outcomes. The discrete random variable x has a uniform probability distribution if its probability function is

$$f(x) = \frac{1}{n} \qquad (8.3.13)$$

The roll of a die with six sides obeys a uniform probability distribution with $n = 6$.

Example 8.3.3: Exponential Probability Distribution

The continuous random variable X has an exponential distribution if its probability density is

$$\rho(x) = \frac{\exp(-x/m)}{m}, \quad x > 0 \qquad (8.3.14)$$

where m is the mean (or expectation value $E(X)$) of the distribution. The exponential probability distribution can be used to describe the decay of a radioactive substance.

8.4   NONRELATIVISTIC SCHROEDINGER EQUATION

The Born representation, probability theory, and basic definitions of physical operations are at the heart of the derivation of an equation for calculating nonrelativistic quantum mechanical fields. We begin the derivation of a nonrelativistic field equation for a single particle by assuming a conditional probability density $\rho(\vec{x}\,|\,t)$ exists. The symbol $\vec{x}$ for the position vector denotes a set of three coordinates for a particle in the space volume L^3 for which $\rho(\vec{x}\,|\,t)$ has nonzero values. The jth component of the position vector may be written as x^j where the index $j = 1, 2, 3$ signifies space components. The index is written as a superscript for technical reasons, and the expanded form of $\rho(\vec{x}\,|\,t)$ is the conditional probability density $\rho\left(x^1, x^2, x^3\,|\,t\right)$.

A word about notation is in order here. In Euclidean space, the length of a vector can be written in several equivalent ways, thus

$$x \cdot x = \vec{x} \cdot \vec{x} = x_1^2 + x_2^2 + x_3^2 = \sum_{j=1}^{3} x^j x_j = \sum_{i=1}^{3}\sum_{j=1}^{3} g_{ij} x^i x^j, \qquad x^j = x_j$$

(8.4.1)

The elements of the matrix $\{g_{ij}\}$ are

$$g_{ij} = \begin{bmatrix} 1 & 0 & 0 \\ 0 & 1 & 0 \\ 0 & 0 & 1 \end{bmatrix}$$

(8.4.2)

in Euclidean space. The matrix $\{g_{ij}\}$ is called the *metric tensor*. Its elements depend on the coordinate system we are working in. The index j can be lowered or raised by using the metric tensor, thus

$$\sum_{j=1}^{3} g_{ij} x^j = x_i, \qquad \sum_{j=1}^{3} g^{ij} x_j = x^i$$

(8.4.3)

In the case of the metric tensor for Euclidean space shown in Equation (8.4.2), there is no difference between x^j and x_j. This is not true in other spaces, as we see in Chapter 9. A vector with components $\{x^j\}$ is called a *contravariant vector*, and a vector with components $\{x_j\}$ is called a *covariant vector*. Contravariant vectors are the vectors you were introduced to in first year physics.

It is useful to rewrite Equations (8.4.1) and (8.4.3) using Einstein's summation convention. Einstein introduced the notation that a sum is implied over an index if the index appears twice in a term. Thus Equations (8.4.1) and (8.4.3) are written as

$$x \cdot x = \vec{x} \cdot \vec{x} = x_1^2 + x_2^2 + x_3^2 = x^j x_j = g_{ij} x^i x^j, \qquad x^j = x_j \qquad (8.4.4)$$

and

$$g_{ij} x^j = x_i, \qquad g^{ij} x_j = x^i \qquad (8.4.5)$$

where Einstein's summation convention has simplified the notation. The mathematical apparatus described in Equations (8.4.1) through (8.4.5) is more sophisticated than it needs to be for our discussion in this chapter, but the mathematics gains importance when considering particles traveling close to the speed of light c in relativistic systems. It is therefore worthwhile to introduce the concepts now so we can use them in a more familiar setting before considering more difficult applications in Chapter 9.

The symbol t is a nonrelativistic evolution parameter that conditions the probability density $\rho(\vec{x} \mid t)$. The nonrelativistic evolution parameter t is interpreted as time in the sense of Newton, that is, it increases monotonically and is an absolute time. The interpretation of the meaning of the nonrelativistic evolution parameter t becomes important in the next chapter when we consider the relativistic extension of quantum mechanics.

The conditional probability distribution $\rho(\vec{x} \mid t)$ represents the probability of observing a particle at position x in space at time t. The field equations described next must be solved to determine the probability distribution.

According to probability theory, $\rho(\vec{x} \mid t)$ must be positive definite

$$\rho(\vec{x} \mid t) \geq 0 \qquad (8.4.6)$$

and normalizable over the volume integral

$$\int_{L^3} \rho(\vec{x} \mid t) \, d^3 x = 1 \qquad (8.4.7)$$

where

$$d^3x = dx^1 dx^2 dx^3 = \prod_{j=1}^{3} dx^j \tag{8.4.8}$$

When applied to a particle, Equation (8.4.7) says that the particle is observed somewhere in space at all times. We now know that many particles are short-lived; they can disappear in a process known as particle decay. A revision to Equation (8.4.7) is discussed in Chapter 9.

Conservation of probability implies that $\rho(\vec{x} \mid t)$ obeys the continuity equation

$$\frac{\partial \rho(\vec{x} \mid t)}{\partial t} + \sum_{j=1}^{3} \frac{\partial \rho(\vec{x} \mid t) v^j}{\partial x^j} = 0 \tag{8.4.9}$$

where ρv^j is the jth component of probability flux of the particle, that is, probability flux is the probability density times the jth component of its velocity. Equation (8.4.9) can be written using Einstein's summation convention for Latin indices in the form

$$\frac{\partial \rho(\vec{x} \mid t)}{\partial t} + \partial_j \rho(\vec{x} \mid t) v^j = 0; \qquad \frac{\partial \rho(\vec{x} \mid t) v^j}{\partial x^j} \equiv \partial_j \rho(\vec{x} \mid t) v^j \tag{8.4.10}$$

where we sum over the Latin index j and we have introduced the notation $\partial_j \equiv \partial/\partial x^j$.

The probability distribution $\rho(\vec{x} \mid t)$ satisfies the positive-definite requirement $\rho(\vec{x} \mid t) \geq 0$ if we write $\rho(\vec{x} \mid t)$ in the form

$$\rho(\vec{x} \mid t) = \psi^*(\vec{x}, t)\, \vec{\psi}(x, t) \tag{8.4.11}$$

where function ψ^* is the complex conjugate of ψ. The function ψ is called many things: wave function, eigenfunction, state function, "psi" function, and probability amplitude are some of the most common names for ψ. Max Born first suggested an interpretation of ψ in the context of probability theory. The formulation of quantum theory presented here highlights the relationship between quantum theory and probability theory.

The function ψ can be written in the form

$$\psi(\vec{x}, t) = [\rho(\vec{x} \,|\, t)]^{1/2} \exp[i\xi(\vec{x}, t)] \tag{8.4.12}$$

where the phase $\xi(\vec{x}, t)$ is a real scalar function. For our purposes, we note that a scalar is a quantity that does not have a directional component. Physical examples of scalars include temperature and mass. By contrast, vectors such as velocity and force have both magnitude and direction.

Quantum mechanical field equations can be derived from the probability formalism by expressing the velocity of the particle as

$$v^j(\vec{x}, t) = \frac{1}{m} \left[\hbar \frac{\partial \xi(\vec{x}, t)}{\partial x_j} - \frac{e}{c} A^j(\vec{x}, t) \right] \tag{8.4.13}$$

where $\hbar = h/2\pi$ is Planck's constant divided by 2π. The vector $\vec{A}$ in Equation (8.4.13) is the vector potential introduced in Chapter 2. Given these assumptions, the field equation for a particle interacting with an electromagnetic field has the form

$$i\hbar \frac{\partial \psi(\vec{x}, t)}{\partial t} = \left[\frac{\pi^j \pi_j}{2m} + U_\mathrm{I} \right] \psi(\vec{x}, t) \tag{8.4.14}$$

where U_I is the interaction potential. The differential operator

$$\pi^j = \frac{\hbar}{i} \frac{\partial}{\partial x_j} - \frac{e}{c} A^j = \frac{\hbar}{i} \partial^j - \frac{e}{c} A^j \tag{8.4.15}$$

is called the momentum operator and has units of momentum. Equation (8.4.14) together with Equation (8.4.15) is called the Schroedinger equation. In this formulation, the definition of expectation value of an observable Ω is

$$\langle \Omega \rangle \equiv \int_{L^3} \psi^* \Omega \psi \, d^3x \tag{8.4.16}$$

An example of an observable is the position vector $\vec{x}$. We illustrate how to use these equations by considering the relatively simple problem of a noninteracting particle.

Example 8.4.1: Free Particle in Quantum Theory

A noninteracting system corresponds to the conditions $\{A^\mu\} = 0$ and $U_I = 0$ so that Equation (8.4.14) becomes

$$i\hbar \frac{\partial \psi_f}{\partial t} = -\frac{\hbar^2}{2m} \nabla^2 \psi_f \qquad (8.4.17)$$

where we have introduced the Laplacian operator

$$\nabla^2 \psi_f = \left[\frac{\partial^2}{\partial (x^1)^2} + \frac{\partial^2}{\partial (x^2)^2} + \frac{\partial^2}{\partial (x^3)^2} \right] \psi_f = \left[\frac{\partial^2}{\partial x^2} + \frac{\partial^2}{\partial y^2} + \frac{\partial^2}{\partial z^2} \right] \psi_f \qquad (8.4.18)$$

in a Cartesian coordinate system. Equation (8.4.17) has the solution

$$\psi_f(\vec{x}, t) = \eta^{1/2} \exp\left[-\frac{i\hbar}{2m} \left(k^j k_j \right) t + i k_j x^j \right] \qquad (8.4.19)$$

where η is the constant calculated from the probability normalization requirement, and we are using Einstein's summation convention. The magnitude of the constant vector $\vec{k}$ is the wave number and is inversely proportional to wavelength. The vector $\vec{k}$ is related to particle momentum below. If Equation (8.4.19) is compared with Equation (8.4.12), you can identify the real scalar function $\xi(\vec{x}, t)$. The probability density $\rho(\vec{x} \mid t) = \psi^*(\vec{x}, t)\psi(\vec{x}, t)$ is a uniform probability density that depends on the limits of the integral in the normalization condition Equation (8.4.7).

The expectation value of particle velocity is

$$\langle v^j \rangle = \int_{L^3} \psi_f^* \pi^j \psi_f \, d^3x = \frac{1}{m} \int_{L^3} \psi_f^* \left[\frac{\hbar}{i} \frac{\partial}{\partial x_j} \right] \psi_f \, d^3x \qquad (8.4.20)$$

Substituting Equation (8.4.19) into (8.4.20) gives

$$\frac{d \langle x^j \rangle}{dt} = \langle v^j \rangle = \frac{\hbar \langle k^j \rangle}{m} = \frac{\langle p^j \rangle}{m} \qquad (8.4.21)$$

where we have related particle momentum to wave number. The relationship is

$$\langle p^j \rangle = \hbar \langle k^j \rangle \tag{8.4.22}$$

Equation (8.4.22) shows that the mechanical properties of a particle can be related to wavelike properties. Integrating Equation (8.4.21) from t to $t + \delta t$ gives

$$\delta \langle x^j \rangle = \frac{\hbar \langle k^j \rangle}{m} \delta t = \frac{\langle p^j \rangle}{m} \delta t \tag{8.4.23}$$

Equation (8.4.23) expresses the trajectory of a free particle in terms of expectation values of the components of the position vector and the components of the momentum vector.

QUANTIZED ENERGY

The concept of quantized energy arises when we solve Schroedinger's equation using a solution of the form

$$\psi(\vec{x}, t) = \eta^{1/2} \psi_E(\vec{x}) \exp\left[-\frac{iEt}{\hbar} \right] \tag{8.4.24}$$

Substituting Equation (8.4.24) into (8.4.14) gives

$$E\psi_E(\vec{x}) = \hat{H}\psi_E(\vec{x}) \tag{8.4.25}$$

where $\hat{H}$ (pronounced "H hat") is the Hamiltonian operator

$$\hat{H} = \frac{\pi^j \pi_j}{2m} + U_I = \frac{1}{2m}\left(\frac{\hbar}{i}\frac{\partial}{\partial x_j} - \frac{e}{c}A^j \right)\left(\frac{\hbar}{i}\frac{\partial}{\partial x^j} - \frac{e}{c}A_j \right) + U_I \tag{8.4.26}$$

The operator $\hat{H}$ is the quantum mechanical analog of the Hamiltonian introduced in Chapter 1. The value of E calculated from Equation (8.4.25) is the energy of the system. Energy may have either discrete or continuous values depending on the interaction potentials $\{A^j, U_I\}$ in Equation (8.4.26). Examples of physical systems that have discrete energy values include the harmonic oscillator and the hydrogen atom.

Example 8.4.2: Harmonic Oscillator in Quantum Theory

The energy of a harmonic oscillator with spring constant k_{HO} in one space dimension and potential energy $U_I = k_{HO} x^1 = k_{HO} x$ is

$$E_n = \left(n + \frac{1}{2}\right) \hbar \omega_{HO}, \qquad \omega_{HO} = \sqrt{\frac{k_{HO}}{m}}, \qquad n = 0, 1, 2, 3, \ldots \quad (8.4.27)$$

The allowed values of energy are determined by the integer n, which is called a quantum number. Each value of the quantum number corresponds to a quantum state. The set of discrete energy values gives a discrete energy spectrum. The difference in energy between two harmonic oscillator states is

$$E_{n+1} - E_n = \left(n + 1 + \frac{1}{2}\right) \hbar \omega_{HO} - \left(n + \frac{1}{2}\right) \hbar \omega_{HO} = \hbar \omega_{HO} \quad (8.4.28)$$

The energy difference in Equation (8.4.28) is a quantum of energy. The frequency ω_{HO} in Equation (8.4.28) is angular frequency with the unit radians/second or rad/s. Angular frequency ω with the unit radians per second is related to frequency f with the unit Hertz by the equation

$$\omega = 2\pi f \qquad\qquad (8.4.29)$$

The quantum energy of a harmonic oscillator E_{HO} can be expressed in terms of frequency and angular frequency by

$$E_{HO} = \hbar \omega_{HO} = \hbar (2\pi f_{HO}) = h f_{HO} \qquad\qquad (8.4.30)$$

The harmonic oscillator can change energy states only by absorbing or releasing (emitting) a quantum of energy $\hbar \omega_{HO}$. Notice that the ground state of the harmonic oscillator, that is, the state with quantum number $n = 0$, has energy $E_0 = \hbar \omega_{HO}/2$. The existence of a non-zero ground state energy E_0 for a harmonic oscillator is a purely quantum mechanical effect.

8.5   PATH INTEGRAL FORMALISM

American physicist Richard Feynman presented a path integral formulation of quantum mechanics in his 1942 Princeton University Ph.D.

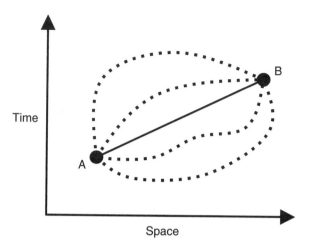

Figure 8-1. Feynman paths.

Dissertation [Feynman, 1942]. Feynman related the motion of a particle to probability amplitude. An example of the probability amplitude is the function ψ introduced in Equation (8.4.11). Feynman postulated that the probability of a particle following a path between two points in a region of space is the absolute square of the sum of probability amplitudes representing all possible paths in the region. The paths contribute equally in magnitude, but the phase of their contribution is the classical action introduced in Section 1.5. Figure 8-1 illustrates a few paths, including the classical path drawn as a solid line, connecting points A and B on a space-time diagram. The Feynman path integral approach is illustrated in the following example. It provides another perspective on the relationship between classical and quantum physics.

Example 8.5.1: Path Integral Analysis of a Free Particle

We consider the motion of a free particle in one space dimension. The probability amplitude ϕ at a space point x_{i+1} and time $t + \varepsilon$ for an infinitesimal duration ε is calculated from the probability amplitude at x_i, t using the relation

$$\phi(x_{i+1}, t + \varepsilon) = \frac{1}{A} \int_{-\infty}^{\infty} e^{iS(x_i)/\hbar} \phi(x_i, t)\, dx_i \qquad (8.5.1)$$

where $dx_i = x_{i+1} - x_i$ and A is a normalization constant. The action S is expressed in terms of the Lagrangian L by

$$S(x) = \int L(\dot{x})\, dt \tag{8.5.2}$$

The Lagrangian for a nonrelativistic free particle with mass m is

$$L(\dot{x}) = \frac{m}{2}\dot{x}^2 = \frac{m}{2}\left(\frac{dx}{dt}\right)^2 \tag{8.5.3}$$

We use the trapezoidal rule to approximate the integral in Equation (8.5.1). The result is

$$S(x_i) = \varepsilon\frac{m}{2}\left(\frac{x_{i+1} - x_i}{\varepsilon}\right)^2 \tag{8.5.4}$$

Defining the variables

$$\delta_x \equiv x_{i+1} - x_i, \qquad x \equiv x_{i+1} \tag{8.5.5}$$

lets us write Equation (8.5.1) in the form

$$\phi(x, t + \varepsilon) = \frac{1}{A}\int_{-\infty}^{\infty} \exp\left[i\frac{m}{2\hbar\varepsilon}\delta_x^2\right]\phi(x - \delta_x, t)\, d\delta_x \tag{8.5.6}$$

Performing Taylor series expansions of the probability amplitudes to first order in ε and second order in δ_x gives

$$\phi(x, t) + \varepsilon\frac{\partial\phi}{\partial t} = \frac{1}{A}\int_{-\infty}^{\infty} \exp\left[i\frac{m}{2\hbar\varepsilon}\delta_x^2\right]\left[\phi(x, t) - \delta_x\frac{\partial\phi}{\partial x} + \frac{\delta_x^2}{2}\frac{\partial^2\phi}{\partial x^2}\right]d\delta_x \tag{8.5.7}$$

The integral on the right hand side of Equation (8.5.7) is evaluated using the integrals

$$\int_{-\infty}^{\infty} \exp\left[-a^2z^2\right] dz = \frac{\sqrt{\pi}}{a}$$

$$\int_{-\infty}^{\infty} z \exp\left[-a^2z^2\right] dz = 0 \qquad (8.5.8)$$

$$\int_{-\infty}^{\infty} z^2 \exp\left[-a^2z^2\right] dz = \frac{\sqrt{\pi}}{2a^3}$$

Equation (8.5.7) becomes

$$\phi(x,t) + \varepsilon\frac{\partial\phi}{\partial t} = \frac{1}{A}\frac{\sqrt{\pi}}{a}\phi(x,t) + \frac{1}{4A}\frac{\sqrt{\pi}}{a^3}\frac{\partial^2\phi}{\partial x^2}, \quad a = \sqrt{\frac{-im}{2\hbar\varepsilon}} \qquad (8.5.9)$$

To assure continuity of the probability amplitudes when $\varepsilon \to 0$, we set $A = \sqrt{\pi}/a$ so that Equation (8.5.9) becomes

$$\phi(x,t) + \varepsilon\frac{\partial\phi}{\partial t} = \phi(x,t) + \frac{1}{4a^2}\frac{\partial^2\phi}{\partial x^2} = \phi(x,t) - \frac{\hbar\varepsilon}{2im}\frac{\partial^2\phi}{\partial x^2} \qquad (8.5.10)$$

Equation (8.5.10) is satisfied for all ε when the coefficients of the ε terms are equal, thus

$$\frac{\partial\phi}{\partial t} = -\frac{\hbar}{2im}\frac{\partial^2\phi}{\partial x^2} \qquad (8.5.11)$$

Multiplying Equation (8.5.11) by $i\hbar$ gives

$$i\hbar\frac{\partial\phi}{\partial t} = -\frac{\hbar^2}{2m}\frac{\partial^2\phi}{\partial x^2} \qquad (8.5.12)$$

Equation (8.5.12) is the Schroedinger equation for a free particle.

8.6   TUNNELING: A QUANTUM MECHANICAL PHENOMENON

Tunneling is one of the most fascinating and unique predictions of quantum theory.[3] Imagine tossing a tennis ball at a wall. When the ball hits the wall, it bounces back. According to quantum theory there is a chance, though extremely small, that the ball will actually pass through the wall. For an object the size of a tennis ball, the chance for it to tunnel through the wall is so small that we would never expect to see it happen. If we consider a much smaller object, such as an atom or molecule, impinging on a much thinner barrier than a wall, such as the thin membrane of a living cell, then the chance for an atom or molecule to tunnel through the cell membrane becomes large enough for us to expect the event to occur and be observable. This tunneling phenomenon is pictorially represented in Figure 8-2.

An incident beam of projectiles is directed from the left of Figure 8-2 toward the rectangular barrier. A beam, or collection, of projectiles is used to simplify our interpretation of the tunneling effect. The projectiles could be tennis balls or atoms. From a practical point of view, projectiles the size of atoms or smaller must be used if we hope to observe tunneling.

The barrier shown in Figure 8-2 is a potential energy barrier. For example, a physical wall may be viewed as a potential energy barrier. Imagine a bullet is fired at a wall. The bullet can penetrate the wall if the bullet has enough kinetic energy to push the molecules of the wall out of the way. The molecular bonds holding the molecules of the wall together create a potential energy barrier. As another example, an atomic electron may be

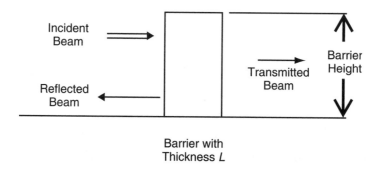

Figure 8-2. Quantum mechanical tunneling.

bound in the electrostatic potential well of a nucleus unless the electron has enough kinetic energy to escape the well. The height of the barrier in Figure 8-2 represents the magnitude of the potential energy barrier, and the height of the arrow for the incident beam represents the magnitude of the kinetic energy of the beam. If the arrow representing the kinetic energy of the incident beam of particles is above the top of the potential energy barrier, the barrier will not prevent the incident beam of particles from passing to the right in the figure. On the other hand, if the arrow is below the top of the barrier, two things can happen.

In classical physics, the particles would all be reflected at the barrier and the spatial position of the left-hand side of the barrier would be called a *turning point*. In quantum physics, when the incident beam of projectiles strikes the barrier, some projectiles are reflected while the rest are transmitted, or tunnel, through the barrier into the classically forbidden region. The precise number will depend on such factors as the thickness of the barrier, the height of the barrier, and the speed of the incident beam. We can calculate the probabilities for the projectile being reflected and transmitted.

The probability that a projectile with mass m and energy E will be transmitted through a barrier with width L and potential energy U is

$$P_T = \left\{ 1 + \frac{1}{4} \left[\frac{U^2}{E(U-E)} \right] \sinh^2 \left(\frac{\sqrt{2m(U-E)}}{\hbar} L \right) \right\}^{-1} \qquad (8.6.1)$$

The probability that the particle will be reflected is

$$P_R = 1 - P_T = 1 - \left\{ 1 + \frac{1}{4} \left[\frac{U^2}{E(U-E)} \right] \sinh^2 \left(\frac{\sqrt{2m(U-E)}}{\hbar} L \right) \right\}^{-1} \qquad (8.6.2)$$

because probability conservation requires that $P_R + P_T = 1$.

If we use a beam of projectiles, the fraction transmitted is equal to the probability of transmission. Likewise, the fraction reflected equals the probability of reflection. On the other hand, we could use one projectile and repeat the experiment many times. In this case the fraction of experiments in which the projectile is transmitted equals the transmission probability, and similarly for the reflection probability. Either approach, using a beam of projectiles or one projectile many times, will show that part of the incident beam is reflected back away from the barrier and part of the incident beam is transmitted through the barrier. These results are a consequence of quantum theory.

Tunneling lends experimental support to the quantum theory. It has been used in the development of transistors, and as a tool—such as the scanning tunneling microscope—for studying the surface of structures on an atomic level. From the point of view of energy, tunneling plays a role in nuclear reactions.

8.7   INTERPRETATION OF QUANTUM THEORY

The most fundamental quantity of a quantum theoretic calculation is the probability amplitude or state vector. Many interpretations[4] of the state vector have been suggested, but two in particular have had the greatest historical impact. They are the statistical interpretation and the Copenhagen interpretation. We consider both. It should be noted, however, that acceptance of one interpretation rather than another is not necessary for useful application of the quantum theory.

Any theory can be characterized as the combination of formalism and its interpretation. Scientists can use the formalism of quantum theory (the calculation procedure) without fully understanding what every aspect of the formalism means. In other words, an interpretation of a mathematically well-defined formalism does not have to be entirely correct before we can get useful information from it. Attempts have been made to show that a difference in interpretation can lead to calculational differences that can be tested experimentally. One such attempt, first proposed by Austrian physicist Erwin Schroedinger, highlights the difference between the statistical and Copenhagen interpretations.

Schroedinger imagined the following scenario. Suppose we place a live cat in a closed chamber. In addition to the cat, we also put a bottle of cyanide in the chamber, a radioactive atom, and a device that breaks the bottle when the radioactive atom decays. The decay of the atom is a chance occurrence; we know it will happen, but we do not know when. If we look in the chamber periodically, we do not know beforehand whether we will find a live cat or a dead cat. We can, in principle, construct a state vector S(live) describing the chamber with a live cat and an undecayed atom. We can also, in principle, construct a state vector S(dead) describing the dead cat and the decayed atom. The state vector for the system as a whole

S(system) is a combination of the state vectors S(live) and S(dead). How should we interpret the state vector S(system)? This question is the essence of Schroedinger's cat paradox.

To proponents of the Copenhagen interpretation, S(system) completely describes everything we can know about Schroedinger's cat. If we look in the chamber when there is a fifty percent chance the atom has decayed, the state vector S(system) will contain equal parts of S(live) and S(dead). Does that mean the cat is half dead and half alive? Copenhagen proponents answer no. They say the act of looking into the chamber forced the state vector S(system) into one of the two state vectors S(live) or S(dead). This is known as reduction of the state vector and is the mechanism Copenhagen proponents use to avoid a nonsensical answer to Schroedinger's cat paradox.

Proponents of the statistical interpretation take a different view. They envision a very large number of cats in the predicament devised by Schroedinger. The state vector S(system) applies to this very large number of systems. The number of systems must be large enough to provide statistically meaningful information. If we look in all of the chambers when there is a fifty percent chance the atoms in each chamber have decayed, we will find half of the chambers contain dead cats and the other half contain live cats. In the statistical interpretation, the state vector S(system) does not apply to a single system, but to a very large number of similarly prepared systems known as an ensemble. The state vector S(system) is making a statistical statement about an ensemble of systems. The statistical interpretation does not need the state vector reduction mechanism required by the Copenhagen interpretation.

Quantum theory is an indeterminate theory in the sense that the result of a measurement is not predictable with certainty; only the probability of various possible results can be calculated. The founders of quantum theory recognized that the concepts of determinism and causality (cause and effect) were changing from indisputable laws describing nature to approximations of nature. Niels Bohr, for example, realized causality might have limits:

> **Q8-4.** Just as freedom of will is an experimental category of our psychic life, causality may be considered as a mode of perception by which we reduce our sense impressions to order. At the same time, however, we are concerned in both cases with idealizations whose natural limitations are open to investigation. [Sambursky, 1975, pg. 535]

Regarding particle collisions, Max Born wrote:

Q8-5. From the standpoint of our quantum mechanics there is no quantity that causally fixes the effect of a collision in an individual event ... I myself am inclined to renounce determinism in the atomic world. [Pais, 1982, pg. 900]

The emergence of quantum theory has changed our views of causality, determinism, and the nature of matter and energy. Determinism, the seemingly unshakable foundation of classical physics, has failed. Taking its place is a theory based on chance, a theory inherently lacking predestination for any individual system.

Quantum mechanics and special relativity were two of the great achievements of physics in the first half of the twentieth century. Quantum mechanics explained the behavior of submicroscopic systems, and special relativity described the behavior of objects moving at speeds approaching the speed of light. The two theories were first combined in the 1920s to form a theory now known as relativistic quantum theory. It is discussed in the next chapter.

8.8 PHOTOVOLTAICS

Solar electric technologies are designed to convert light from the sun directly into electrical energy.[1] Some of the most important solar electric processes are the photoelectric effect and photovoltaics. To understand these processes, we must understand how sunlight can provide the energy to make electrons move in some types of materials. On a more fundamental level, we must understand how electrons move in materials. We begin this section with a discussion of the free electron gas model of conductivity. This gives us the necessary background to provide a modern discussion of the photoelectric effect and photovoltaics.

FREE ELECTRON THEORY OF CONDUCTIVITY

Paul Drude introduced a classical model of conductivity in metals in 1900 [Bernstein, et al., page 394]. Drude's model is the basis of the modern theory of electron conduction in metals. The classical model treats metals as materials that contain immobile positive charges and electrons that are loosely bound valence electrons. *Valence electrons* are the outer electrons, or electrons in the highest energy levels, of elements in the main group of the

Periodic table. Valence electrons can move like a free gas through the metal. Interactions between valence electrons and other objects are negligible between collisions. The velocity of an electron changes instantaneously when the electron does experience a collision. If we call τ the average time between collisions, then the probability of an electron participating in a collision during an interval dt is dt/τ. If the metal is subjected to a temperature change, electrons reach thermal equilibrium by participating in collisions.

The quantum mechanical model of metal conductivity adopts many of the ideas of the classical model, but expresses them in a nonrelativistic quantum mechanical framework. We begin with the three-dimensional Schroedinger equation for a free particle, namely

$$-\frac{\hbar^2}{2m}\nabla^2 \psi(x,y,z) = E\psi(x,y,z) \tag{8.8.1}$$

where m is the mass of the valence electron, E is its energy, $\hbar$ is Planck's constant divided by 2π, and $\psi(x, y, z)$ is the wave function in Cartesian coordinates. We envision a metal as a collection of atoms or molecules occupying the corners of cubes in a cubic lattice with side L. A valence electron occupies each of the corners of the lattice, and we assume Equation (8.8.1) describes the behavior of the valence electron. The wave function for describing this symmetric system must satisfy the boundary condition

$$\psi(x,y,z) = \psi(x+L,y,z) = \psi(x,y+L,z) = \psi(x,y,z+L) \tag{8.8.2}$$

The boundary condition in Equation (8.8.2) is called a *periodic boundary condition*. The periodic boundary condition represents a cubic lattice in k-space, that is, a three-dimensional space with coordinates $\{k_x, k_y, k_z\}$. The spacing of points in k-space is π/L, and a vector in k-space can be written as

$$\vec{k}_F = \frac{\pi}{L}\left(n_x, n_y, n_z\right) \tag{8.8.3}$$

for the integers $n_x = 0, 1, 2, \ldots$; $n_y = 0, 1, 2, \ldots$; $n_z = 0, 1, 2, \ldots$. The vector $\vec{k}_F$ is called the *Fermi vector*. A solution of Equation (8.8.1) that satisfies the periodic boundary condition in Equation (8.8.2) is

$$\psi(x,y,z) \sim \sin\left(\frac{\pi}{L}n_x\,x\right)\sin\left(\frac{\pi}{L}n_y\,y\right)\sin\left(\frac{\pi}{L}n_z\,z\right) \tag{8.8.4}$$

subject to the condition

$$\frac{2mE}{\hbar^2} = \frac{\pi^2}{L^2}\left(n_x^2 + n_y^2 + n_z^2\right) = \vec{k}_F \cdot \vec{k}_F = k_F^2 \tag{8.8.5}$$

where the integers $n_x = 0, 1, 2, \ldots$; $n_y = 0, 1, 2, \ldots$; $n_z = 0, 1, 2, \ldots$ are now considered quantum numbers.

If we recognize that the number of electrons in the ground state (or unexcited state) of a metal is large, then the occupied region of k-space can be viewed as a sphere with radius $k_F = \|\vec{k}_F\|$ where k_F is the magnitude of the Fermi vector. The sphere is called the *Fermi sphere* and contains all occupied, one-electron quantum states. Since each of the quantum numbers $\{n_x, n_y, n_z\}$ is zero or positive, we must consider only one octant of the sphere. A surface called the *Fermi surface* can be defined for the Fermi sphere. The Fermi surface separates the occupied levels from unoccupied levels.

The number of electrons in the Fermi sphere is

$$N = 2\frac{k_F^3}{6\pi^2}V = \frac{k_F^3}{3\pi^2}V \tag{8.8.6}$$

where the volume $V = L^3$. Equation (8.8.6) includes a factor of 2 because electrons obey the Pauli exclusion principle. The Pauli exclusion principle states that two indistinguishable particles cannot occupy the same quantum state, and applies to fermions like electrons, protons, and neutrons. The factor of 2 recognizes that electrons with the same energy can be in different quantum states because they can have spin up or spin down. Electron density in the Fermi sphere is found from Equation (8.8.6) to be

$$n = \frac{N}{V} = \frac{k_F^3}{3\pi^2} \tag{8.8.7}$$

The magnitude of the momentum of an electron at the Fermi surface is the Fermi momentum

$$p_F = \hbar k_F \tag{8.8.8}$$

The corresponding Fermi velocity has the magnitude

$$v_F = \frac{p_F}{m} = \frac{\hbar k_F}{m} \tag{8.8.9}$$

The kinetic energy of the electron is the Fermi energy

$$E_F = \frac{p_F^2}{2m} = \frac{\hbar^2 k_F^2}{2m} = \frac{1}{2}mv_F^2 \qquad (8.8.10)$$

The probability that an electron is in a state with energy E is given by the distribution function

$$f(E) = \frac{1}{\exp[(E - E_F)/(k_B T)] + 1} \qquad (8.8.11)$$

with Boltzmann's constant k_B and a material at temperature T. Equation (8.8.11) is called the *Fermi-Dirac distribution function*. It provides a physical meaning of Fermi energy: an electron has a 50% probability of having the energy $E = E_F$ at any temperature. The Fermi energy can be used to define the Fermi temperature

$$T_F = \frac{E_F}{k_B} \qquad (8.8.12)$$

Fermi energy plays an important role in determining the energy an electron needs to begin moving.[5] Figure 8-3 illustrates the energy levels involved

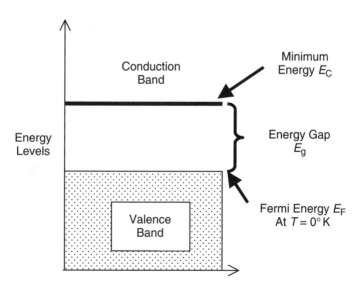

Figure 8-3. Energy levels in a metal.

in electrical conduction in a metal at $0°$ K. The Fermi energy E_F represents the maximum energy of an electron in a noninteracting material. All energy levels below the Fermi energy are filled with electrons, and all levels above the Fermi energy are empty. The lower band filled with electrons is the valence band. The upper band of energy levels is the conduction band. If the metal is at a temperature $T > 0°$ K, a few electrons will have enough thermal energy to exceed the Fermi energy barrier. The distribution of electrons is given by the Fermi-Dirac distribution function.

The energy of conduction E_C is the lowest energy level that must be attained by an electron before it can move through the material. The difference in energy between the conduction energy E_C and the Fermi energy E_F in a metal at $0°$ K is the energy of the gap E_g between the valence band and the conduction band, thus

$$E_g = E_C - E_F \qquad (8.8.13)$$

The energy gap E_g is also called the band gap energy because it is the energy difference between the conduction band with minimum conduction energy E_C and the valence band of energy levels. In a metal at $0°$ K, the maximum energy level in the valence band is the Fermi energy E_F. Energy levels with energies less than the Fermi energy E_F are valence band levels. They correspond to energy levels for bound electrons. Energy levels with energies greater than the conduction energy E_C are free particle energy levels in the conduction band. An electron can move through a material if it has enough kinetic energy KE_e to leap across the band gap and enter the conduction energy levels of the material.

The band gap energy can be used to classify the electrical properties of materials. Figure 8-4 shows the energy levels associated with insulators and semiconductors. The Fermi energy is at the midpoint between the valence band and conduction band. An insulator does not have any energy levels above the valence band or below the conduction band. Electrons must be excited with enough energy to cross the band gap.

A semiconductor has some partially filled energy levels above the valence band and below the conduction band. This effectively reduces the energy needed to excite electrons into the conduction band. A semiconductor doped with electron donor atoms, such as arsenic with five valence electrons, is an n-type semiconductor. A semiconductor doped with electron acceptor atoms, such as boron with three valence electrons, is a p-type semiconductor. The energy levels of the donor atoms are just below the conduction band energy levels, and the energy levels of the acceptor atoms are just above the valence band energy levels.

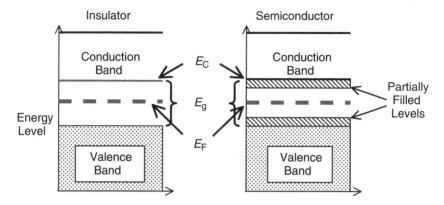

Figure 8-4. Energy levels in an insulator and a semiconductor.

A conductor such as a metal has a relatively small band gap E_{gC}. An insulator such as plastic has a relatively large band gap E_{gI}. Semiconducting materials have an energy gap E_{gS} in the range $E_{gC} < E_{gS} < E_{gI}$. The band gap of an insulator is approximately 10 eV (1.6×10^{-18} J), and a semiconductor has a band gap on the order of 1 eV (1.6×10^{-17} J). For comparison, the average thermal energy of an electron at room temperature is about 0.025 eV (0.04×10^{-17} J).

PHOTOELECTRIC EFFECT

We saw that an electron can move in a metal when its kinetic energy exceeds the Fermi energy, thus $KE_e > E_F$. It was known in the late 1800s that electrons could be ejected from a metal exposed to electromagnetic radiation, but the effect depended on the frequency of the radiation. The effect was called the *photoelectric effect*. Albert Einstein used the concept of quantized energy to explain the photoelectric effect in 1905, the same year he published his special theory of relativity.

An electron cannot be ejected from a metal unless it has enough energy to overcome the work function W of the metal. The work function is the smallest energy needed to extract an electron from the metal, and typically ranges from 2 eV (3.2×10^{-19} J) to 8 eV (12.8×10^{-19} J). The work function depends on the type of material and the condition of its surface. Einstein postulated that a collision between a photon and an electron could transfer enough energy from the photon to the electron to eject the electron from the metal. The photon must have enough energy to overcome the work function. If an electron collides with a photon with frequency ν, the electron will be ejected from the metal if the kinetic energy of the electron satisfies

$$KE_e = h\nu - W > 0 \qquad (8.8.14)$$

where $h\nu$ is the quantum energy of the photon, and h is Planck's constant. An electron ejected from a metal because of a collision with a photon is called a *photoelectron*.

PHOTOVOLTAICS

Photovoltaics can be described as the use of light to generate electrical current.[6] We can make a photovoltaic cell, or photocell, by placing two semiconductors in contact with each other. The semiconductors in a photocell are solid-state materials with atoms arranged in lattice structures. When sunlight shines on the photocell, it can eject a valence electron from an atom in the lattice and leave behind a positive ion. The positive ion is called a *hole* in the lattice of the cell. An electron from a neighboring neutral atom can move from the neutral atom to the positively charged hole. The original neutral atom becomes a hole when the electron moves, and the original hole becomes a neutral atom when the positively charged ion accepts the moving electron. The current of electrons moving from one hole to another can be considered a current of holes moving in the opposite direction as the current of electrons.

A photocell is a diode. A diode can conduct current in one direction but not the opposite direction. The semiconducting material that receives valence electrons and retains positively charged holes is called a p-type semiconductor. It is the positively charged anode shown in Figure 8-5. The semiconducting material that receives valence electrons and becomes negatively charged is called an n-type semiconductor. It is the negatively charged cathode shown in Figure 8-5.

Electrons are ejected by the photoelectric effect when sunlight strikes a p-type semiconductor. An electrical current will flow from the p-type semiconductor to the n-type semiconductor when the electrons have enough energy to leap across the band gap, enter the conduction band, and cross the p-n junction. The *p-n junction* is the interface between the p-type semiconductor and the n-type semiconductor.

Sunlight interacts with electrons in photocells and produces electron-hole pairs. The sunlight must transfer enough energy to the valence electron to enable the electron to leap across the band gap. If the frequency of light

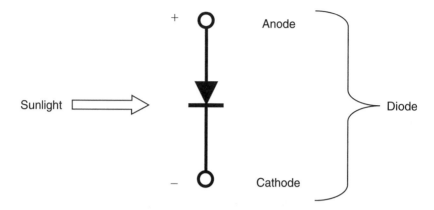

Figure 8-5. Photovoltaic cell.

does not provide enough energy to stimulate the electron leap, the light will be ineffective. On the other hand, if the frequency of light is too large, the electron will receive enough energy to make the quantum leap across the band gap and the rest of the energy will generate heat. Most of the photons in sunlight have energy between 1 eV (1.6×10^{-17} J) and 3 eV (4.8×10^{-17} J) [Serway, et al., 1997, page 471]. The band gap that maximizes the conversion of solar power to electrical power is approximately 1.5 eV (2.4×10^{-17} J).

Photocells are not sources of energy, and they do not store energy. Photocells transform sunlight into electrical energy. When sunlight is removed, the photocell will stop producing electricity. If a photocell is used to produce electricity, an additional system is also needed to provide energy when light is not available. The additional system can be an energy storage system that is charged by sunlight transformed into electrical energy and then stored, or it can be a supplemental energy supply provided by another source.

ENDNOTES

1. Sources on modern physics include Lawrie [1990], Serway, et al. [1997], and Bernstein, et al. [2000].
2. Sources on the standard formulation of quantum theory include Peebles [1992], Weinberg [1995], Serway, et al. [1997], and Bernstein, et al. [2000].
3. Tunneling is discussed in standard quantum mechanics textbooks, such as Peebles [1992], Serway, et al. [1997], and Bernstein, et al.

[2000]. Merzbacher [2002] provides a history of the early days of quantum tunneling.

4. There is extensive literature on the interpretation of quantum mechanics. Some useful references include d'Abro [1951], Jammer [1966], Ballentine [1970], Sambursky [1975], Newton [1980], Bell [1987], Bohm and Hiley [1993], Fanchi [1993], Omnes [1994], Wick [1995], Greenstein and Zajonc [1997], and Tegmark and Wheeler [2001].

5. See Serway, et al. [1997, Chapter 11] for more details about Fermi energy and electrical conduction in metals.

6. The primary references used for solar energy are Sørensen [2000], Goswami, et al. [2000], Bernstein, et al. [2000], and Serway, et al. [1997].

EXERCISES

8-1. Find the phase factor $\xi(\vec{x}, t)$ for a free particle by comparing Equation (8.4.19) with Equation (8.4.12).

8-2. Derive Equation (8.5.9) from (8.5.7).

8-3. Free particle momentum p is related to quantum mechanical wavenumber k by the relation $p = \hbar k = 2\pi\hbar/\lambda$ where λ is the wavelength of the particle. Estimate the wavenumber and wavelength of a proton that is moving at a speed equal to 0.001 c. Express your answer in SI units.

8-4. A. A bowling ball rolls a distance of 30 m in 10 s and has a mass of 5 kg. What is its kinetic energy?
B. Suppose we treat the bowling ball like a quantum mechanical free particle. What is the magnitude of its wave number?
C. Calculate the wavelength of the bowling ball using the relation wave number $k = 2\pi/\lambda$ where λ is wavelength.

8-5. Suppose the Lagrangian for a nonrelativistic particle is $L(\dot{x}, x) = (1/2)m\dot{x}^2 - V(x)$ where V is the potential energy. Use Feynman's path integral analysis to derive Schroedinger's equation for the particle.

8-6. A. Calculate the transmission coefficient P_T and the reflection coefficient P_R for quantum mechanical tunneling through the potential energy barrier shown in Figure 8-2 using the values of physical parameters U, L, E presented in the table below. The parameters U,

L, E are defined in Section 8.6. Assume U, L, E are finite, non-zero values unless stated otherwise.

Case	Parameter	P_T	P_R
1	$U \to 0$		
2	$U \to \infty$		
3	$L \to 0$		
4	$L \to \infty$		
5	$E \to 0$		
6	$E \to \infty$		

B. Which case(s) correspond to tunneling through an impenetrable barrier?
C. Which case(s) correspond to an incident beam with negligible momentum?

8-7. Derive Equation (8.8.5) from Equations (8.8.1) to (8.8.3).

8-8. A. Plot the Fermi-Dirac distribution as a function of energy for a metal at room temperature ($T = 20°$ C). Assume the Fermi energy is 1 eV and let the range of energy be 0 eV $\leq E \leq$ 2 eV. The energy on the horizontal axis of the plot should be expressed in eV, and the scale of the vertical axis should range from 0 to 1.
B. What is the energy value (in eV) that corresponds to a Fermi-Dirac distribution value of 0.5?
C. Estimate the energy interval (in eV) between the Fermi-Dirac distribution values 0.98 and 0.02?

8-9. A. The work function for a metal is approximately 4 eV. What is the minimum frequency of light needed to eject an electron from the metal by the photoelectric effect?
B. What is the corresponding wavelength (in m)?

8-10. A. Suppose we have a joint probability density $\rho(x, y) = 2(x+2y)/3$ that is valid in the intervals $0 \leq x \leq 1, 0 \leq y \leq 1$. Calculate the marginal probability distribution $\rho(y)$.
B. Calculate the conditional probability $\rho(x \mid y) = \rho(0.3 \mid 0.2)$.

8-11. A. The half-life $t_{1/2}$ of carbon-14 is about 5600 years. Half-life is related to the probability of particle decay per unit time λ by $\lambda = \ln 2 / t_{1/2}$. What is λ for carbon-14?
B. If the decay of carbon-14 is described by the exponential probability distribution $\rho(t) = \lambda \exp(-\lambda t)$, what is the mean of the distribution?

CHAPTER NINE

Mass–Energy Transformations

The notion of quantized energy introduced in the previous chapter is useful
for objects moving at speeds that are much less than the speed of light in
vacuum. In such a nonrelativistic system, it is possible to treat matter and
energy as distinct physical quantities. Scientists have learned, however, that
objects moving close to the speed of light obey a different set of rules. The
quantum theory had to be extended to the relativistic domain: a domain
where the distinction between matter and energy disappears.

The most widely accepted theory of elementary particle physics is relativistic quantum theory, and the standard model of elementary particle
physics is the quark model. The quark model is the modern day equivalent
of Democritus' atomic theory. An observation that is central to relativistic
quantum theory is that energy can be converted into mass, and mass can
be converted into energy. We encountered mass–energy transformations in
our discussion of solar energy. Solar energy is generated by mass–energy
transformations that occur in nuclear fusion reactions inside a star. The conversion of mass to energy is a mechanism at work in stars, nuclear weapons,
and nuclear reactors. In this chapter, we acquire a more sophisticated
understanding of mass–energy transformations and nuclear reactions by
becoming familiar with relativistic quantum theory and the quark model.
First, we begin with an introduction to relativity.

9.1 EINSTEIN'S RELATIVITY

The concept of relativity was around long before Albert Einstein published his first paper on relativity in 1905. Relativity is concerned with how
the motion of an observer influences the observer's determination of the
relationships between physical quantities. The idea is quite simple.

Suppose we are moving in a spaceship at half the speed of light relative
to Max, an observer standing still on the ground. The speed of light is

approximately 3×10^8 m/s. Now suppose we are sitting beside each other in the spaceship. How fast am I moving relative to you? You can easily answer this question just by looking at me and deciding if the distance between us is getting larger, smaller, or not changing. If the distance between us is not changing, I am not moving relative to you and you would say my speed is zero. Max disagrees.

To Max, we are both moving at half the speed of light. Which observer is correct? My momentum relative to you is zero, but my momentum relative to Max is non-zero. Which momentum should we use in Newton's laws of motion? If the results obtained by applying Newton's laws depend on the motion of the observer, how can we justifiably say Newton's laws are universally true?

The principle of relativity asserts that the laws of physics should not depend on the relative motion of observers. The laws of physics govern the relationships between measured quantities and should be the same for all observers, even if the numerical values of the measured quantities differ from one observer to the next. The numerical values depend on the observer's frame of reference. We define a reference frame as the coordinate system used to describe a physical system. The principle of relativity represents a "democratic" view of different observers: none of us occupies a preferred reference frame. One person's reference frame is in principle equivalent to the reference frame of anyone else.

Let us return to our imaginary scenario: we are traveling in a spaceship at half the speed of light relative to Max, a motionless observer. Notice that Max is motionless by convention. We could just as easily consider ourselves motionless and Max is moving at half the speed of light. Either point of view is acceptable in principle. For our purposes, it is more convenient to think of Max as being motionless.

Suppose we observe Alice approaching us in another spaceship. Alice's speed relative to Max is two-thirds the speed of light. According to Newton's view of space and time, Max should observe us approaching Alice at a speed in excess of the speed of light. If Max actually measured the speed with which we are approaching Alice as observed from his stationary reference frame, he would find the speed of approach of the two spaceships to be less than the speed of light. In other words, Max's measurement does not agree with Newton's theory. Something must be wrong!

The problem resides in Newton's conception of space and time. He thought space and time were absolute and independent quantities. Einstein resolved this problem by solving another: why is the value of the speed of light independent of the motion of the observer?

Experimentalists had discovered that the speed of light was the same whether you were measuring it while you were at rest or while you were moving at any speed not exceeding the speed of light. Indeed, it appears to be physically impossible for any object to move faster than the speed of light. In his 1905 paper entitled "On the Electrodynamics of Moving Bodies" Einstein presented his explanation of this extraordinary phenomenon.

Before Einstein's 1905 paper, physicists had devised a set of rules for transforming the dynamical description of an object in one uniformly moving (nonaccelerating) reference frame to that in another. The set of rules is called the Galilean coordinate transformation. A simple expression for the Galilean coordinate transformation can be obtained by considering the relative motion of the two frames of reference shown in Figure 9-1. The primed frame F′ is moving with a constant speed v in the x-direction relative to the unprimed frame F. The origins of the two frames F and F′ are at $x = x' = 0$ at time $t = t' = 0$. The Galilean coordinate transformation is

$$
\begin{aligned}
x' &= x - vt \\
y' &= y \\
z' &= z \\
t' &= t
\end{aligned}
\tag{9.1.1}
$$

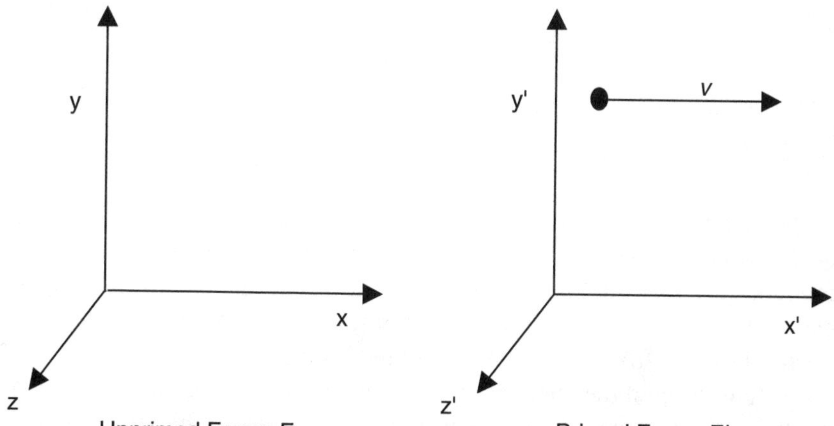

Unprimed Frame F Primed Frame F′

Figure 9-1. Relative motion.

The spatial coordinate x' depends on the values of x and t in the unprimed frame, while the time coordinates t and t' are equal. The equality of t and t' implies that the rate of change of time measured by clocks in two different reference frames does not depend on the relative motion of the two frames. The equality of t and t' expresses Newton's view of time as absolute and immutable.

The Galilean coordinate transformation does not change the form of Newton's laws, but it does change the form of Maxwell's equations. The measurement of the speed of light by Michelsen and Morley showed that Maxwell's equations should not depend on the motion of an inertial observer. Inertial observers are observers whose relative motion is uniform, or nonaccelerating, as in Figure 9-1. Einstein proposed using a new set of transformation rules that made the form of Maxwell's equations independent of the uniform motion of an observer. These new transformation rules, known collectively as the Lorentz transformation after their founder, Dutch physicist Henrik Anton Lorentz, introduced a mutual dependence between space and time coordinates in the primed and unprimed frames. The relationship between the unprimed and primed space and time coordinates for the unprimed and primed frames in Figure 9-1 is given by the Lorentz transformation

$$x' = \frac{x - vt}{\sqrt{1 - \left(\frac{v}{c}\right)^2}}$$

$$y' = y$$

$$z' = z$$

$$t' = \frac{t - x\frac{v}{c^2}}{\sqrt{1 - \left(\frac{v}{c}\right)^2}}$$

$$(9.1.2)$$

The speed v is the speed of the primed frame F' moving relative to the unprimed frame F as measured in the unprimed frame F. Unlike Equation (9.1.1), Equation (9.1.2) shows that space and time depend upon one another in both the primed and unprimed frames. We can recover Equation (9.1.1) from (9.1.2) by taking the limit $c \to \infty$.

Einstein rejected the concept of "absolute time" and adopted the notion of relative time. Time, like space, is a physically measurable quantity and its numerical value depends on the reference frame of the observer. According to Einstein's relativity, the complete description of the trajectory of a moving object requires that we specify the space and time coordinates of each point along the trajectory in a given reference frame. The space and time coordinates are written as space–time coordinates in relativity to show that they are no longer independent variables. A point along the trajectory of a moving object is referred to as an event and is defined by its space–time coordinates in a given reference frame.

Newton's concepts of space and time, though still useful at speeds much less than the speed of light, could no longer be considered fundamental. Furthermore, it became meaningless to specify the duration of an event without specifying the reference frame in which the duration was measured. Thus, to claim an event lasted one day is no better than saying the same event lasted millions of years if the reference frame of the observer is not known.

In addition to replacing Newton's "absolute time" with relative time, Einstein also changed the physics community's understanding of simultaneity. Einstein showed that we should regard the notion of simultaneity as a concept that depends on the motion of the observer. Two events that occur simultaneously in reference frame A may not occur simultaneously from the point of view of an observer in reference frame B that is in motion relative to reference frame A. Thus, as a simple example, if we see the hood and trunk of our moving car pop up at the same time, an observer standing on the ground may correctly claim the hood popped up before the trunk. This dependence of simultaneity on the motion of the observer becomes significant only for speeds comparable to the speed of light; otherwise Newton's ideas are acceptable as a special case of Einstein's more general simultaneity concept.

The theory presented by Einstein in his 1905 paper is known today as the *special theory of relativity*. It is concerned with reference frames moving uniformly with respect to other reference frames. These uniformly moving reference frames are known as nonaccelerating, or inertial, reference frames. As a part of the theory of relativity, Einstein showed that mass and energy are but different manifestations of the same physical quantity. The equivalence of mass and energy is the basis of nuclear power and nuclear weapons.

Einstein was unable to present a theory of relativity in 1905 for reference frames that were accelerating with respect to one another. Twelve years later, in 1917, he published a relativity theory for handling accelerating reference frames. The 1917 theory is called the *general theory of relativity*. The general theory applies to a broad range of reference systems.

The special theory of relativity is widely accepted, whereas the general theory is still considered to be in the verification stage. Much more experimental evidence is available for supporting the special theory than the general theory. Experimental tests of the general theory have been successful, but are less numerous and encompassing than tests of the special theory. Although experimental tests of the general theory are not yet conclusive, most physicists accept the tenets of the general theory, particularly when applied to such macroscopic phenomena as the motion of heavenly bodies, because both the special theory and the general theory have successfully explained data and phenomena that could not be explained as convincingly using alternative theories. The general theory provides a geometric interpretation of the gravitational interaction and predicts, among other things, the existence of black holes (massive, collapsed stars) and the evolution of the universe.

Different assumptions regarding the evolution of the universe result in different mathematical predictions of physical behavior. Each set of assumptions and the results derived from the general theory of relativity for that set of assumptions may be considered a cosmological model. Determining which cosmological model is supported by the physical evidence provides a means of testing the validity of the assumptions underlying each cosmological model and provides insight into the origin of important energy sources, as we discuss in more detail later.

9.2   INVARIANCE, SYMMETRY, AND RELATIVITY

Physical theories consist of a mathematical formalism with a set of rules for relating mathematical quantities and operations to physical quantities and measurements. Relationships between data measurements are described by equations. One approach to constructing the mathematical formalism is to specify how physical variables should behave when observed from different frames of reference. The principle of relativity says that relationships between physically significant quantities should remain invariant with respect to transformations between different reference frames.

Invariance requirements in relativity may be represented mathematically as transformations between coordinate systems. The mathematical objects that represent coordinate transformations are groups. A group is a nonempty set G of elements that has the following properties:

- The elements a, b of the set G can form a new element c of G from the binary operation $c = ab$. Addition and multiplication are examples of binary operations.

- The elements of G obey the associative law $(ab)c = a(bc)$.
- The set G contains an identity element e that satisfies the equality $ae = ea = a$ for every element a in G.
- The set G contains an inverse element a^{-1} for each element a in the set.

As an example, suppose the binary operation is multiplication and the set G is the set of all positive, non-zero real numbers. In this case, G is a group with the identity element 1 and the inverse a^{-1} is $1/a$. Group theory is the field of mathematics that is used to describe symmetry in nature.[1]

Coordinate transformations include three-dimensional rotations and translations in Euclidean space. A translation in the context of coordinate transformations refers to a separation between the origins of the reference frames. For our purposes, Euclidean space is the three-dimensional space in which the length of a vector $\vec{r}$ is given by

$$r^2 = (x^1)^2 + (x^2)^2 + (x^3)^2 \qquad (9.2.1)$$

The Cartesian coordinate components x, y, z are represented by the symbols x^1, x^2, x^3 respectively. The index is written as a superscript to agree with the more sophisticated notation in the modern literature. For future reference, it is useful to rewrite this equation in the form

$$r^2 = \sum_{i=1}^{3} \sum_{j=1}^{3} g_{ij} x^i x^j \quad \text{for} \quad i, j = 1, 2, 3 \qquad (9.2.2)$$

where the coefficient $g_{ij} = 1$ if $i = j$, and $g_{ij} = 0$ if $i \neq j$. The set of coefficients $\{g_{ij}\}$ are called *metric coefficients* and may be written in matrix form as

$$[g_{ij}] = \begin{bmatrix} 1 & 0 & 0 \\ 0 & 1 & 0 \\ 0 & 0 & 1 \end{bmatrix} \qquad (9.2.3)$$

The Euclidean group describes rotations and translations in Euclidean space. The Euclidean group satisfies the requirement of Galilean relativity: the equations of physics should be the same for two observers at rest with respect to each other. Equations that are invariant with respect to Galilean relativity are not necessarily invariant if observers are moving relative to one another. If we want the equations of physics to be valid whether we are standing on land, traveling in sea, or moving through space, then the Euclidean group and Galilean relativity are too restrictive.

EINSTEIN AND RELATIVITY

Albert Einstein's principle of relativity extends Galilean relativity by requiring that the equations of physics be invariant for inertial observers in relative motion. An inertial observer is an observer in a reference frame that is not accelerating. Special relativity requires that inertial observers in equivalent reference frames measure the same speed for the propagation of light in vacuum. Coordinates in the special theory of relativity are no longer defined as components in Euclidean space and an absolute time, but are defined as components of a four-dimensional space–time called *Minkowski space–time*. The length of a four-vector $\{s\} = \{x^0, x^1, x^2, x^3\}$ in Minkowski space–time is

$$s^2 = \sum_{\mu=0}^{3} \sum_{\nu=0}^{3} g_{\mu\nu} x^{\mu} x^{\nu} \quad \text{for} \quad \mu, \nu = 0, 1, 2, 3 \tag{9.2.4}$$

with a metric

$$[g_{\mu\nu}] = \begin{bmatrix} 1 & 0 & 0 & 0 \\ 0 & -1 & 0 & 0 \\ 0 & 0 & -1 & 0 \\ 0 & 0 & 0 & -1 \end{bmatrix} \tag{9.2.5}$$

The 0th component of the space–time four-vector is the time coordinate $x^0 = ct$, where c is speed of light in vacuum, and the three remaining components represent the three space coordinates. The length of the space–time four-vector is

$$s^2 = (x^0)^2 - (x^1)^2 - (x^2)^2 - (x^3)^2$$
$$= c^2 t^2 - (x^1)^2 - (x^2)^2 - (x^3)^2 \tag{9.2.6}$$

The four-vector is called timelike if $s^2 > 0$ and it is called spacelike if $s^2 < 0$. If an object is moving at the speed of light, then the space–time four-vector has length $s^2 = 0$.

ENERGY-MOMENTUM FOUR-VECTOR

Another four-vector that has special importance is the energy-momentum four-vector $\{p^\mu\} = \{p^0, p^1, p^2, p^3\} = \{\frac{E}{c}, p^1, p^2, p^3\}$. The 0th component of the energy-momentum four-vector is the total energy E divided by the speed of light in vacuum. The remaining three spatial components are the components of the momentum vector. The length of the energy-momentum four-vector for a relativistic particle with mass m is given by Equation (9.2.4) with space–time four-vector replaced by the energy-momentum four-vector, thus

$$
\begin{aligned}
m^2 c^2 &= (p^0)^2 - (p^1)^2 - (p^2)^2 - (p^3)^2 \\
&= \frac{E^2}{c^2} - (p^1)^2 - (p^2)^2 - (p^3)^2
\end{aligned}
\tag{9.2.7}
$$

The four-vector is called timelike if $m^2 c^2 > 0$ and it is called spacelike if $m^2 c^2 < 0$. An object moving slower than the speed of light is called a *bradyon* and has a positive mass. It is mathematically possible for an object to be moving faster than the speed of light. Faster-than-light objects are called *tachyons*. In the context of classical special relativity given by Equation (9.2.7), the mass of a faster-than-light object would have to be imaginary, an unphysical result. There are other ways to treat tachyons, but that discussion is beyond the scope of this book.[2] If an object is moving at the speed of light, then the space–time four-vector has length $m^2 c^2 = 0$.

Equation (9.2.7) can be rearranged to let us calculate total energy from mass and momentum:

$$
E^2 = m^2 c^4 + (p^1)^2 c^2 + (p^2)^2 c^2 + (p^3)^2 c^2
\tag{9.2.8}
$$

If the momentum is negligible compared to mass, Equation (9.2.8) reduces to the famous result

$$
E \approx mc^2
\tag{9.2.9}
$$

If our frame of reference is moving with the object, the object is at rest and Equation (9.2.9) becomes the equality

$$
E = mc^2 = m_0 c^2
\tag{9.2.10}
$$

where m_0 is called the rest mass of the object.

Equation (9.2.9) is the basis of the observation that mass can be converted into energy, and vice versa. This observation is put to use in the generation of nuclear energy, and is discussed in more detail later.

9.3 AN ILLUSTRATION FROM PARTICLE PHYSICS

To illustrate the ideas presented in the preceding section, let us consider an application of the scientific method to a particle physics experiment. The purpose of the experiment is to trace the motion of a particle, an electron, in space and time. We are especially interested in measuring the world-line of the electron. This requires some explanation.

The arena of nature is four-dimensional. Three dimensions are spatial (height, length, and width), and the fourth dimension is time. A length is one dimension, an area or surface requires two dimensions, the volume occupied by this book requires three dimensions, and the evolution of a three-dimensional object takes place in the time dimension. Because it is difficult to draw four-dimensional figures on two-dimensional pages, people often draw two-dimensional figures in which one of the dimensions is time and the other is a space dimension. A coordinate system with time on the vertical axis and space on the horizontal axis is the coordinate system for a space–time diagram. The point of intersection of the space and time axes represents the present time and location of an object such as an electron. A space–time diagram is shown in Figure 9-2.

Suppose we wish to trace the motion of an electron in one space dimension. The electron can move forward, backward, or not at all. If the electron does not move at all as time passes, we would trace the "time motion" of the electron up the vertical axis. We would trace down the vertical axis if the electron were traveling into the past. The lines we have traced are called *world-lines*. They represent the motion of the electron through space–time. Every individual point of the world-line is called an *event*. Thus the present time and position of the electron, the origin of the space–time diagram, is an event on the world-line of the electron. These terms are displayed in Figure 9-2 for an electron moving away from its original location as time proceeds into the future. The dashed lines in Figure 9-2 denote the world-line of an object moving at the speed of light. The area bounded by the dashed lines in the future and past resembles a cone in shape and is known as the *light cone*.

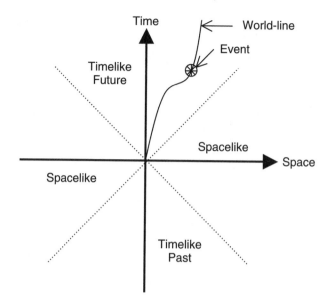

Figure 9-2. Events, world-lines, and the light cone.

Albert Einstein pioneered a theory that forbids any massive object such as the electron from entering the regions of the space–time diagram marked "spacelike." To enter these regions, the object would have to move faster than the speed of light. Since Einstein postulated that no object could attain a speed in excess of the speed of light, a measurement of the electron's world-line can test Einstein's theory. If the electron leaves the light cone, which is forbidden by formal deductions using Einstein's postulates, doubt would be cast on his theory.

Suppose we build a detection system consisting of two detectors and two clocks (Figure 9-3). Detectors 1 and 2 measure the location of the electron along the space axis. Clocks 1 and 2 mark the times when the electron passes through Detectors 1 and 2 respectively.

All location and time measurements are uncertain to some extent. Although small, our detectors have a finite width that obscures the precise location of the electron. Our clocks can be read accurately but not exactly. It is important to realize that sources of uncertainty always exist in physical measurement. To minimize the effects of uncertainty, we repeat the measurements many times. We average our results to obtain reliable results, where "reliable" simply means the final values used to trace the electron's world-line are reproducible within a calculable uncertainty.

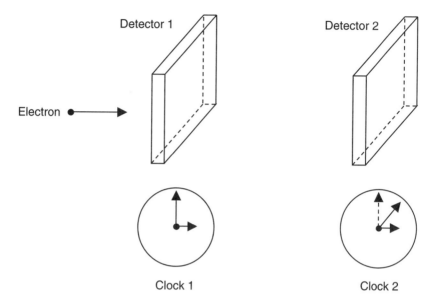

Figure 9-3. Measuring the electron world-line.

Circled dots in Figure 9-4 show the average values of our measurements. The size of each circle represents the uncertainty: the larger the circle, the larger the uncertainty. A straight line marked "Classical" is drawn between the two measurements. This is the simplest world-line we can draw for the electron. It is possible that the electron took the more circuitous path represented by the dotted line, however, the most probable path is the Classical path. We learned from the path integral formulation of nonrelativistic quantum mechanics that paths other than the Classical path play an important role in the modern scientific view of nature. For now it is enough to consider only the straight path. Doing so, we conclude that our measurement of the electron's world-line supports Einstein's theory because the world-line remains within the light cone.

Had the world-line entered the "spacelike" region of the space–time diagram, our experiment would not have supported the theory. Before rejecting the theory, however, we would first examine the experimental arrangement and procedure to make sure that no extraneous factors were present. After checking the operational aspects of the experiment and finding no errors, we would check the formal procedure used to construct the light cone. Again finding no errors, our only recourse would be to revise the theory. This can be done by altering the original hypotheses or starting

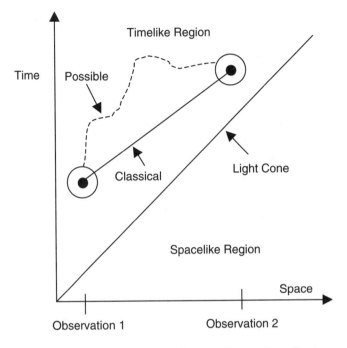

Figure 9-4. The electron world-line on a space–time diagram.

with completely new ones. Scientific progress is made when a new theory explains all of the phenomena explained by an old theory, provides an explanation for unexplained observations, and makes experimentally testable predictions.

From a quantum mechanical perspective, the world-line experiment is designed to measure the world-line of an electron by measuring two points, or events, along the world-line. The path of the electron between the two events is unknown. Many paths between the events are possible. For each path, quantum theory lets us calculate the probability that an electron would take that path. If only ten paths are possible, and each path is equally likely, then there is one chance in ten that the electron will take any one of the ten paths we might arbitrarily choose. In many cases the paths are not probabilistically equal—that is, one path is more likely, or more probable, than the other paths. The most probable path is the path determined by classical theory. Although classical theory tells us which path is most likely, it does not account for alternative paths. Quantum theory takes into consideration all possible paths and makes a statement about how likely each one is.

9.4   SCATTERING

Much of what we know about the submicroscopic world has been determined from the analysis of particle scattering experiments. Scattering is actually a very simple concept with many everyday examples. One familiar example is the case of a ball striking a bat. The collision between a ball and a bat causes a change in the initial direction of the ball. It is even possible to send the ball back along its original path. This change in direction of a moving object due to an interaction or collision with another object is an example of scattering. Another is the case of an automobile grazing the side of a guardrail. An astronomical scattering process occurs when a meteor comes near the earth without being trapped by the gravitational pull of the earth. The path of the meteor is altered by the influence of the earth's gravitational field.

Submicroscopic particles often behave in a manner analogous to that of the meteor–Earth scattering system. Although a gravitational interaction is also present when two massive submicroscopic particles interact with each other, gravity is usually not the dominant interaction. Three other interactions may be present when submicroscopic scattering occurs. These interactions are the strong, electromagnetic, and weak interactions. They are discussed in more detail later.

Scattering processes can be very complex. To better understand the factors involved in any particular scattering process, we consider space–time diagrams. Time is drawn along the vertical axis and a spatial dimension is drawn along the horizontal axis of a space–time diagram. Arrows depict the paths of particles. Figure 9-5 shows some simple cases. Case A is a typical example of a motionless object with its clock running. Case B is the same object with its clock running in reverse. Cases C and D represent objects moving forward and backward in space without any passage of time. To date, there is no known physical phenomenon with the characteristics of cases C and D. The two cases E and F are routinely observed in scattering experiments. Case E shows a particle moving forward in space and time and is familiar from everyday life. Case F, on the other hand, is not familiar from everyday life because it shows a particle moving backward in space and time.

The possibility of motion backward in time is a physical consequence of the relativistic mass–energy relationship

$$m^2 c^4 = E^2 - p^2 c^2$$

(9.4.1)

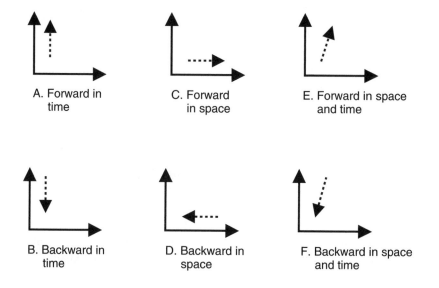

A. Forward in
time

C. Forward
in space

E. Forward in space
and time

B. Backward in
time

D. Backward in
space

F. Backward in space
and time

Figure 9-5. Space–time diagrams.

for a free particle with mass m, energy E, and magnitude of momentum p. Solving for energy gives

$$E = \pm\sqrt{m^2 c^4 + p^2 c^2} \qquad (9.4.2)$$

Energy can be either positive or negative. A particle can be in either a positive energy state or a negative energy state. A particle in a negative energy state is interpreted as a particle in a positive energy state going backward in time. This interpretation, which is due to physicists E.C.G. Stueckelberg and Richard Feynman [Fanchi, 1993; Weinberg, 1995], is possible because the probability amplitude for the motion of a relativistic free particle includes the product of energy E and coordinate time $x^0 = ct$. The product Ex^0 is negative if either $E < 0$, $x^0 > 0$ or $E > 0$, $x^0 < 0$.

Another point should be noted about Equation (9.4.1). Equation (9.4.1) presents mass as a derived concept: mass is calculated from energy and momentum. By contrast, Equation (9.4.2) presents energy as the derived concept: energy is calculated from mass and momentum. In practice, particle mass is not measured, but is inferred from measurements of energy in calorimeters and momentum from particle tracks. These inferred masses are not fundamental measurements because they depend on the theory used to establish the relationship between energy and momentum, in this case special relativity is the theory underlying Equation (9.4.1). If we assumed that a particle was nonrelativistic so that $m^2c^4 \gg p^2c^2$, and we were only interested in positive values of energy, Equation (9.4.2) would become

$$E = \sqrt{m^2c^4 + p^2c^2} = mc^2\sqrt{1 + \frac{p^2c^2}{m^2c^4}} \approx mc^2 + \frac{p^2}{2m} \qquad (9.4.3)$$

Equation (9.4.3) shows that the total energy of a nonrelativistic free particle with positive energy is the sum of the energy due to the mass of the particle and the kinetic energy of the particle.

Although scattering experiments study the behavior of very small submicroscopic systems, the machines needed to perform these experiments are often huge and expensive. The chief function of scattering machines is to guide a collimated beam of particles into either a stationary material target or into another beam of particles. It is usually necessary to accelerate the incident beam of particles until its particles have attained a desired velocity. For this reason machines built to perform particle scattering experiments are called *particle accelerators*. To justify their high cost relative to other types of experiments, particle accelerators and associated particle detectors have been valuable in studying many fundamental natural processes, and their development has yielded spin-off technology in commercially important areas such as magnet design and superconductivity.

Figure 9-6 displays some examples of processes and interactions that have been observed during scattering experiments. Space–time diagrams illustrating scattering processes are known as Feynman diagrams in honor of their originator, American physicist Richard P. Feynman. The top diagram illustrates the notation. Once again an arrow represents the motion of a particle in space–time. A dot, or vertex, denotes the space–time event when an interaction between two or more particles occurs.

Pair production and pair annihilation are shown in the second and third diagrams in Figure 9-6. The wavy line denotes a photon γ, or particle of light. The symbols e^- and e^+ denote an electron and its antiparticle, the positron, respectively. Pair production is the process in which a particle and its antiparticle are created when sufficient energy is available. The energy takes the form of electromagnetic radiation in the pair production process. Conversely, a particle and its antiparticle may annihilate each other in the pair annihilation process. The energy of the particle–antiparticle pair is converted into electromagnetic radiation following the scattering.

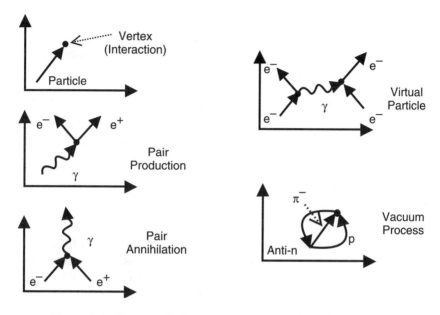

Figure 9-6. Example Feynman diagrams for scattering processes.

The "virtual particle" diagram in Figure 9-6 shows what may happen when two electrons pass near each other. From a simple point of view, we say the two electrons repel each other because they have the same electrical charge. This repulsion has been established by innumerable experiments. On a more fundamental, and modern, level we say the electrons exchange a photon, or particle of light. The photon exists for a very short time only and is called a *virtual particle.* The duration the virtual particle exists is controlled by the energy-time uncertainty principle

$$\Delta E \Delta t \geq \frac{\hbar}{2} \tag{9.4.4}$$

where ΔE is the uncertainty in the energy measurement and Δt is the uncertainty in the time it takes to measure the energy. Equation (9.4.4) says that the measurements of energy and time must be uncertain by a non-zero, finite amount that is proportional to Planck's constant. If ΔE is known, the uncertainty in the time measurement may be estimated from the relation

$$\Delta t \approx \frac{\hbar}{(2\Delta E)} \tag{9.4.5}$$

Thus, an accurate energy measurement will yield a small value of ΔE, which implies a large uncertainty Δt in the time measurement. A virtual particle can exist during the duration Δt. From another perspective, we cannot measure the energy to an accuracy greater than

$$\Delta E \approx \frac{\hbar}{(2\Delta t)} \qquad (9.4.6)$$

during the time interval Δt. Particles that appear and disappear during this time are called *virtual particles*.

An especially interesting process is the vacuum process shown in Figure 9-6. Calculations performed to quantitatively analyze some scattering processes have had to include an effect known as the vacuum process to be correct. The vacuum process represents the physical situation in which virtual particles arise from the vacuum, interact, and then return to the vacuum, all within a short time duration satisfying Heisenberg's uncertainty principle. The particular example in Figure 9-6 shows the creation, interaction, and annihilation of a proton (p), antineutron (n), and a negatively charged pion (π). Some people believe the vacuum process was the seed from which the entire universe began. The vacuum process is an example of an interaction between virtual particles that might serve as a future energy source.

In the modern view, the force between two particles is due to the exchange of a particle between the two particles. According to this view, the interactions between particles require the exchange of a particle. For example, the change in trajectory, or scattering, of two electrons is caused by the interaction with the virtual photon. The modern view is a local theory. An alternative and more historical view of electron repulsion is an example of an action-at-a-distance theory. The action-at-a-distance concept is that two electrons repulse each other because one electron interacts with the electric field generated by the other electron, and vice versa. Modern physics favors the local concept of interactions, although action-at-a-distance theories are often easier to apply and yield the same results.

One of the apparent tenets of particle physics is that anything that is not explicitly forbidden can occur [Bernstein, et al., 2000, pg. 483]. There are some constraints, however. The constraints restrict what is possible and are known as *conservation laws*. The search for conservation laws is an important task of contemporary high-energy physics. A conservation law can be surmised if a physical quantity does not change as a result of an interaction. For example, conservation of electric charge says that total electric charge does not change during an interaction, and conservation of energy says that the energy of a closed system does not change following an interaction. Conservation of electric charge and conservation of energy are part of a set of conservation laws. Our set of conservation laws may change as new observations are made. Conservation of energy was considered a valid conservation law until mass–energy transformations were discovered. Experiments have shown that the correct conserved quantity is not energy, but energy and momentum. The observation that mass can be converted into energy and vice versa means that the number of particles in a system can no longer be considered a conserved quantity. The number of particles will change when the mass of a particle is converted to energy.

In the processes considered so far, and this is only a small sampling, six particles have been introduced: the electron, the positron, the photon, the proton, the pion, and the antineutron. These particles are but six residents of the particle zoo.

9.5   THE PARTICLE ZOO

Scattering experiments have led to the discovery of a large number of particles. Most of these particles are unstable; they do not exist very long. Unlike virtual particles, the lifetimes of unstable particles are not determined by the uncertainty principle. Rather, the instability of these particles is a consequence of interactions with other particles. The interaction associated with the decay of unstable particles is called the *weak interaction*. It is one of four recognized interactions. In order of decreasing strength, the interactions are:

1. Strong interaction
2. Electromagnetic interaction
3. Weak interaction
4. Gravitational interaction

The strong interaction is the force that binds nuclei together. All atoms consist of a nucleus, which contains most of the mass of the atom, and an electron cloud surrounding the nucleus. Positively charged protons and particles without an electrical charge—neutrons—are the principal building blocks within the nucleus. Protons and neutrons are also called nucleons. Negatively charged electrons occupy regions of space surrounding the nucleus. The actual location of the electrons is not known because of Heisenberg's uncertainty principle. Instead, we can calculate the probability of finding an electron at a particular spatial location. The term *electron cloud* is used to denote both our ignorance of exact spatial locations and our knowledge of probability distributions. The volume of space occupied by the nucleus is very small in comparison to the volume of space occupied by the electron cloud. The electromagnetic attraction between protons and electrons is primarily responsible for maintaining the electron cloud around the nucleus. By contrast, electromagnetic repulsion between protons in the small volume of the nucleus would lead to nuclear disintegration if electromagnetic repulsion between protons was not countered by another force. The strong interaction is that other force. One expression of the strong force is the Yukawa potential energy introduced in Section 5.1. Both protons and neutrons are carriers of the strong force. Nuclear power, both peaceful and military, depends on breaking the strong force that binds nucleons in the nucleus of an atom.

Both the strong and the weak interactions are short-range forces. Their influence is dominant only when the interacting particles are very close to each other: at distances comparable to the diameter of a nucleus. The electromagnetic and gravitational interactions are long-range interactions. They exert an influence on particles separated by distances ranging from the diameter of an atom—on the order of 100,000 times the diameter of the nucleus—to the distance between stars. All of the interactions can be viewed as an exchange of particles. Figure 9-7 presents Feynman diagrams for interactions. The myriad of known particles can be classified according to the types of interactions they participate in.

There are two general categories of particles besides photons. Particles in these categories are called either leptons or hadrons. Hadrons can participate in the strong interaction; leptons cannot. Examples of leptons are the neutrino, the electron, and the positron. The majority of known particles are not leptons, however, but hadrons. Hadrons are further organized into subgroupings. The two hadron subgroupings are mesons and baryons.

Mesons, such as the pion we encountered earlier, can be their own antiparticle, whereas baryons exist in distinct particle–antiparticle pairs. An example of a baryon pair is the proton–antiproton combination. Another common baryon is the neutron. Examples of different types of particles are depicted in Table 9-1. Four classification levels are shown and are interpreted as follows. Consider the proton in level 4. The proton is a nucleon, nucleons are baryons, and baryons are hadrons.

The particle classification scheme shown in Table 9-1 is based on the strong and weak interactions. The electromagnetic and gravitational interactions transcend the classification lines. Any electrically charged particle, such as the electron or the proton, can participate in

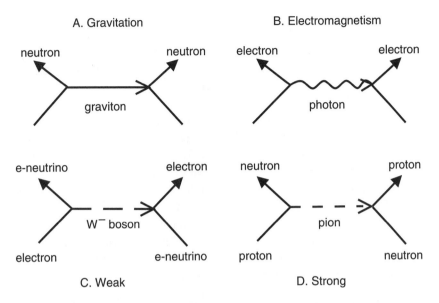

Figure 9-7. Fundamental particle interactions.

Table 9-1
Elementary particles

Level 1	Level 2	Level 3	Level 4
Photons			
Leptons (no strong interaction)	Neutrinos		
	Electron e⁻ Positron e⁺		
Hadrons (strongly interact)	Mesons (interact weakly)	Pion π Kaon K Eta η	
	Baryons (distinct particle-antiparticle pairs)	Nucleons Lambda Λ Sigma Σ	Proton p Neutron n

electromagnetic interactions. All particles can participate in gravitational interactions. Because of the relatively broad applicability of the gravitational and electromagnetic interactions, they are not as well suited for categorizing particles as the relatively less applicable strong and weak interactions. Consequently the classification scheme outlined in Table 9-1 has proven to be most useful in attempting to understand the particle zoo.

QUARKS

Historically, the treatment of matter as an aggregate of tiny, indivisible particles has been a useful approach. It lets us model objects that have behavior that may otherwise be too paradoxical for us to understand. Because of its conceptual usefulness and proven success, people searching for an understanding of unexplained phenomena often grasp the notion that elementary, indivisible particles exist. As technology progresses, objects once thought to be elementary are found to be composed of even more fundamental objects. Hadrons are an example.

Prior to the beginning of the twentieth century, atoms seemed to be the fundamental particles of nature. By the turn of the century a number of observations existed that could best be explained if atoms were thought to be objects comprised of smaller objects. The electron, proton, and neutron were born. The proton and neutron were constituents of the nucleus of an atom, and negatively charged electrons orbited the positively charged nucleus.

For three decades the electron, proton, and neutron were sufficient for explaining the results of most experiments. Among the unexplained observations was the existence of stable nuclei. The stability of nuclei was eventually explained in 1935 when the Japanese physicist Hideki Yukawa postulated the existence of a new particle called the pi meson. In 1936 a particle having many of the properties of Yukawa's pi meson was discovered. Further analysis of the properties of this particle showed it did not interact strongly with other matter. Yukawa's pi meson was not actually discovered until 1947, when British physicist Cecil F. Powell observed it as a constituent of cosmic radiation. By then people realized that the number of elementary particles was greater than originally thought.

The existence of so many particles prompted people to look for particles of a more elementary nature. Thus was born the quark model. The quark model, conceived independently by Murray Gell-Mann and George Zweig, appeared in 1964. Gell-Mann borrowed the word "quark" from James Joyce's *Finnegan's Wake*. In terms of the quark model, mesons such as the pi meson, or pion for short, are composed of a quark and an anti-quark. Baryons such as the proton, on the other hand, are composed of three quarks. By postulating the existence of a relatively small number of quarks, it was possible in the late 1960s and 1970s to classify the many observed hadrons. Thus the number of "truly elementary" particles was reduced to the number of quarks and leptons. Today the number of quarks and leptons has grown from the original two quarks and two leptons to six "flavors" of

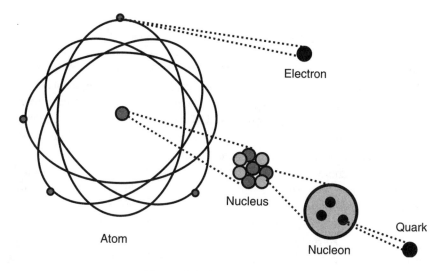

Figure 9-8. Particle hierarchy.

quarks and six kinds of leptons. Figure 9-8 illustrates the particle hierarchy. The nucleon in the figure can be either a proton or a neutron.

Some people have speculated on the existence of yet another level of structure more fundamental than quarks and leptons. One class of particles at the new level refers to particles as squarks and sleptons.[3] Physicist O.W. Greenberg [1985] has suggested that another class of particles consisting of quarks and leptons be dubbed *dirons* because he thought English physicist P.A.M. Dirac's name should be associated with fundamental particles at the level of the electron. This honor is a tribute to the achievements of Dirac, who was instrumental in developing a mathematical structure for calculating the effects of interactions between electrons and photons. The new levels of particles are speculative. The fact that several other types of particles have been postulated at a more fundamental level than the quark indicates that the theory of elementary particles may still be evolving.

One of the most interesting properties of quarks is the postulated value of their electrical charge. The charge of every observed particle, if it has a charge at all, has the same magnitude as the charge of the electron. This experimental fact is the basis for the historical practice of calling the electrical charge of the electron the elementary charge. The electrical charge of quarks is one third or two thirds of the elementary charge in the standard quark model. This difference in the magnitude of the electrical charge provides a means for searching for and identifying quarks. Such a search has been underway since 1964. To date, indisputable experimental proof of the existence of particles having an electrical charge less than the elementary charge is not available. Does this mean quarks do not exist?

The evidence for quarks is much more comprehensive than just a search for fractional elementary charge. Highly energetic scattering experiments have shown that nucleons—protons and neutrons—exhibit an internal structure. The internal structure can be understood if we think of nucleons as composed of smaller point particles: quarks.

Attempts to create quarks have been made at particle accelerators around the world. Theory has predicted that the collision of a beam of electrons and a beam of positrons should produce quarks. Such attempts to produce free quarks have resulted in the production of jets of hadrons that are composites of quarks. No free quarks have been observed. Jets are taken as evidence for quarks. In the initial collision between an electron and its antiparticle, the positron, the electron and positron annihilate each other and form a highly energetic photon. If the photon has enough energy, it can transform into quarks by a pair creation process. When the transformation occurs we should have free quarks. In practice, when the two beams collide, so many quarks are produced in such close proximity that they immediately acquire partners because of the strong interaction between them. The subsequent groupings of quarks appear to us as jets—or groupings—of hadrons. If quarks exist, as seems necessary based on a wealth of experimental data, why have free quarks not been observed?

The theoretical appeal of the quark concept—contemporary atomism— has led several people to search for reasons to explain why free quarks may not be observable. The most plausible explanation is a mechanism known as asymptotic freedom. Simply stated, *asymptotic freedom* means the strength of the interaction between quarks increases as the quarks move away from each other. Conversely, the interaction between quarks should decrease as bound quarks approach each other. If the asymptotic freedom concept is correct, experiments capable of scattering a particle from the quark system should show quarks behaving like free particles within the region of space they occupy. This type of behavior is analogous to the behavior of two balls bound together by a rubber band. If the balls are near each other, the rubber band is slack and the balls behave as if they are free. Were we to pull the balls apart, the rubber band would lengthen and tighten. There would be a resistance to separating the balls completely. If the balls were released, they would be pulled together again. The only way to completely separate the balls would be to break the rubber band: the force binding the balls. A similar requirement is suggested for quarks; to separate quarks, it is necessary to break the strong force binding them.

To date, people have been unable to split a hadron into its quark constituents and isolate the quark debris, although they have shown that quarks behave like free particles inside the hadron. Thus, although considerable experimental evidence supporting the quark concept exists, we do not yet have indisputable proof of quark existence. People must prove that free quarks and anti-quarks do not exist because they have all combined to form hadrons. Such a proof depends on the origin of quarks and anti-quarks, which depends in turn on the origin of the universe. Once again we have encountered a problem of cosmological importance.

9.6 TIME

Twentieth-century experiments have shown that the description of microscopic and submicroscopic objects moving at or near the speed of light requires the synthesis of two distinct theories: special relativity and quantum theory. Much of the experimental evidence described in the preceding sections was not available to physicists in the early twentieth century, which is when crucial choices were being made. Physicists had to modify their notions of space, time, mass, and energy. They were faced with questions of fundamental importance. We can better understand the consequences of the choices that were made if we first review some of the options that were available for a critical concept: time.

Time t played the role of a monotonically increasing evolution parameter in classical Newtonian mechanics, as in the force law $\vec{F} = d\vec{p}/dt$ for a nonrelativistic, classical object with momentum $\vec{p}$. To Newton, time was an "arrow" that parametrized the direction of evolution of a system. It has been known since the nineteenth century that entropy, or the disorder of a system, can be constant or increase, but never decrease. This is the content of the second law of thermodynamics. Ludwig Boltzmann (1844–1906) demonstrated that entropy could only be constant or increase with respect to a monotonically increasing evolution parameter. Boltzmann used this observation to establish a temporal direction that distinguished between future and past.[4]

Einstein rejected the Newtonian concept and identified t as the fourth coordinate of a space–time four-vector. Einstein's view of time requires a physical equivalence between coordinate time and coordinate space. In this view, time should be a reversible coordinate in the same manner as space.

Table 9-2
Temporal hypotheses

#	Hypothesis
I	Assume t is Einsteinian time and reject Newtonian time.
II	Introduce two temporal variables:
	• A coordinate time ($x^0 = ct$) in the sense of Einstein
	• An invariant evolution parameter (s) in the sense of Newton

Evidence for such reversibility has come from particle physics, where antiparticles are interpreted as particles moving backward in time.

The development of nonrelativistic quantum mechanics in the early twentieth century preserved the Newtonian concept of time in the Schroedinger equation. Attempts to understand the physical meaning of the wave function Ψ led to a probability conservation equation

$$\frac{\partial \rho}{\partial t} + \nabla \cdot \rho \vec{v} = 0, \quad \rho = \Psi^* \Psi \tag{9.6.1}$$

where ρ is the probability density and $\rho \vec{v}$ represents probability flux.

The success of nonrelativistic quantum mechanics and special relativity motivated efforts to extend quantum concepts to the relativistic domain. The role played by time was a key difference between Einsteinian and Newtonian views of classical theory. What role should time play in relativistic quantum theory? Two hypotheses that were consistent with special relativity are listed in Table 9-2.

Hypothesis I led to a relativistic probability conservation equation

$$\sum_{\mu=0}^{3} \partial_\mu \rho v^\mu = 0 \tag{9.6.2}$$

where the index $\mu = \{0, 1, 2, 3\}$. The integer 0 denotes the time component, and the integers 1,2,3 denote three space components. Equation (9.6.2) is essentially a re-statement of the nonrelativistic continuity equation shown in Equation (9.6.1) with the time (0th) component of probability flux given by $\rho v^0 = \rho c$, $\partial_0 \rho v^0 = \partial \rho c / \partial ct = \partial \rho / \partial t$ so that

$$\sum_{\mu=0}^{3} \partial_\mu \rho v^\mu = \partial_0 \rho v^0 + \sum_{j=1}^{3} \partial_j \rho v^j = \frac{\partial \rho}{\partial t} + \nabla \cdot \rho \vec{v} = 0 \tag{9.6.3}$$

Time in the relativistic probability conservation equation is Einstein's time and is a consequence of adopting Hypothesis I. Most physicists today say that time is a coordinate in the Einsteinian sense, not an arrow in the Newtonian sense. From the perspective of Hypothesis I, time must be both an irreversible arrow tied to entropy and a reversible coordinate in the Einsteinian sense. By adopting Hypothesis I, physicists retained a temporal paradox: motion relative to a single temporal variable must be reversible even though the second law of thermodynamics establishes an "arrow of time" for evolving systems, including relativistic systems.

A few physicists have worked with Hypothesis II. Twentieth-century American physicist Richard Feynman used the evolution parameter concept to develop the relativistic path integral formalism outlined in the following section.[5] Hypothesis II employs two temporal variables: a coordinate time, and an evolution parameter. The early theories treated the evolution parameter as a convenient mathematical variable that had no physical significance. More recently, the evolution parameter has been viewed as a physically measurable quantity and a procedure has been presented for designing evolution parameter clocks.[6]

By recognizing the existence of a distinct coordinate time and a distinct parametrized time, Hypothesis II does not have a conflict between a universal direction of time s and a time x^0 that may proceed as readily from future to past as from past to future. The two times are entirely different temporal concepts: the former time s is a parameter like Newton's time, and the latter time x^0 is a coordinate like Einstein's time. Unlike Newton's time, the evolution parameter s with time units in the relativistic domain must be a scalar—that is, the relativistic evolution parameter s is invariant with respect to inertial frames of reference. The two different times in Hypothesis II make it possible to have an arrow of (evolution parameter) time s and equivalence between (coordinate) time x^0 and spatial coordinates. The difference between Hypothesis I and Hypothesis II is illustrated in Figure 9-9.

Figure 9-9 shows a system with one spatial coordinate x^1, a temporal coordinate x^0 corresponding to Einsteinian time, and an evolution parameter s. The x^0-x^1 plane is a space–time diagram. Hypothesis I applies within the space–time plane defined by the x^0-x^1 axes. Point A is an event in the x^0-x^1 plane. Relativistic dynamics allows events to evolve from one x^0-x^1 plane to another. The curve connecting events A and B in Figure 9-9 illustrates one such evolutionary path. A mathematical formalism governing the evolution is introduced using Feynman's path integral approach in the next section. It is a natural extension of the nonrelativistic theory

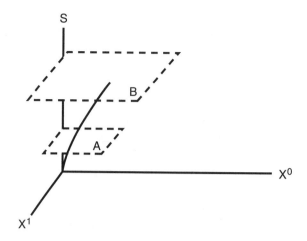

Figure 9-9. Temporal evolution.

presented previously and contains the theory associated with Hypothesis I as a special case.

9.7 RELATIVISTIC PATH INTEGRAL FORMALISM

Feynman's path integral formalism introduced previously can be extended to relativistic systems. The extension of the nonrelativistic path integral technique to relativistic systems highlights many of the issues discussed in the preceding section. The relativistic path integral formalism is introduced by analyzing the behavior of a relativistic free particle.

Example 9.7.1: Path Integral Analysis of a Relativistic Free Particle

We want to find the equation that describes the motion of a relativistic free particle in one space dimension. The probability amplitude Φ at a space–time point x_{i+1}, t_{i+1} and parameter $s + \varepsilon$ for an infinitesimal duration ε is calculated from the probability amplitude at x_i, t_i, s using the relation

$$\Phi(x_{i+1}, t_{i+1}, s + \varepsilon) = \frac{1}{A} \int e^{iS(x_i,t_i)/\hbar} \, \Phi(x_i, t_i, s) \, dx_i d(ct_i) \qquad (9.7.1)$$

where $dx_i = x_{i+1} - x_i$, $dt_i = t_{i+1} - t_i$ and A is a normalization constant. The action S is expressed in terms of the Lagrangian L by

$$S(x_i, t_i) = \int L(\dot{x}_i, \dot{t}_i) ds \qquad (9.7.2)$$

where the dot over the letter denotes differentiation with respect to the parameters. The Lagrangian for a relativistic free particle with mass m is

$$L(\dot{x}, \dot{t}) = \frac{m}{2}\left[\left(c\frac{dt}{ds}\right)^2 - \left(\frac{dx}{ds}\right)^2\right] \qquad (9.7.3)$$

We use the trapezoidal rule to approximate the integral in Equation (9.7.1). The result is

$$S(x_i, t_i) = \varepsilon\frac{m}{2}\left[c^2\left(\frac{t_{i+1} - t_i}{\varepsilon}\right)^2 - \left(\frac{x_{i+1} - x_i}{\varepsilon}\right)^2\right] \qquad (9.7.4)$$

Defining the variables

$$\delta_x \equiv x_{i+1} - x_i, \quad \delta_t \equiv t_{i+1} - t_i, \quad x \equiv x_{i+1}, \quad t \equiv t_{i+1} \qquad (9.7.5)$$

lets us write Equation (9.7.1) in the form

$$\Phi(x, t, s + \varepsilon) = \frac{1}{A}\int e^{i\frac{m}{2\hbar\varepsilon}[c^2\delta_t^2 - \delta_x^2]} d\,\delta_x d(c\delta_t)$$
$$\cdot\Phi(x - \delta_x, t - \delta_t, s) \qquad (9.7.6)$$

Performing Taylor series expansions of the probability amplitudes to first order in ε and second order in δ_x gives

$$\Phi(x, t, s) + \varepsilon\frac{\partial\Phi}{\partial s} = \frac{1}{A}\int e^{i\frac{m}{2\hbar\varepsilon}[c^2\delta_t^2 - \delta_x^2]}d\delta_x d(c\delta_t)$$
$$\cdot\left[\Phi(x, t, s) - \delta_t\frac{\partial\Phi}{\partial t} - \delta_x\frac{\partial\Phi}{\partial x} - \delta_x\delta_t\frac{\partial^2\Phi}{\partial x\partial t} + \frac{\delta_t^2}{2}\frac{\partial^2\Phi}{\partial t^2} + \frac{\delta_x^2}{2}\frac{\partial^2\Phi}{\partial x^2}\right]$$
$$(9.7.7)$$

The integral on the right-hand side of Equation (9.7.7) is evaluated using the integrals in Equation (8.5.8). Equation (9.7.7) becomes

$$\Phi(x, t, s) + \varepsilon \frac{\partial \Phi}{\partial s} = -\frac{2\varepsilon\hbar\pi}{mA} \Phi(x, t, s)$$
$$- \frac{i\varepsilon\hbar}{2m} \left(\frac{2\varepsilon\hbar\pi}{mA} \right) \left[\frac{1}{c^2} \frac{\partial^2 \Phi}{\partial t^2} - \frac{\partial^2 \Phi}{\partial x^2} \right] \tag{9.7.8}$$

To assure continuity of the probability amplitudes when $\varepsilon \to 0$, we set $A = -2\varepsilon\hbar\pi/m$ so that Equation (9.7.8) becomes

$$\Phi(x, t, s) + \varepsilon \frac{\partial \Phi}{\partial s} = \Phi(x, t, s) + \frac{i\hbar\varepsilon}{2m} \left[\frac{1}{c^2} \frac{\partial^2 \Phi}{\partial t^2} - \frac{\partial^2 \Phi}{\partial x^2} \right] \tag{9.7.9}$$

Equation (9.7.9) is satisfied for all ε when the coefficients of the ε terms are equal, thus

$$\frac{\partial \Phi}{\partial s} = \frac{i\hbar}{2m} \left[\frac{1}{c^2} \frac{\partial^2 \Phi}{\partial t^2} - \frac{\partial^2 \Phi}{\partial x^2} \right] \tag{9.7.10}$$

Multiplying Equation (9.7.10) by $i\hbar$, introducing four-vector notation and imposing the Einstein summation convention gives

$$i\hbar \frac{\partial \Phi}{\partial s} = -\frac{\hbar^2}{2m} \partial^\mu \partial_\mu \Phi \tag{9.7.11}$$

We can find an equation that does not include the parameter s by writing Φ in the form

$$\Phi(x, t, s) = \Phi_{KG}(x, t) \exp\left[-i \frac{M^2 c^2 s}{2m\hbar} \right] \tag{9.7.12}$$

where Φ_{KG} depends only on x, t and M is a constant with the unit of mass. Equation (9.7.12) is considered a solution that is stationary with respect to s. Substituting Equation (9.7.12) into Equation (9.7.11) gives

$$M^2 c^2 \Phi_{KG} = -\hbar^2 \partial^\mu \partial_\mu \Phi_{KG} \tag{9.7.13}$$

Equation (9.7.13) is the Klein-Gordon equation for a relativistic free particle. The value of M is determined from Equation (9.7.13) as a function of momentum and energy. An example calculation is presented in the following section.

9.8   RELATIVISTIC QUANTUM THEORY

We highlight the importance of probability theory to our understanding of relativistic quantum theory in this section by combining the ideas outlined in the preceding section with the procedure presented in Section 8.4. The result is another formulation of relativistic quantum theory that begins with the assumption that a conditional probability density $\rho(y|s)$ exists. The symbol y denotes a set of four coordinates for a particle in the space–time volume D for which $\rho(y|s)$ has non-zero values. The μth component of the position four-vector is written as y^μ where $\mu = 0, 1, 2, 3$. Indices 1, 2, 3 signify space components and 0 signifies the geometric time component. The symbol s is an invariant evolution parameter that conditions the probability density $\rho(y|s)$.

The conditional probability distribution $\rho(y|s)$ is capable of representing the disappearance or reappearance of particles in space–time. We illustrate this capability by considering a single particle. In this case, the conditional probability density can be expressed as the product $\rho(y^1, y^2, y^3|y^0, s)\rho(y^0|s)$ for a single particle. The distribution $\rho(y^0|s)$ is the marginal probability density in time and is conditioned by the evolution parameter s. When $\rho(y^0|s)$ is zero, the probability of observing a particle at time y^0 given parameter s is zero. In other words, the particle cannot be detected anywhere in space when $\rho(y^0|s)$ is zero. By contrast, when $\rho(y^0|s)$ is non-zero, there is a non-zero probability of observing a particle at time y^0 given parameter s. The field equations described next must be solved to determine the probability distributions.

According to probability theory, $\rho(y|s)$ must be positive definite and normalizable, thus

$$\rho(y|s) \geq 0 \qquad (9.8.1)$$

and

$$\int_D \rho(y|s)dy = 1 \qquad (9.8.2)$$

where

$$dy = d^4y = dy^0 dy^1 dy^2 dy^3 = \prod_{\mu=0}^{3} dy^\mu \tag{9.8.3}$$

Conservation of probability implies the continuity equation

$$\frac{\partial \rho}{\partial s} + \frac{\partial \rho V^\mu}{\partial y^\mu} = 0 \tag{9.8.4}$$

where Einstein's summation convention for Greek indices is assumed and the term ρV^μ represents the μth component of probability flux of a particle.

For ρ to be differentiable and non-negative, its derivative must satisfy $\partial \rho / \partial s = 0$ if $\rho = 0$. The positive-definite requirement for $\rho(y|s)$ is satisfied by writing $\rho(y|s)$ in the Born representation:

$$\rho(y|s) = \psi^*(y,s)\,\psi(y,s) \tag{9.8.5}$$

The scalar eigenfunctions ψ can be written as

$$\psi(y,s) = [\rho(y|s)]^{1/2} \exp[i\xi(y,s)] \tag{9.8.6}$$

where $\xi(y,s)$ is a real scalar function.

Field equations can be derived from the probability formalism by expressing the four-velocity of the particle as

$$V^\mu(y,s) = \frac{1}{m}\left[\hbar\frac{\partial \xi(y,s)}{\partial y_\mu} - \frac{e}{c}A^\mu(y,s)\right] \tag{9.8.7}$$

Given these assumptions, the field equation for a particle interacting with an electromagnetic field has the form

$$i\hbar\frac{\partial \psi(y,s)}{\partial s} = \left[\frac{\pi^\mu \pi_\mu}{2m} + V_I\right]\psi(y,s) \tag{9.8.8}$$

where V_I is the interaction potential. The differential operator

$$\pi^\mu = \frac{\hbar}{i}\frac{\partial}{\partial y_\mu} - \frac{e}{c}A^\mu \tag{9.8.9}$$

is called the four-momentum operator and represents the four-momentum, or energy-momentum, of the particle. The four-vector A^μ contains the scalar and vector potentials of electromagnetism introduced in Section 2.2. Equation (9.8.8) together with Equation (9.8.9) is called the Stueckelberg equation. In this formulation, the definition of expectation value of an observable Ω is

$$\langle\Omega\rangle \equiv \int_{D^N} \psi^+\Omega\psi\,dy \tag{9.8.10}$$

where ψ^+ is the conjugate transpose of ψ. We illustrate next how to use Equations (9.8.8) through (9.8.10) by considering the relatively simple problem of a noninteracting particle.

Example 9.8.1: Relativistic Free Particles

A noninteracting system corresponds to the conditions $\{A^\mu\} = 0$ and $V_I = 0$ so that Equation (9.8.8) becomes

$$i\hbar\frac{\partial\psi_f}{\partial s} = -\frac{\hbar^2}{2m}\frac{\partial^2\psi_f}{\partial y^\mu\partial y_\mu} \tag{9.8.11}$$

Equation (9.8.11) has the solution

$$\psi_f(y,s) = \eta^{1/2}\exp\left[-\frac{i\hbar}{2m}\left(k^\mu k_\mu\right)s + ik_\mu y^\mu\right] \tag{9.8.12}$$

where η is the constant calculated from the probability normalization requirement.

The expectation value of the particle four-velocity is

$$\langle V^\mu\rangle = \frac{1}{m}\int \psi_f^*\left[\frac{\hbar}{i}\frac{\partial}{\partial y_\mu}\right]\psi_f\,dy \tag{9.8.13}$$

Substituting Equation (9.8.12) into (9.8.13) gives

$$\langle V^\mu \rangle = \frac{d \langle y^\mu \rangle}{ds} = \frac{\hbar \langle k^\mu \rangle}{m} \tag{9.8.14}$$

The four-momentum of the relativistic free particle is

$$m \langle V^\mu \rangle = \hbar \langle k^\mu \rangle \tag{9.8.15}$$

Integrating Equation (9.8.14) from s to $s + \delta s$ gives

$$\delta \langle y^\mu \rangle = \frac{\hbar \langle k^\mu \rangle}{m} \delta s \tag{9.8.16}$$

Equation (9.8.16) expresses the expectation value of the components of the position four-vector for the relativistic free particle.

The observable world-line of the particle is found from Equation (9.8.16) by calculating

$$\delta \langle y^\mu \rangle \, \delta \langle y_\mu \rangle = \frac{\hbar^2 \langle k^\mu \rangle \langle k_\mu \rangle}{m^2} \delta s^2 \tag{9.8.17}$$

In the classical limit of negligible dispersion, we have

$$\delta \langle y^\mu \rangle \, \delta \langle y_\mu \rangle = \frac{\hbar^2 \langle k^\mu k_\mu \rangle}{m^2} \delta s^2 \tag{9.8.18}$$

The observable mass of the particle is

$$\langle p^\mu p_\mu \rangle = \hbar^2 \langle k^\mu k_\mu \rangle \equiv m^2 c^2 \tag{9.8.19}$$

Using Equation (9.8.19) in Equation (9.8.18) gives the light cone constraint of classical special relativity:

$$\delta \langle y^\mu \rangle \, \delta \langle y_\mu \rangle = c^2 \delta s^2 \tag{9.8.20}$$

Equation (9.8.20) shows that the classical world-line of the particle from special relativity is its expected trajectory.

9.9   GUT AND TOE

Physicists recognize four fundamental interactions: strong, electromagnetic, weak, and gravitational. Many physicists believe a theory can be formulated that will unify these four seemingly disparate interactions into a single type of interaction. Several people have tried and continue to try to formulate such an interaction. Einstein, for example, attempted to unify the gravitational and electromagnetic interactions in his formulation of general relativity. He failed. No one, in fact, has succeeded in unifying all four interactions. Some progress has been made in this area, however.

Steven Weinberg, Abdus Salam, and Sheldon Lee Glashow were awarded the 1979 Nobel Prize in Physics for their work in developing a unified theory of the weak and electromagnetic interactions. The combination of interactions is now known as the electroweak interaction and has spurred efforts to unify the remaining two interactions. Scientists are searching for a "grand unified theory" of the strong, electromagnetic, and weak interactions. A widely accepted theory that can combine the strong, electromagnetic, and weak interactions with the gravitational interaction remains elusive. Although existing theories have important limitations, there are striking similarities among the modern mathematical descriptions of the four fundamental interactions to encourage further unification efforts.

Feynman diagrams illustrating the four fundamental interactions are shown in Figure 9-7. In every case, the interaction between two particles occurs by the exchange of a third particle. The exchanged particle, represented by a horizontal line in each of the Feynman diagrams, is said to mediate the interaction. Thus a pion mediates the interaction between a neutron and a proton in the hadron example of the strong interaction. The other interactions are thought to behave similarly in this standard formulation of particle physics. A question now arises: How can we determine if nature actually behaves as depicted in Figure 9-7? The Feynman diagrams provide a clue to the answer.

One way to verify the validity of the particle exchange view of interactions is to verify the existence of the intermediate exchange particles. The mediating particles of the strong interaction diagram for hadrons (pion) and the electromagnetic interaction diagram (photon) have both been observed. In 1983, the Nobel Prize in Physics was awarded to Carlo Rubbia and Simon van der Meer for their experimental detection of the W boson, the mediating particle of the weak interaction. No one has directly observed the mediating particle of the gravitational interaction (graviton). In the remaining case,

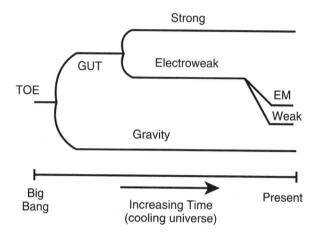

Figure 9-10. The theory of everything.

the strong interaction diagram for quarks, not only have gluons not yet been observed, but free quarks have evaded the watchful eye of experimentalists. The standard theoretical view depicted in Figure 9-7 has been partially confirmed experimentally, but contains important predictions that must still be verified by experiment.

The search for a theory of everything continues. The theory of everything will combine the four fundamental interactions and is expected to include a grand unified theory and a theory of gravity as special cases. The Grand Unified Theory (GUT) describes the strong and electroweak interactions. The electroweak interaction appears as the electromagnetic (EM) and weak interactions at low enough temperatures. Figure 9-10 presents one scenario that relates the dominant interaction in the universe, and the corresponding theory that best explains it, to the history of the universe. The dominant interaction is expected to change as the universe cools from the moment of the Big Bang to the present. The Big Bang is considered the moment of conception of the universe. It is discussed in more detail in the next chapter.

One of the most promising approaches to a theory of everything is string theory. String theory is based on the assumption that particles are not points but one-dimensional lines in space–time. The one-dimensional line is called a string. Strings can be open on both ends or closed into a loop. A string represents a particle, and the mass of the particle is related to the number of nodes of a vibrating string. So far, the predictions of string theories have not been easy to test experimentally.

ENDNOTES

1. Symmetry can be related to conservation laws such as conservation of energy and momentum. For further details, see textbooks such as Feynman, et al. [1963, Volume I], Goldstein [1980], Lawrie [1990], McCauley [1997], and Singer [2001].
2. For a discussion of tachyons and additional references, see Fanchi [1990, 1993] and Pavsic [2001].
3. See Greenberg [1985], the discussion of supersymmetry by Börner [1993, Section 6.6], Weinberg [1999], Greene [2000], and Kane [2000].
4. The thermodynamic arrow of time proposed by Boltzmann is now called Boltzmann's H-theorem (pronounced "eta-theorem" because of the Greek letter capital eta "H"). For further discussion, see D'Abro [1951, pp. 419–420], Fanchi [1993, Chapter 17], Greiner, et al. [1995, pg. 43], and Weinberg [1995, pg. 151].
5. Richard Feynman is the most recognizable physicist to pioneer use of the evolution parameter in relativistic quantum theory. Papers by V.A. Fock in 1937 and E.C.G. Stueckelberg in 1941 predated Feynman's work in 1948. L.P. Horwitz and C. Piron rekindled interest in relativistic dynamics with an evolution parameter in 1973. For a sketch of the history of the evolution parameter in relativistic quantum theory, see Fanchi [1993, Chapter 1].
6. See Fanchi [1993, Chapter 11] for further discussion.

EXERCISES

9-1. Derive Equation (9.1.1) from (9.1.2) by taking the limit $c \to \infty$.

9-2. A. Suppose a primed reference frame F' is moving at a speed $v = 0.5\,c$ in the x-direction of an unprimed reference frame F. Write the Galilean and Lorentz transformation equations for x', t' using the numerical values of v, c.

B. Fill in the following table.

		Galilean		Lorentz	
$x\,(m)$	$t\,(s)$	$x'\,(m)$	$t'\,(s)$	$x'\,(m)$	$t'\,(s)$
0	0				
1	10^{-10}				

9-3. Verify that Equation (9.8.12) is a solution of (9.8.11).

9-4. Find the phase factor $\xi(\vec{x}, t)$ for a relativistic free particle by comparing Equation (9.8.12) with Equation (9.8.6).

9-5. An electron and a positron annihilate each other. Assume the kinetic energy of the two particles was negligible relative to their rest mass energy. Estimate the amount of electromagnetic energy generated in the annihilation process. Express your answer in J and MeV.

9-6. Use the uncertainty principle for energy-time to estimate the length of time a virtual electron-positron pair can exist. Neglect the momentum of the electron and positron.

9-7. A. The classical speed of a free electron is 0.5 c. What is the magnitude of the momentum of the electron?
B. Free particle momentum p is related to quantum mechanical wavenumber k by the relation $p = \hbar k = 2\pi \hbar / \lambda$ where λ is the wavelength of the particle. What is the wavenumber of the electron in Part A.

9-8. What is the relativistic energy of an electron in its rest frame? Express your answer in J and MeV.

9-9. What is the rest mass energy of the hydrogen nucleus? Express your answer in J and MeV.

CHAPTER TEN

Nucleosynthesis

The formation and decay of nuclei are the central mechanisms of nuclear fusion and nuclear fission. We can obtain a deeper understanding of nuclei and nuclear reactions by becoming familiar with the science of nuclear formation: nucleosynthesis. Modern theories of nuclear formation are based on observations from astronomy and particle physics, and depend on cosmological models of the origin and evolution of the universe.[1] Many cosmological models have been proposed. These models must explain the observation that the universe is expanding. In this chapter, we present the evidence for an expanding universe and then discuss the observable parameters that influence our selection of a cosmological model. We complete the chapter with a discussion of nucleosynthesis and the emergence of matter in the universe as we see it today.

10.1   THE EXPANDING UNIVERSE

The twentieth century saw a significant increase in the amount of observational evidence that could be used to test the diverse set of cosmological concepts that have appeared throughout human history. A particularly significant observation is the cosmological red shift, which is an electromagnetic example of the Doppler effect.

Imagine standing next to a railroad crossing as a train approaches. When the train gets close to the crossing the engineer blasts the train horn in warning. The sound of the horn begins before the train enters the crossing and continues until the train exits the crossing. To the engineer on the train, the sound of the horn never changes. To those of us standing next to the crossing, the sound of the horn rises in pitch as the train approaches, and then falls as the train recedes.

The sound of the train horn is the transmission of vibrational energy through the atmosphere. Air molecules are forced to vibrate by an oscillating membrane in the train horn. The molecular vibrations propagate as sound waves from the horn and through the atmosphere until the energy of vibration is dissipated or until it reaches our ears. Because sound is periodic, hence wavelike, we can define a sound wavelength and sound frequency (or pitch). If the wavelength of the sound changes with respect to an observer, so also will its pitch. Changes in wavelength or pitch can result from the relative motion of a source of wave motion and an observer. Austrian physicist and mathematician Christian Johann Doppler was the first to explain the change in pitch. The change in pitch that results from the relative motion of the wave source and an observer is called the *Doppler effect*. The Doppler effect for the train example is shown in Figure 10-1.

Figure 10-1A shows a train moving relative to the observer, and Figure 10-1B shows the train at rest relative to the observer. The train engineer does not hear a change in pitch because the sound of the horn and the engineer are not moving relative to each other. By contrast, the train is moving relative to us. The change in pitch we hear is the Doppler effect for sound. As the train approaches, the number of sound waves reaching us in a given time interval increases. This is the cause of the increase in pitch we hear. When the train passes and recedes, the number of waves reaching

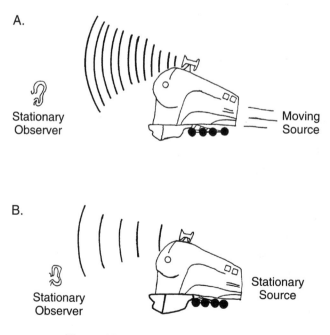

Figure 10-1. Doppler effect for sound.

us in the given time interval decreases with the result that we hear a falling pitch.

The Doppler effect applies not only to sound waves, but to any wave phenomenon. Suppose ν_S is the frequency of a light wave emitted from a source and measured in the frame of reference of the source. The Doppler effect modifies the frequency ν_S so that a receiver at rest relative to the source (Figure 10-2) measures the apparent frequency ν_R

$$\nu_R = \nu_S \frac{\sqrt{1 - \beta^2}}{1 + \beta \cos \theta} \tag{10.1.1}$$

where $\beta = \nu/c$, ν is the magnitude of the velocity of the source, and θ is the angle shown in Figure 10-2 [Garrod, 1984, Section 1.11].

An optical example of the Doppler effect is the cosmological red shift of stars. Stars are sources of light. Many mechanisms at work in a star can generate starlight. The mechanism of interest to us is the release of energy from an atom that has been energetically excited. As the atom relaxes to a less energetic state, it releases energy in the form of electromagnetic quanta (light). Each kind of atom generates a light spectrum that can serve as an identifying characteristic of that atom. For example, the spectrum of one hydrogen atom is the same as the spectrum of another hydrogen atom, but differs from the helium spectrum. By determining the spectra of atoms on

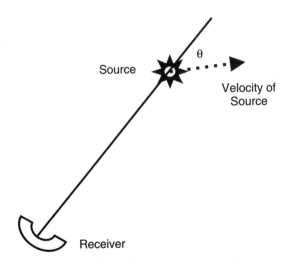

Figure 10-2. Doppler effect between source and receiver.

Earth, we can identify the types of atoms present on stars by comparing their spectra.

Suppose a star is moving away from the earth. Each succeeding wave of light from the star has to travel farther than the preceding waves because the star is moving away from us. The wavelength of the light appears longer than the wavelength would be if the star were at rest relative to the earth because of the Doppler effect. This phenomenon is sketched in Figure 10-3.

Figure 10-3A shows a star at rest relative to the observer, and Figure 10-3B shows the star moving relative to the observer. In the case of a receding source, the angle θ in Equation (10.1.1) is 0 degrees and Equation (10.1.1) becomes

$$\nu_{\text{receding}} = \nu_S \frac{\sqrt{1-\beta^2}}{1+\beta} = \nu_S \left| \frac{1-\beta}{1+\beta} \right|^{1/2} \tag{10.1.2}$$

The corresponding Doppler effect for the wavelength of the receding source is the reciprocal of Equation (10.1.2), or

$$\lambda_{\text{receding}} = \lambda_S \sqrt{\frac{1+\beta}{1-\beta}} \tag{10.1.3}$$

The wave frequency of light from a moving source decreases because the number of waves we count in any fixed time interval has decreased relative to the frequency of light from a stationary source. This frequency decrease makes the spectra of atoms on the moving star shift to less-energetic values. If the spectra are in the visible light range, the shift is towards the red. Consequently, the shift of a spectrum to less-energetic values is called a *red shift*. The red shift is often reported as the parameter

$$Z = \left(\frac{\lambda_{\text{receding}}}{\lambda_S} \right) - 1 \tag{10.1.4}$$

The ratio β, the speed of recession divided by the speed of light in vacuum, can be calculated by combining Equations (10.1.3) and (10.1.4) to obtain

$$\beta = \frac{Z^2 - 1}{Z^2 + 1} \tag{10.1.5}$$

For comparison, suppose a star was moving toward us. The wave frequency would increase and the atomic spectra would shift toward more

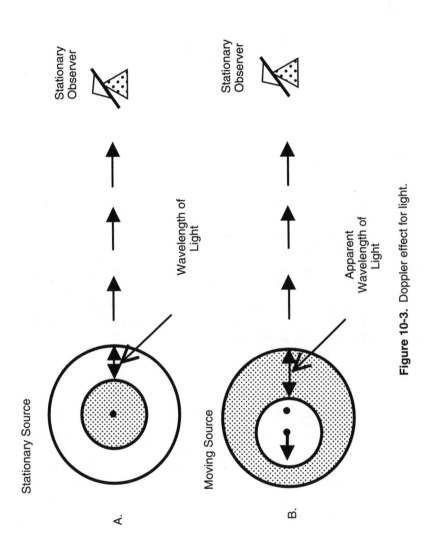

Stationary Observer

Stationary Observer

Wavelength of Light

Apparent Wavelength of Light

Stationary Source

Moving Source

A.

B.

Figure 10-3. Doppler effect for light.

energetic values. In the case of an approaching source, the angle θ in Equation (10.1.1) is 180 degrees and Equation (10.1.1) becomes

$$
\nu_{\text{approaching}} = \nu_S \frac{\sqrt{1 - \beta^2}}{1 - \beta} = \nu_S \left| \frac{1 + \beta}{1 - \beta} \right|^{1/2} \tag{10.1.6}
$$

This is a shift toward the blue for visible light, and is therefore referred to as a *blue shift*. The corresponding Doppler effect for the wavelength of an approaching object is

$$
\lambda_{\text{approaching}} = \lambda_S \sqrt{\frac{1 - \beta}{1 + \beta}} \tag{10.1.7}
$$

In 1929 the American astronomer Edwin Powell Hubble was examining visible light from the distant Bootes galaxy. He observed a red shift in atomic spectra as compared with light from nearer, similar galaxies. Interpreting the red shift as the Doppler effect, Hubble concluded that the Bootes galaxy was traveling away from the earth at one-eighth the speed of light. Prompted by his discovery, astronomers studied other distant galaxies and found that light from each of the galaxies was red shifted. Astronomers have found that the speed of recession ν_{receding} of galaxies is proportional to the distance d to the galaxy, or

$$
\nu_{\text{receding}} = H_0 d \tag{10.1.8}
$$

The proportionality constant H_0 is called *Hubble's constant* and has units of reciprocal time. The value $1/H_0$ is an estimate of the age of the universe, and the distance $d_{\text{max}} = c/H_0$ is an estimate of the size of the universe if we assume an object cannot travel faster than the speed of light in vacuum. If Hubble's interpretation is correct, galaxies throughout the universe are receding from the earth. In some cases, the speed of recession is almost the speed of light. How can the red shifts be explained within a cosmological context?

One reasonable explanation is based on a relatively simple observation. Suppose we inflate a black balloon with two white dots painted on its surface. What happens to the distance separating the two dots as the balloon expands? The two dots move away from each other. They continue to separate as long as we inflate the balloon without popping it. Galaxies in the universe behave much as the dots on the balloon behave. If the universe is expanding, each galaxy moves away from every other galaxy just like the white dots separated when the balloon was inflated. Figure 10-4 illustrates the expansion. The relative position of each object remains the same, but their separation increases as time increases and the overall size of the universe expands. Figure 10-4 illustrates a uniform expansion. Light emanating from the galaxies is red shifted in an expanding universe because the galaxies are moving away from each other.

An expanding universe model explains the red shift observed by astronomers, but it raises new questions. What caused the expansion is the most obvious. Less obvious, but perhaps more important from a scientific point of view, is the implication in the balloon analogy that the universe may be curved and finite. Indeed, mathematical descriptions of an expanding universe treat the universe as both curved and finite, but the curvature and finiteness are in four-dimensional space–time. Modern cosmological models exist in the arena of four-dimensional space–time. This arena would be quite unacceptable to an advocate of Newton's concepts of absolute space and time. It is built from the concepts of Einstein's general theory of relativity.

Modern models of cosmology describe the expanding universe as a relic of an explosion. According to the most popular explanation of universal

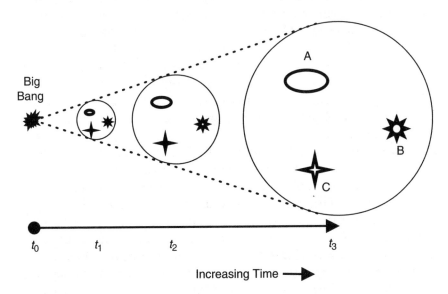

Figure 10-4. Expanding universe.

expansion, all of the matter and energy of the early universe was located at a single point in space at a specific time. The explosion of the matter-energy is known as the Big Bang. In open universe models the Big Bang is the beginning of the universe. By contrast, the Big Bang initiates the expansion phase of an infinite succession of expansion and contraction cycles in oscillating universe models.

Perhaps you have noticed that we have not discussed the cause of the Big Bang. Although many have speculated on this issue, a consensus does not exist and speculation in this area is beyond the scope of this book. It is enough for now to note the cosmological red shift is experimental justification of an expanding universe.

10.2   COSMIC RADIATION

Open universe and oscillating universe cosmological models assume the universe is expanding as a result of a primordial explosion. If such an explosion did take place, is it possible that remnants of the explosion, some type of debris, may still exist as a relic of the cataclysm?

Americans Arno A. Penzias and Robert W. Wilson made a discovery of cosmological significance in 1964. These two men were developing an extremely sensitive instrument for measuring radiation having wavelengths on the order of seven to 20 centimeters (about 0.07 m to 0.20 m). They opted to use this instrument—known as a horn reflector or microwave antenna—to observe any microwave radiation not originating on the earth. Such cosmic radiation, which presumably originated at the center of our Milky Way galaxy, could interfere with satellite communications. Penzias and Wilson's employer, Bell Laboratories, wanted to determine if such interference exists.

Prior to 1964 no one had successfully managed to isolate the cosmic microwave frequencies in a study of radiation streaming to the earth from outer space. Radiation from space covers a relatively broad range of spectral frequencies and can be loosely characterized as noise. Its effect on electronic communications systems is familiar to anyone who owns a television and has picked up "snow" or static. Some of this interference could be inherent to the electrical device, it could arise as interference from Earth-based radiating systems such as television or radio transmitters, or it could originate in the cosmos. The challenge for Penzias and Wilson was to remove all sources of microwave radiation other than the cosmos.

When Penzias and Wilson performed their observations after a painstaking effort to eliminate all unwanted sources of radiation, they did not expect to find any signal. Instead, they detected a signal—a hiss—coming from every direction in which they aligned their antenna. The hiss was present everywhere in the sky and in all of the seasons. The energy of the radiation they observed was equivalent to an absolute temperature of about 3.5° Kelvin, or about −453° Fahrenheit. What is the meaning of this ubiquitous radiation?

Contemporary physicists have interpreted the microwave radiation observed by Penzias and Wilson as cosmic background radiation. It is a relic of the primordial explosion that initiated the cosmological red shifts we attribute to an expanding universe. Measurements of the microwave background radiation are being used to determine the isotropy of matter in the universe. The distribution of matter in the universe is isotropic if the distribution does not depend on which direction we look. If there is more matter in one direction than another, the universe is anisotropic. The isotropy of the universe is one of the fundamental parameters of cosmological models that can be measured. The ability to measure the microwave background radiation has helped make cosmology an experimental science.

10.3 ASTRONOMICAL DISTANCES

Our choice of a cosmological model depends on three observable parameters: the rate of expansion of the universe (the Hubble expansion rate), a dimensionless deceleration parameter, and the critical density of the universe. All three parameters depend on measurements of astronomical distances.

Astronomical distances are measured using several methods.[2] Distances to stars near the earth are determined by triangulation. Triangulation is illustrated in Figure 10-5. It is a method that was developed by surveyors for measuring distances on Earth. The triangulation method for measuring astronomical distances to nearby stars is most effective when it uses the baseline of the orbit of the earth. The earth follows an elliptical orbit around the sun similar to the orbit shown in Figure 10-5. The semi-major axis of the ellipse is the distance a and the distance from the sun to the center of the ellipse is the distance ea. The variable e is a number between 0 and 1 called the eccentricity of the orbit. If the orbit was circular, the value of eccentricity would be $e = 0$.

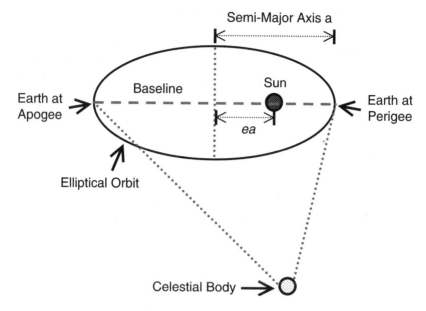

Figure 10-5. Triangulation.

The triangulation method uses a triangle that is formed with the star at one apex and observations from two points of the earth's orbit at the remaining two apexes. The greatest accuracy is achieved by making observations of the star at six-month intervals. The longest baseline is formed between the point of the earth's orbit closest to the sun (perigee) and the point farthest from the sun (apogee). From the baseline length and the angles of measurements to the stars we can determine the distance of the star from the earth. Triangulation using the earth's orbit to form a baseline works well for distances on the order of 150 light-years, where a light-year is the distance light travels in one year. A similar method using the apparent motion of stars to form a baseline can yield estimates of astronomical distances up to a few hundred light-years.

Triangulation is not effective for distances of greatest interest to the study of the origin of the universe: distances stretching over billions of light-years. Measurements of such large distances are needed because we have instruments capable of observing light emanating from astronomical objects at such distances. Light from these objects began the journey to the earth billions of years ago. By observing distant astronomical objects using such sophisticated technology as the Cosmic Background Explorer (COBE) satellite and the Hubble Space Telescope, we are getting a glimpse of the universe as it appeared billions of years ago, near the time of the Big Bang.[3] How do we measure such vast distances?

A discovery by Henrietta Leavitt at the Harvard Observatory in 1912 is the basis for one method of measuring distances on the order of a few tens of millions of light-years. She observed a pulsation in the visible light originating from some stars: the light dims and brightens periodically. These pulsating stars are called *Cepheid variables*. The period of the pulsation is related to the luminosity, or total energy radiated each second, by a Cepheid variable. Thus, the period of pulsation is a "fingerprint" or "signature" of a Cepheid variable. If we determine the period of pulsation of a distant Cepheid variable, we can compare its observed luminosity to the luminosity we would expect based on studies of nearby Cepheid variables. The difference between the observed and expected luminosities of distant Cepheid variables is a measure of their distance from the earth. Once we know the distance to a Cepheid variable, we have an estimate of the distance to its neighbors. Edwin Hubble used Cepheid variables to measure distances in his study of red shifted stellar spectra. He originally compared velocity relative to the sun with the distance from the sun. The Hubble expansion rate can be determined from the linear relationship between the red shift of a galaxy and its distance from us.

Cepheid variables are inadequate to establish the distances of the most remote objects we can detect with our telescopes. Another astronomical discovery was needed, and it came in the form of quasars. Quasars, or quasi-stellar radio sources, were first observed in the late 1950s when radiotelescopes began to mature. Radiotelescopes are essentially radiowave antennae that are used to collect electromagnetic radiation in the radiowave range of the electromagnetic spectrum. Quasars are intense emitters of radiowaves—comparable, in fact, to the strongest galactic emitters. Dutch astronomer Maarten Schmidt provided an explanation of quasar emissions in 1963. He interpreted quasar spectra as highly red shifted, so much so that the source of the spectra must be moving at a significant fraction of the speed of light. Speeds of recession of some quasars have been calculated to be in excess of 90% of the speed of light.

A quasar is an example of a standard candle. A *standard candle* in the astronomical sense is a light source with a known luminosity. By comparing the magnitudes of electromagnetic emission of nearby standard candles with the apparent magnitudes of distant standard candles, we can establish a correlation between the magnitudes and red shifts of standard candles. Historically, standard candles have included quasars, selected bright galaxies, and supernovae. Supernovae are exploding stars. The value of quasars as standard candles is limited because the magnitudes of quasars are highly irregular. A better standard candle is the brightest galaxy in a cluster of galaxies with a recognizable structure.

Galaxies typically exist in clusters. Our own galaxy, for example, is part of a cluster of galaxies called the Local Group. Galaxies also have recognizable shapes. The Milky Way Galaxy is a spiral galaxy. Many galaxies are elliptical. Elliptical galaxies are said to be E-type, and spiral galaxies are S-type. The brightest members of E-type galaxies are so similar to one another that they make excellent standard candles. The deceleration parameter discussed in Section 10.5 is estimated from the relationship between the magnitude of E-type galaxies and their red shifts.

A problem with using the brightest galaxy in a cluster of galaxies as a standard candle is that galaxies can change. For example, their brightness can increase if they merge with other galaxies. Another standard candle for measuring astronomical distances is the type Ia supernova. A supernova which does not exhibit hydrogen features in its spectrum is a type I supernova. If the type I supernova spectrum includes a silicon absorption feature at 6150 Angstroms, it is called a type Ia supernova. Type Ia supernovae have been found to work well as standard candles [Perlmutter, 2003].

Astronomical distance measurements provide data that can be used to determine the parameters in cosmological models. Both the Hubble expansion rate and the deceleration parameter are necessary values for determining the most appropriate cosmological model of our universe. They are relatively well known, however, in comparison to the critical density of the universe. The cosmological parameters are discussed in more detail following the introduction to the standard cosmological model presented in the next section.

10.4   THE STANDARD COSMOLOGICAL MODEL

The standard cosmological model is a model of a universe that evolves after the Big Bang. Two important and simplifying assumptions are made in the standard cosmological model: (1) the distribution of matter in the universe is homogeneous, and (2) the distribution of matter in the universe is isotropic. The assumption of homogeneity says that matter is uniformly distributed throughout space. The assumption of isotropy says that the distribution of matter is independent of direction. These two assumptions are expressed together in a principle that is known as the *cosmological*

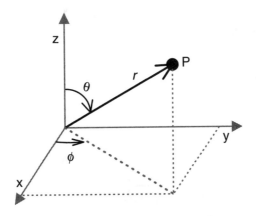

Figure 10-6. Spherical coordinates.

principle. The cosmological principle asserts that the distribution of matter in the universe is homogeneous and isotropic.

The four-dimensional geometry of the standard cosmological model is based on a metric called the Robertson-Walker metric and uses spatial coordinates expressed as spherical coordinates $\{r, \theta, \phi\}$. A set of spherical coordinates is shown in Figure 10-6. The z-axis is sometimes called the polar axis, and spherical coordinates are sometimes called spherical polar coordinates. The radius r is the radius of a sphere, θ is the polar angle measured clockwise from the z-axis, and ϕ is the azimuthal angle around the z-axis and measured counterclockwise from the x-axis.

The length of a four-vector in the standard cosmological model is defined by the Robertson-Walker metric:

$$ds^2 = -dt^2 + R^2(t)\left[\frac{dr^2}{1 - K_0 r^2} + r^2\left(d\theta^2 + \sin^2\theta\, d\phi^2\right)\right] \qquad (10.4.1)$$

The variable K_0 is an arbitrary constant and $R(t)$ is a time dependent scale factor with the initial condition

$$R_0 \equiv R(0) = 1 \qquad (10.4.2)$$

The curvature of three-space is given in terms of $R(t)$ and K_0 as

$$K(t) = \frac{K_0}{R^2}, \quad K(0) \equiv K_0 \qquad (10.4.3)$$

where the initial condition of $K(t)$ is derived from Equation (10.4.2).

Substitution of Equation (10.4.1) in the Einstein field equations from general relativity leads to the Lemaître equations for the scale factor:

$$\left[\frac{1}{R}\frac{dR}{dt}\right]^2 = \frac{8}{3}\pi G\rho(t) - c^2\frac{K_0}{R^2} + \frac{\Lambda}{3} \tag{10.4.4}$$

and

$$\frac{d\rho(t)}{dt} = -3\left[\rho(t) + \frac{p(t)}{c^2}\right]\left(\frac{1}{R}\frac{dR}{dt}\right) \tag{10.4.5}$$

The speed of light in vacuum is c, $\rho(t)$ is the mass density including the mass equivalent of all energy present, $p(t)$ is pressure, G is the gravitational constant, and Λ is the cosmological constant. The Lemaître equations together with an equation of state relating p and ρ, determine the behavior of $R(t)$.

A model of an evolving universe is obtained by finding the scale factor $R(t)$. The standard model of an expanding universe uses a scale factor that was first obtained by Russian meteorologist and mathematician Alexsandr Alexandrovich Friedmann in 1922. The Friedmann scale factor is the function $R(t)$ that satisfies Equation (10.4.4) with the assumption $p \ll \rho c^2$. In this case Equation (10.4.5) has the solution

$$\rho(t) = \frac{\rho_0}{R^3}, \quad \rho(0) \equiv \rho_0 \tag{10.4.6}$$

The Friedmann equation for the scale factor $R(t)$ is derived from Equations (10.4.4) through (10.4.6) by imposing the assumptions $p \ll \rho c^2$ and $\Lambda = 0$. We do not impose the assumption $\Lambda = 0$ here because the cosmological constant is an important parameter, as we see later. Equation (10.4.4) takes on the simplified form

$$\left[\frac{1}{R}\frac{dR}{dt}\right]^2 = \frac{8}{3}\pi G\frac{\rho_0}{R^3} - c^2\frac{K_0}{R^3} + \frac{\Lambda}{3} \tag{10.4.7}$$

The Friedmann scale factor $R(t)$ is the solution of Equation (10.4.7) with $\Lambda = 0$.

We now introduce the Hubble parameter

$$H(t) = \frac{1}{R}\frac{dR}{dt}, \quad H_0 \equiv H(0) \tag{10.4.8}$$

and the dimensionless density (or closure) parameter

$$\Omega(t) = \frac{8}{3}\pi G\frac{\rho(t)}{H^2(t)}, \quad \Omega_0 \equiv \Omega(0) = \frac{8}{3}\pi G\frac{\rho_0}{R_0^3 H_0^2} \tag{10.4.9}$$

The Friedmann equation with a cosmological constant can be written in terms of these parameters as

$$c^2 K(t) = \frac{8}{3}\pi G\rho(t) + \frac{\Lambda}{3} - H^2(t) \tag{10.4.10}$$

Introducing the dimensionless time

$$t_D \equiv H_0 t \tag{10.4.11}$$

and substituting Equations (10.4.8) through (10.4.11) in Equation (10.4.7) gives

$$\left[\frac{1}{R}\frac{dR}{dt_D}\right]^2 = \Omega_0 \frac{(1-R)}{R^3} + \frac{1}{R^2} + \frac{1}{3}\left[\frac{\Lambda}{H_0^2}\right]\frac{(R^2-1)}{R^2} \tag{10.4.12}$$

Equation (10.4.12) can be solved numerically for the scale factor $R(t)$.

The scale factor $R(t)$ is sometimes called the *scale factor of the universe* [Liddle, 1999, page 21]. It is a measure of the universal expansion rate. We can use $R(t)$ to change from a coordinate system that is fixed with respect to the expanding universe to a coordinate system that is moving with the expanding universe. We call the latter coordinate system a *co-moving coordinate system*. A co-moving coordinate system is sketched in Figure 10-7.

Suppose we define $\vec{r}_{fix}$ as the spatial distance to an object from the origin of the fixed coordinate system, and we define $\vec{r}_{co\text{-moving}}$ as the spatial

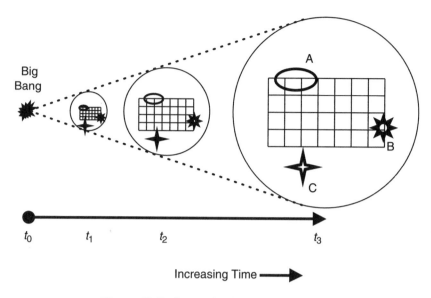

Figure 10-7. Co-moving coordinate system.

distance to the object from the origin of the co-moving coordinate system. The spatial distances are related by the transformation

$$\vec{r}_{\text{fix}} = R(t)\,\vec{r}_{\text{co-moving}} \tag{10.4.13}$$

The homogeneity assumption of the cosmological principle is used to justify making the scale factor $R(t)$ a function of time only. The distance between two objects does not change as the result of the expansion of the universe if we measure the distance in the co-moving coordinate system. The fixed distances are illustrated by the unchanging locations of objects A, B, C relative to the co-moving grid shown in Figure 10-7. The distances between objects measured relative to the co-moving grid do not change because the co-moving grid expands as the universe expands.

It is instructive to express the length of a four-vector in our four-dimensional coordinate system. The square of the length of a four-vector ds with components $ds = \{dx^0 = dt,\ dx^1 = dr,\ dx^2 = d\theta,\ dx^3 = d\phi\}$ may be written as

$$ds^2 = \sum_{\mu=0}^{3}\sum_{\nu=0}^{3} g_{\mu\nu}\,dx^\mu dx^\nu \quad \text{for} \quad \mu, \nu = 0, 1, 2, 3 \tag{10.4.14}$$

The metric corresponding to the Robertson-Walker metric is obtained by comparing Equation (10.4.1) and (10.4.14) to find

$$
[g_{\mu\nu}] = \begin{bmatrix} -1 & 0 & 0 & 0 \\ 0 & \dfrac{R^2(t)}{1 - K_0 r^2} & 0 & 0 \\ 0 & 0 & R^2(t)r^2 & 0 \\ 0 & 0 & 0 & R^2(t)r^2 \sin^2\theta \end{bmatrix}
\tag{10.4.15}
$$

If the elements of the metric were constant, the four-dimensional coordinate system would be called flat, as it is for special relativity. The four-dimensional coordinate system in the standard cosmological model is called curved because of the dependence of the diagonal elements of the metric on the spatial coordinates $\{r, \theta, \phi\}$. The model universe is closed, open, or spatially flat if K_0 is $+1$, -1, or 0 respectively. Two of the cosmological models are illustrated in Figure 10-8.

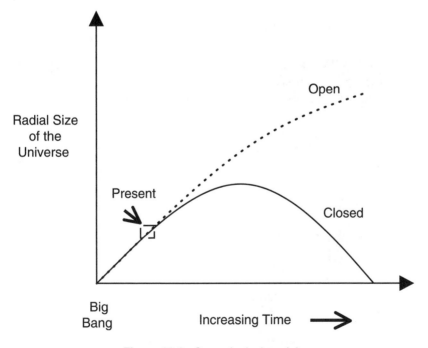

Figure 10-8. Cosmological models.

10.5   COSMOLOGICAL PARAMETERS

We have identified three important cosmological parameters: the Hubble expansion rate or Hubble parameter, the deceleration parameter, and the critical density. The Hubble parameter can be measured using the Doppler effect and the spectra of standard candles, as we have discussed previously. We focus here on the deceleration parameter and the critical density.

DECELERATION PARAMETER

The rate of expansion of the universe can be described in terms of the deceleration parameter as follows [Liddle, 1999, Section 6.3]. We first use the Taylor series to expand the scale factor as

$$R(t) = R(t_0) + \dot{R}(t_0)\,(t - t_0) + \frac{1}{2}\ddot{R}(t_0)\,(t - t_0)^2 + \cdots \qquad (10.5.1)$$

If we divide by $R(t_0)$, we obtain

$$\frac{R(t)}{R(t_0)} = 1 + \frac{\dot{R}(t_0)}{R(t_0)}(t - t_0) + \frac{1}{2}\frac{\ddot{R}(t_0)}{R(t_0)}(t - t_0)^2 + \cdots \qquad (10.5.2)$$

The ratio $R(t)/R(t_0)$ is a normalized scale factor. The universe is expanding if $R(t)/R(t_0) > 0$ and it is contracting if $R(t)/R(t_0) < 0$.

We now use Equation (10.4.8) to replace the coefficient of $(t - t_0)$ with the Hubble parameter:

$$\frac{R(t)}{R(t_0)} = 1 + H_0(t - t_0) + \frac{1}{2}\frac{\ddot{R}(t_0)}{R(t_0)}(t - t_0)^2 + \cdots \qquad (10.5.3)$$

If we introduce the term

$$q_0 = -\frac{\ddot{R}(t_0)}{R(t_0)}\frac{1}{H_0^2} = -\frac{R(t_0)\,\ddot{R}(t_0)}{[\dot{R}(t_0)]^2} \qquad (10.5.4)$$

in Equation (10.5.3), we can write

$$\frac{R(t)}{R(t_0)} = 1 + H_0(t - t_0) - \frac{q_0}{2}H_0^2(t - t_0)^2 + \cdots \qquad (10.5.5)$$

Keeping only terms to second order in $(t - t_0)$, we calculate the time rate of change of the normalized scale factor. The result is

$$\frac{d}{dt}\frac{R(t)}{R(t_0)} = H_0 - q_0 H_0^2 (t - t_0) \qquad (10.5.6)$$

The second order time derivative of the normalized scale factor is

$$\frac{d^2}{dt^2}\frac{R(t)}{R(t_0)} = -q_0 H_0^2 \qquad (10.5.7)$$

The term q_0 is called the *deceleration parameter*. Distance measurements using type Ia supernovae as standard candles indicate that $q_0 < 0$, which implies that the expansion of the universe is accelerating [Liddle, 1999, page 51].

CRITICAL DENSITY

Critical density is a measure of the mass contained in the universe. It is the density that would make the universe cosmologically flat when the cosmological constant Λ is zero. In this case Equation (10.4.10) gives

$$0 = \frac{8}{3}\pi G \rho_c - H^2 \qquad (10.5.8)$$

or

$$\rho_c = \frac{3H^2}{8\pi G} \qquad (10.5.9)$$

By comparing the observed density of the universe to the critical density, we can decide what type of cosmological model provides the best description of the evolution of our universe. We must therefore try to determine the density of the universe. If the observed density, defined as the observable mass of the universe divided by the volume of the universe, is greater than the critical density, then the universe is closed. On the other hand, an observed density less than or equal to the critical density implies the universe is open and will continue to expand indefinitely. To decide, all we have to do is measure the mass in the universe, divide by the volume of the universe, and compare this observed density to the critical density.

Answers to two problems must be found: how much mass is in the universe, and how big is the universe? A somewhat reliable estimate of the size of the universe is obtained by using the cosmological red shift to determine the rate of expansion of the universe. Observations of quasars and supernovae are particularly valuable here. With this information, we use the speed of light—the greatest known speed of any physical object—to estimate the size of the universe.

The amount of mass in the universe is estimated by simply adding up all of the mass we observe. Masses of astronomical assemblages, such as galaxies and nebulae, are determined by studying the motion of astronomical candles (such as quasars or star clusters) within the assemblage. Evaluating the motion of an astronomical candle in terms of gravitational laws leads to an estimate of the mass of the assemblage. We must correct for the amount of mass we cannot observe due to the limitations of our instruments. Dividing the corrected amount of mass in the universe by the size of the universe gives us an observed density to compare with the critical density of our cosmological models.

As you may have surmised, the process is not quite this easy. There are a number of uncertainties and assumptions built into the procedure outlined above. What if we do not know the correct rate of expansion of the universe? Many astronomers now believe the Hubble expansion rate may have changed with time. The change in the Hubble expansion rate is called *inflation*.[4] Figure 10-9 illustrates an inflationary period during the expansion of the universe. The time scale is the same as in Figure 10-4, but there is a period of rapid expansion immediately after the Big Bang. The "Old Size" refers to the size of the universe in Figure 10-4. The inflationary expansion sketched in Figure 10-9 yields a bigger universe than the expansion without inflation.

Inflation is included in cosmological models by adding a term called the *cosmological constant* to a general relativistic equation that relates matter and the geometry of space–time. Einstein first introduced the cosmological constant in 1917 in an effort to explain a universe that was then considered unchanging, only to learn later from Hubble's measurements that the universe appeared to be expanding. Einstein viewed the cosmological constant as a major blunder,[5] but the cosmological constant has been resurrected to help explain the change in the expansion rate of the universe. Inflation is part of the standard model of cosmology, yet the theoretical explanation for inflation is an open area of research.

Another source of error in our calculation of the cosmic density is the estimate of the amount of mass in the universe. Calculations of the universal

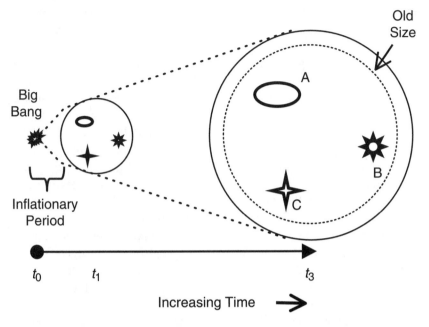

Figure 10-9. Inflation.

mass ordinarily assume the distribution of mass in the universe is uniform (homogenous) and independent of the direction in which we look (isotropic). This assumption is so important it has been given a name: the cosmological principle. The cosmological principle says that the universe is homogenous and isotropic. If we look at our solar system, at our Milky Way galaxy, or even at the Local Group of galaxies, we do not see a uniform distribution of mass. The cosmological principle does not apply on this scale because the gravitational influence of astronomical objects on one another in the Local Group disrupts the expected uniformity. We must look at an even larger region of space before we begin to observe a uniform and homogeneous distribution of mass. For the region of the universe we can observe, the cosmological principle is quite good. What about elsewhere in the universe? Does the cosmological principle actually apply everywhere? The search for answers to these questions has been the basis for research programs[6] using the COBE satellite, a microwave interferometer at the South Pole, and the microwave anisotropy probe spacecraft.

One possible source of error in our estimate of the size of the universe is our assumption that we have accurately measured the masses of particles comprising the universe. This assumption has been challenged by recent observations. One observation in particular, the mass of the neutrino, stands out as a long-standing anomaly of the standard cosmological model. The neutrino was originally introduced into physics to save the laws of conservation of energy and momentum. It is classified as a lepton like the electron. Unlike the electron, the neutrino is supposed to be massless. A massless particle, like the neutrino and the photon, is essentially a quantum of energy. According to Einstein's special theory of relativity, massless particles can only travel at the speed of light. Particles with a rest mass—a non-zero value of mass as measured by an observer moving with the particle—can move at speeds less than the speed of light, but never equal to or greater than light speed. Some people have suggested massive particles can travel faster than the speed of light, but such particles, which are called tachyons, have never been observed. Others have suggested the neutrino, presently thought to be massless, may actually have a small but nonzero rest mass. If they are right, the abundance of neutrinos in the universe could substantially contribute to the cosmic density.[7] Our inability to accurately measure the mass of the neutrino is significant because the neutrino is pervasive in the universe.

Black holes are another possible source of mass that may not be adequately accounted for in the calculation of the cosmic density. Black holes are very difficult to detect. According to theory, the gravitational field of a black hole is so strong that not even light can escape by classical methods from its gravitational pull. Quantum mechanical tunneling is a means by which particles can escape black holes, and provides a mechanism for detecting them. Several astronomical objects have many of the properties people expect a black hole to have.[8] If these objects are black holes, we do not know how many there are. Consequently, their mass, which could be substantial, may be overlooked.

Measurements of the mass of galaxies have indicated that many galaxies behave as if they contain more matter than the matter that shines.[9] The matter that is not visible to telescopes is called *missing matter* or *dark matter*. It ranges from the more conventional mass associated with possibly massive neutrinos to more exotic forms of matter that are being debated by the research community. One possibility for the dark matter is nonbaryonic matter consisting of weakly interacting, massive, stable elementary particles (WIMPS). If dark matter exists, it would be considered a challenge to the standard model of physics that underlies the standard cosmological model.[10]

Measurements of anisotropy in the cosmic microwave background radiation using the Wilkinson Microwave Anisotropy Probe (WMAP) are providing more data for developing improved cosmological models.[11] The WMAP is a space-based instrument that can observe five different microwave frequencies in the full sky. By combining the WMAP measurements with cosmological models, cosmologists are finding evidence to support the view that we live in a flat universe that underwent an inflationary expansion period immediately after the Big Bang. Furthermore, the expansion of the universe may be accelerating. This view, however, depends on a poorly understood type of energy called *dark energy* that is represented by a non-zero cosmological constant. Dark energy has been associated with the energy of the vacuum. Research to understand dark matter and dark energy is continuing.

10.6   THE BIG BANG

Astronomical observations indicate the universe is expanding as a result of a Big Bang, a term coined by astrophysicist George Gamow in the 1940s. In this section the most widely accepted scenario of the evolution of the universe from its beginning to the onset of galaxy formation is described.[12] Table 10-1 presents a summary of the timetable of universal evolution. Cosmic time in the figure refers to the amount of elapsed time between the Big Bang and a particular epoch.

Initially all of the energy of the universe was located at one point in space called a *singularity*. The source of the initial energy is not known and is the subject of both scientific and philosophical speculation. For example, one scientific scenario hypothesizes that the entire universe is

Table 10-1
Timetable of universal evolution

Cosmic time	Epoch	Event
0 second	Singularity	Big Bang
Up to 1 millionth of a second	Hadronic Era	Strong force dominant; Particle–antiparticle annihilation
Up to 1 second	Leptonic Era	Weak force gains importance; beta decay (neutron decays)
Up to 1 week	Radiation Era	Electromagnetism gaining importance; smallest atoms forming (e.g. helium, deuterium, and lithium)
Up to 10,000 years	Matter Era	Matter now dominant energy form; opaque universe
Up to 300,000 years	Decoupling Era	Gravity significant; transparent universe

the outgrowth of a single quantum process that took an improbable path.[13] Quantum fluctuations in the vacuum, such as the vacuum process discussed previously, can distort the four-dimensional geometry of space-time and create a locally dense accumulation of particles. If the fluctuation is self-sustaining, the presence of induced matter can impart a local expansion of space-time, which creates more room for additional fluctuations in a snow-balling creation of matter.

At the moment of the Big Bang, we consider the existence of the energy an experimental fact. The energy density, or energy per unit volume of space, was very large because a finite quantity of energy occupied an infinitesimally small volume of space. Primordial energy was primarily in the form of electromagnetic radiation. The ratio of matter to antimatter in the early, hot universe was almost one to one. The matter asymmetry is sketched in Figure 10-10. Modern estimates suggest there was slightly more matter than antimatter: for every billion grams of antimatter, there were one billion and one grams of matter. This slight difference is considered sufficient for explaining the preponderance of matter in the contemporary universe. Temperature of the primordial fireball was in excess of ten billion degrees Kelvin, which is a thousand times hotter than the interior of our sun.

Our reckoning of cosmic time begins with the explosion of the very dense primordial energy. The explosion propelled the primordial energy away from the singularity. As cosmic time increases, the universe expands.

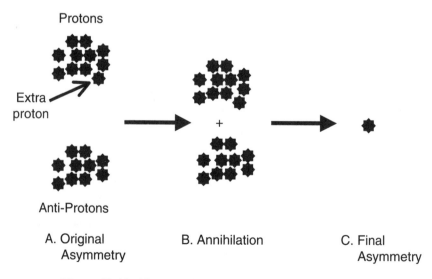

Figure 10-10. Matter–antimatter asymmetry at the Big Bang.

A consequence of an expanding universe is an increase in the distance between particles. The radius of the universe, stretching from the singularity to the furthest bit of primordial energy, increased as the universe expanded. A millionth of a second after the Big Bang the distances between particles was still very short. The dominant interaction between the proximate particles was the short-range strong interaction. Pair creation processes transformed some of the primordial energy from electromagnetic to dironic. The dirons—quarks and antiquarks—were close enough to other dirons to interact strongly and form composite hadrons. This era is dubbed the Hadronic Era. During the Hadronic Era it was virtually impossible to distinguish between hadrons and electromagnetic radiation because of extremely efficient pair creation and annihilation processes.

Continuing expansion of the universe eventually—by one second after the Big Bang—resulted in a sufficiently large separation between particles to enable the weak interaction to gain importance. Protons and neutrons were kept in thermal equilibrium by the weak interaction through such processes as the decay of a neutron into a proton, an electron, and an antineutrino. No longer were hadrons such as the proton and the neutron the only actors in the play. Leptons such as the electron and the neutrino existed and provided this era with an identity: the Leptonic Era.

The expansion of the universe continued. Within a minute after the Big Bang we find ourselves in a new era: the Radiation Era. The electromagnetic interaction has become significant. Strong and weak interactions were still present, but apply only locally within the continually expanding universe much like they do today. Nuclei of the smallest atoms—such as helium, deuterium, and lithium—were forming as a result of the strong interaction between nucleons. The electromagnetic interaction between charged leptons and nucleons has bound electrons to nuclei to complete the atomic structure of the smallest atoms. Hydrogen, being composed of an electron and a proton, is the most abundant. It is followed in decreasing abundance by more complex atoms: deuterium with an electron, proton, and neutron; helium with two electrons, two protons, and two neutrons; and so on. These atoms are not yet stable, however, because of still high temperatures (one million to one hundred million degrees Kelvin).

PRIMORDIAL NUCLEOSYNTHESIS

Primordial matter is formed during the first few minutes after the Big Bang in a process called nucleosynthesis.[14] *Nucleosynthesis* is the formation of complex nuclei from protons and neutrons. Nucleosynthesis may

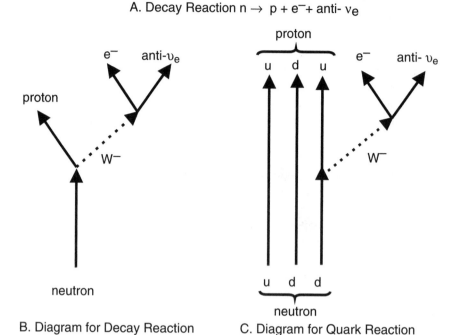

A. Decay Reaction n → p + e⁻+ anti- ν_e

B. Diagram for Decay Reaction   C. Diagram for Quark Reaction

Figure 10-11. Neutron decay.

be classified into primordial nucleosynthesis and stellar nucleosynthesis. *Primordial nucleosynthesis* is the formation of nuclei from protons and neutrons in the early universe. *Stellar nucleosynthesis* is the formation of nuclei in the interior of stars or by the explosion of stars. Stellar nucleosynthesis is an ongoing process that is briefly discussed at the end of this section. Primordial nucleosynthesis occurred early in the life of the universe and was the beginning of nuclear formation.

Two properties of protons and neutrons that are important in nucleosynthesis are particle mass and stability. Proton mass (938.3 MeV in energy units or 1.6726×10^{-27} kg) is slightly less than neutron mass (939.6 MeV in energy units or 1.6749×10^{-27} kg). Free protons are stable, but a free neutron decays into a proton, an electron, and an anti-electron neutrino in the reaction shown in Figure 10-11. The quark content of the neutron includes an up quark labeled "u" and two down quarks labeled "d." One of the down quarks in the neutron decays into an up quark and a W boson. The W boson then decays into an electron and an anti-electron neutrino. The half-life of a free neutron is approximately 940 seconds, so a neutron formed in the first few minutes after the Big Bang would behave as if it were a stable particle until the universe was at least fifteen minutes old.

Free neutrons and protons existed in the first few minutes of the early universe when it was hot enough to prevent the formation of stable nuclei. As the temperature of the universe cooled, the neutrons with mass m_n and protons with mass m_p became nonrelativistic and achieved thermal equilibrium. The distribution of particles is best described by number density N, which is the number of particles per unit volume of space. The ratio of the number density of neutrons N_n to the number density of protons N_p in thermal equilibrium is satisfied by the Maxwell-Boltzmann distribution [Liddle, Section 11.1] and may be written as

$$\frac{N_n}{N_p} = \left(\frac{m_n}{m_p}\right)^{3/2} \exp\left[-\frac{(m_n - m_p)c^2}{k_B T}\right] \tag{10.6.1}$$

where k_B is Boltzmann's constant, c is the speed of light in vacuum, and T is temperature. A nonrelativistic system corresponds to a temperature that satisfies the inequality $k_B T \ll m_p c^2$ or

$$T \ll T_p \equiv \frac{m_p c^2}{k_B} \tag{10.6.2}$$

The temperature T_p is the temperature equivalent of the rest mass energy of the proton.

The formation of primordial helium begins with the reaction of a proton and a neutron to form the nucleus of deuterium and a gamma ray:

$$p + n \rightarrow d + \gamma \tag{10.6.3}$$

The deuterium nucleus contains a proton and a neutron and may be called a deuteron. Two deuterium nuclei combine to form a tritium nucleus in the reaction

$$d + d \rightarrow t + p \tag{10.6.4}$$

The tritium nucleus contains a proton and two neutrons and may be called a triton. Both deuterium and tritium are hydrogen isotopes.

Once deuterium and tritium are available, they can interact to form the helium nucleus in the reaction

$$d + t \rightarrow {}^4He + n \qquad (10.6.5)$$

An alternative process forms the light helium isotope ${}^3He$ instead of tritium in the reaction

$$d + d \rightarrow {}^3He + n \qquad (10.6.6)$$

where a neutron is produced instead of a proton. Helium is produced from light helium by the reaction

$$d + {}^3He \rightarrow {}^4He + p \qquad (10.6.7)$$

The abundance of helium in the early universe was due to these nucleosynthetic reactions.

THE MATTER ERA

Years pass before we enter the next recognized era: the Matter Era. For the first ten thousand years after the Big Bang the dominant energy form was electromagnetic radiation. Beginning with the Matter Era, matter becomes the dominant energy form. The total matter per unit volume of space is large enough to restrict the passage of light from one point of the universe to another in the Matter Era. Since light is absorbed over relatively short distances compared to the size of the expanding universe, the universe appears to be opaque. The presence of significant amounts of matter is accompanied by the presence of an increasingly influential force: the gravitational interaction.

Three hundred thousand years of cosmic time elapsed before gravity became a major force in shaping the evolution of the universe. By then particles were sufficiently separated to allow the passage of light over great distances. The universe was no longer opaque: it had become transparent, much as it is today. Most matter existed as stable elemental hydrogen or helium. Temperatures had cooled to merely thousands of degrees Kelvin. Small perturbations, or disturbances, in the distribution of matter were growing in significance. Gravitational attractions of locally dense matter distributions were gaining strength relative to the other three fundamental interactions as distances between accumulations of matter increased. These regions of relatively strong gravitational forces are believed to be the seeds from which galaxies sprouted some one billion years after the Big Bang.

STELLAR NUCLEOSYNTHESIS

Primordial nucleosynthesis was able to create the first nuclei, such as hydrogen, deuterium, tritium, helium, and, to a lesser extent, lithium. The formation of stars from primordial matter was an important step in building larger nuclei. Stellar nucleosynthesis is the process that has created, and continues to create, the nuclei we see listed in the Periodic Table. Stellar nucleosynthesis is a name that encompasses a set of processes. The first process is the creation of relatively small nuclei by nuclear fusion in the stellar interior. In this process, nuclei with mass number comparable to iron (^{56}Fe) or smaller are created. To create larger nuclei, more energy and more neutrons are needed. This set of conditions occurs when some stars begin to die.

As stars age, the fusion process consumes available fuel such as helium and carbon and eventually reaches a stage when the repulsive pressure associated with nuclear fusion is overcome by gravitational forces associated with the mass of the star. Depending on the size and composition of the star, it will collapse and may experience an explosive fate by becoming a nova or supernova. The explosive gravitational collapse of a star is a stage in the formation of nuclei with mass numbers $A \approx 60$ or greater. Several processes may occur, but their details are beyond the scope of this book. The endnotes[14] provide more information about stellar nucleosynthesis.

ENDNOTES

1. Sources that discuss cosmological parameters include Misner, et al. [1973], Kolb and Turner [1990; 2000], Peebles [1993], Weinberg [1993], Hogan [1998], Bergstrom and Goobar [1999], Liddle [1999], Ludvigsen [1999], Bernstein, et al. [2000], Fukugita and Hogan [2000], and Goldsmith [2000]. For an analysis of the age of the universe using cosmological parameters in the context of an evolution parameter theory, see Fanchi [1988, 1993]. Burbridge, et al. [1999] present a different approach to cosmology that does not rely on a Big Bang.

2. Sources that discuss the measurement of astronomical distances include Trefil [1985], Abell, et al. [1991], Freedman [1992], Börner [1993], Goldsmith [2000], and Hester, et al. [2002].
3. Mather and Boslough [1996] discuss the COBE project. Ferguson, et al. [1997] discuss the Hubble Space Telescope results.
4. For discussions of inflation, see Linde [1987], Weinberg [1992], Peebles [1993], Krauss [1999], Bucher and Spergel [1999], Kolb and Turner [2000], and Hester, et al. [2002]. Peebles [2001] provides additional references and a report card for major theories of cosmology. He gives inflation an incomplete because of a lack of direct evidence and the need to significantly extrapolate beyond the accepted laws of physics.
5. See Pais [1982, Chapter 15e.] and Goldsmith [2000, Chapter 2].
6. Bergstrom and Goobar [1999], Liddle [1999] and Goldsmith [2000, Chapter 11] discuss COBE results, and cosmic microwave observations providing more evidence in support of inflation are reported by Schwarzschild [2003]. Bennett, et al. [2001] discuss the microwave anisotropy probe. Smoot and Scott [2000] summarize measurements of cosmic background radiation.
7. The effect of mass in cosmology is discussed in several sources, such as those in Endnote 1. Inconclusive measurements of neutrino mass are reported by the Particle Data Group [2000]. For an analysis of massive neutrino behavior in the context of an evolution parameter theory, see Fanchi [1998].
8. See Levy [1998], Blandford and Gehrels [1999], Bernstein, et al. [2000], and Hester, et al. [2002] for discussions of black holes.
9. For discussions of dark matter, see Gribbin [1991], Srednicki [2000], Goldsmith [2000], Bernstein, et al. [2000], and Hester, et al. [2002].
10. See Kolb and Turner [2000, pg. 126].
11. See Perlmutter [2003], Turner [2003], and Schwarzschild [2003].
12. The timetable is a synopsis of the timetables presented by Weinberg [1977, 1993], Hogan [1998], and Liddle [1999].
13. Several authors have speculated on the role of quantum fluctuations of the vacuum, including speculation on the creation of the cosmos from vacuum, and quantum foams. For example, see Misner, et al. [1973], Lindley [1987], Linde [1987], Thuan [1995], Liddle [1999], and Greene [2000].
14. For discussions of nucleosynthesis, see such books as Silk [2001], Liddle [1999], and Williams [1991].

EXERCISES

10-1. A. Suppose light from an object has a frequency of 1000 Hz. If the object is approaching us at a speed of $0.1c$, what is the frequency we observe?
B. If the object in part A is receding at a speed of $0.9c$, what is the frequency we observe?

10-2. A quasar is a star that emits a pulsating burst of energy as it rotates. Suppose a quasar in a remote part of the universe has a red shift parameter $Z = 5$. Estimate the ratio β, the speed of recession divided by the speed of light in vacuum.

10-3. Hubble's constant H_0 can be used to estimate the age of the universe and the size of the universe. Fill in the following table for different possible values of Hubble's constant.

H_0 [km/sec/million light years]	Age of Universe [billion years]	Size of Universe [billion light years]
20		
22		
25		

10-4. A. The nuclide chlorine-35 has 17 protons and a mass of 34.97 m_u where the mass unit $m_u = 1.660 \times 10^{-27}$ kg. Calculate the binding energy of chlorine-35 using the mass of chlorine-35. Express your answer in Joules.
B. Estimate the binding energy per nucleon.
C. How does the value calculated in Part B compare with the value read off Figure 7-1?

10-5. Calculate the equivalent temperature T_p of the rest mass energy of the proton in Equation (10.6.2).

10-6. Calculate the ratio of neutron number density to proton number density given in Equation (10.6.1) at the temperatures shown in the following table.

Temperature [°K]	N_n/N_p
10^{11}	
10^{10}	
10^9	

10-7. A. Write the matrix of metric elements in Equation (10.4.14) for the case when the scale factor $R(t) \to 1$ and the parameter $K_0 \to 0$.
B. Use the metric elements in part A to calculate the square of the length of the four-vector $ds = \{dx^0 = dt, dx^1 = dr, dx^2 = d\theta, dx^3 = d\phi\}$.

10-8. The transformation equations for the transformation from spherical polar coordinates to Cartesian coordinates are $x = r \sin\theta \cos\phi$, $y = r \sin\theta \sin\phi$, $z = r \cos\theta$. Use these equations to find the transformation equations for the transformation from Cartesian coordinates to spherical polar coordinates.

10-9. A. Suppose 1 kg of hydrogen with negligible kinetic energy is completely transformed into energy. How much energy is released? Express your answer in Joules.
B. How many kg of hydrogen would need to be converted into energy to provide 380 quads?
C. How many barrels of oil would be needed to provide 380 quads of energy if the energy density of oil is 37,000 MJ/m^3?

CHAPTER ELEVEN

Nuclear Energy

The fundamental science underlying nuclear energy was introduced in the previous two chapters. Nuclear energy can be obtained from two principal types of reactions: fission and fusion. *Fission* is the splitting of one large nucleus into two smaller nuclei; *fusion* is the joining of two small nuclei into one large nucleus. In both reactions, significant amounts of energy can be released. The historical development of nuclear energy and nuclear energy technology is discussed in this chapter.[1]

11.1 HISTORY OF NUCLEAR ENERGY

Physicist Leo Szilard conceived of a neutron chain reaction in 1934. Szilard knew that neutrons could interact with radioactive materials to create more neutrons. If the density of the radioactive material and the number of neutrons were large enough, a chain reaction could occur. Szilard thought of two applications of the neutron chain reaction: a peaceful harnessing of the reaction for the production of consumable energy, and an uncontrolled release of energy (an explosion) for military purposes. Recognizing the potential significance of his concepts, Szilard patented them in an attempt to hinder widespread development of the military capabilities of the neutron chain reaction. This was the first attempt in history to control the proliferation of nuclear technology.

Italian physicist Enrico Fermi (1901–1954) and his colleagues in Rome were the first to bombard radioactive material using low-energy (slow) neutrons in 1935. The spatial extent of the nucleus is often expressed in terms of a unit called the *fermi*, in honor of Enrico Fermi. One fermi is equal to 10^{-15} m, or $1\,\text{fm} = 10^{-15}$ m, and is the range of the nuclear force. The correct interpretation of Fermi's results as a nuclear fission process was provided in 1938 by Lise Meitner and Otto Frisch in Sweden, and Otto Hahn and Fritz Strassmann in Berlin. Hahn and Strassmann observed that neutrons colliding with uranium could cause the uranium to split into smaller elements. This process was called *nuclear fission*. The fission process produces smaller nuclei from the break-up of a larger nucleus, and can release energy and neutrons. If neutrons interact with the nuclei of fissionable material, it can cause the nuclei to split and produce more neutrons, which can then react with more fissionable material in a chain reaction. Fermi succeeded in achieving the first sustained chain reaction on December 2, 1942, in a squash court at the University of Chicago.

The scientific discovery of radioactivity and nuclear fission did not occur in a peaceful society, but in a world threatened by the militaristic ambitions of Adolf Hitler and Nazi Germany. In 1939, Hitler's forces plunged the world into war. Many prominent German scientists fled to the United States and joined an Allied effort to develop the first nuclear weapons. Their effort, known as the Manhattan Project, culminated in the successful development of the atomic bomb. The first atomic bomb was exploded in the desert near Alamogordo, New Mexico, in 1945.

By this time, Hitler's Germany was in ruins and there was no need to use the new weapon in Europe. Japan, however, was continuing the Axis fight against the Allied forces in the Pacific and did not seem willing to surrender without first being defeated on its homeland. Such a defeat, requiring an amphibious assault against the Japanese islands, would have cost many lives, both of combatants and Japanese noncombatants. United States President Harry Truman decided to use the new weapon.

The first atomic weapon to be used against an enemy in war was dropped by a United States airplane on the Japanese city of Hiroshima on August 6, 1945. Approximately 130,000 people were killed and 90% of the city was destroyed. When the Japanese government refused to surrender unconditionally, a second atomic bomb was dropped on the Japanese city of Nagasaki on August 9, 1945. The Japanese government surrendered.

People throughout the world realized that nuclear weapons had significantly altered the potential consequences of an unlimited war. By the early 1950s, the Soviet Union had acquired the technology for building nuclear weapons and the nuclear arms race was on. As of 1972, the United States maintained a slight edge over the Soviet Union in strategic nuclear yield and almost four times the number of deliverable warheads. Yield is a measure of the explosive energy of a nuclear weapon. It is often expressed in megatons, where one megaton of explosive is equivalent to one million tons of TNT.

During the decade from 1972 to 1982, the size of the United States nuclear arsenal did not change significantly, but the quality of the weapons, particularly the delivery systems such as missiles, improved dramatically. By 1982, the Soviet Union had significantly closed the gap in the number of deliverable warheads and had surged ahead of the United States in nuclear yield. The Soviet Union and the United States had achieved a rough nuclear parity. The objective of this parity was a concept called deterrence.

Deterrence is the concept that neither side will risk engaging in a nuclear war because both sides would suffer unacceptably large losses in life and property. This is the concept that underlies the doctrine of Mutual Assured Destruction: the societies of all participants in a nuclear war would be destroyed. The "star wars" concept advanced by United States President Ronald Reagan in the 1980s is a missile defense system that provided an alternative strategy to Mutual Assured Destruction, but threatened the global balance of nuclear power.

The Soviet Union was unable to compete economically with the United States and dissolved into separate states in the late 1980s. Some of the states of the former Soviet Union, notably Russia and Ukraine, retained nuclear technology. Other nations around the world have developed nuclear technology for peaceful purposes, and possibly for military purposes.

NUCLEAR POWER

The first commercial nuclear power plant was built on the Ohio River at Shippingport, Pennsylvania, a city about 25 miles from Pittsburgh. It began operation in 1957 and generated 60 MW of electric power output [Murray, 2001, page 202]. Today, nuclear power plants generate a significant percentage of electricity in some countries. Table 11-1 lists the top ten producers of electric energy from nuclear energy and their percentage of the world's total electric energy production from nuclear energy for the year 2000. The source of these statistics is the website of the Energy Information Administration of the United States Department of Energy. These statistics should be viewed as approximate. They are presented here to indicate the order of magnitude of electric power generated by power plants that use nuclear energy as their primary energy source. According to these statistics, we see that the total amount of electric energy generated from nuclear energy in the world was 2,434 billion kWh in 2000. Total electric energy consumed in the world that same year was 13,719 billion kWh. Electric energy generated from nuclear energy provided approximately 17.7% of the electricity consumed in the world in 2000.

Table 11-1
Top ten producers of electric energy from nuclear energy in 2000

Country	Electric energy from nuclear energy (billion kWh)	% Total electric energy produced from nuclear energy (world = 2,434 billion kWh)
United States	753.9	30.97%
France	394.4	16.20%
Japan	293.8	12.07%
Germany	161.2	6.62%
Russia	122.5	5.03%
South Korea	103.5	4.25%
United Kingdom	81.7	3.36%
Ukraine	71.1	2.92%
Canada	68.7	2.82%
Spain	58.9	2.42%

Source: Table 2.7, EIA website, 2002.

Table 11-2
Dependence of nations on nuclear energy in 2000

Country	Electric energy from nuclear energy (billion kWh)	Total electricity generation (billion kWh)	Nuclear share (% national total)
United States	753.9	3799.9	19.8%
France	394.4	513.9	76.7%
Japan	293.8	1014.7	29.0%
Germany	161.2	537.3	30.0%
Russia	122.5	835.6	14.7%
South Korea	103.5	273.2	37.9%
United Kingdom	81.7	355.8	23.0%
Ukraine	71.1	163.6	43.5%
Canada	68.7	576.2	11.9%
Spain	58.9	211.6	27.8%

Source: Table 6.3, EIA website, 2002.

Some countries are highly dependent on nuclear energy. Table 11-2 shows the percentage of electric energy generated from nuclear energy compared to total electricity generation for the top ten producers of electric energy from nuclear energy for the year 2000. According to the EIA statistics, most of the electricity generated in France was generated from nuclear energy.

11.2   NUCLEAR STABILITY AND DECAY

Nuclear stability and the decay of nuclei have a significant impact on the commercial development and application of nuclear fission. We consider each of these topics in this section.

NUCLEAR STABILITY

Stable nuclei are nuclei that are in a low-energy state. A condition for nuclear stability is determined by finding the ratio Z/A of atomic number Z to mass number A that minimizes the energy of a nucleus. We obtain an estimate of nuclear energy $E(N, Z)$ that depends on atomic number Z and the number of neutrons N by multiplying nuclear mass in Equation (7.1.2) by c^2 where c is the speed of light in vacuum. We then replace the binding energy of the nucleus $B(N, Z)$ with Equation (7.1.3) to find

$$E(N,Z)=m(N,Z)c^2$$

$$=Nm_nc^2+Zm_pc^2-\left[c_1A+c_2A^{2/3}+c_3\frac{Z(Z-1)}{A^{1/3}}+c_4\frac{(N-Z)^2}{A}\right]$$

$$(11.2.1)$$

Values of the coefficients are presented in Equation (7.1.4). The stability condition is obtained by first substituting $N = A - Z$ into Equation (11.2.1) to get

$$E(A,Z) = (A - Z)\, m_nc^2 + Zm_pc^2 - \left[c_1A + c_2A^{2/3}\right.$$

$$\left.+c_3\frac{Z(Z-1)}{A^{1/3}} + c_4\frac{(A-2Z)^2}{A}\right]$$

$$(11.2.2)$$

We minimize the energy by solving the extremum condition $\partial E(A,\,Z)/\partial Z = 0$ for Z/A as a function of A. The extremum condition gives

$$\frac{\partial E(A,Z)}{\partial Z} = -m_nc^2 + m_pc^2 - c_3\frac{2Z-1}{A^{1/3}} - c_4\frac{(A-2Z)(-2)}{A} = 0$$

$$(11.2.3)$$

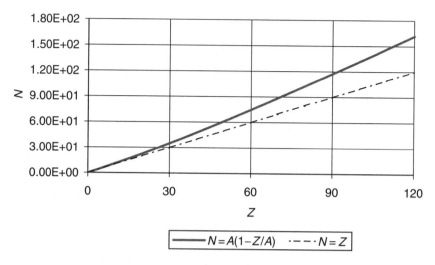

Figure 11-1. Nuclear stability.

Equation (11.2.3) is written in terms of Z/A as

$$-\left(m_n - m_p\right)c^2 - 2c_3\frac{Z}{A}A^{2/3} + \frac{c_3}{A^{1/3}} + 2c_4 + 4c_4\frac{Z}{A} = 0 \qquad (11.2.4)$$

or

$$\frac{Z}{A} = \frac{1}{2c_3A^{2/3} + 4c_4}\left[-\left(m_n - m_p\right)c^2 + \frac{c_3}{A^{1/3}} + 2c_4\right] \qquad (11.2.5)$$

The mass difference between the neutron and proton yields an energy difference

$$\left(m_n - m_p\right)c^2 \approx 1.3 \text{ MeV} \approx 2.1 \times 10^{-13}\text{J} \qquad (11.2.6)$$

Figure 11-1 is a plot of the number of neutrons $N = A[1 - (Z/A)]$ in a nucleus with minimized energy versus the number of protons Z and the condition $N = Z$ versus Z. The curve $N = A[1 - (Z/A)]$ does a good job of identifying the number of neutrons N in a stable nucleus with Z protons. Nuclei that are not on the curve tend to be unstable. Figure 11-1 also shows that light nuclei are stable when the number of neutrons equals the number of protons.

NUCLEAR DECAY PROCESSES

A general nuclear decay process can be written as the decay series

$$A \rightarrow B \rightarrow C \rightarrow \cdots \qquad (11.2.7)$$

The original nucleus A is called the parent. The symbol B represents the nuclei that are formed by the decay of the parent nucleus. The B nuclei are called daughter nuclei, the C nuclei are called granddaughter nuclei, and so on. Nuclei formed in each step of the decay series are called decay products. The amount of energy released in a nuclear decay process is called the disintegration energy Q. If disintegration energy Q is positive, the decay process releases energy as kinetic energy of the decay products.

Unstable nuclei can decay in a variety of ways. The three most common decay processes are α decay, β decay, and γ decay. The α decay process is the emission of an α particle (helium nucleus) in the general process

$$^A_Z X_N \rightarrow ^{A-4}_{Z-2} Y_{N-2} + ^4_2 He_2 \qquad (11.2.8)$$

The parent nucleus is $^A_Z X_N$ and the daughter nucleus is $^{A-4}_{Z-2} Y_{N-2}$. The disintegration energy for α decay is

$$Q = (m_X - m_Y - m_\alpha)c^2 \qquad (11.2.9)$$

where m_X is the mass of the parent nucleus, m_Y is the mass of the daughter nucleus, and m_α is the mass of the α particle. The α particle escapes the nucleus by quantum mechanical tunneling. An example of α decay is the decay of uranium-238 to produce thorium-234 and a helium nucleus:

$$^{238}_{92} U_{146} \rightarrow ^{234}_{90} Th_{144} + ^4_2 He_2 \qquad (11.2.10)$$

The β decay process is the emission of a β particle (an electron) in the general process

$$^A_Z X_N \rightarrow ^A_{Z+1} Y_{N-1} + e^- + \bar{\nu} \qquad (11.2.11)$$

where $\bar{\nu}$ is an antineutrino that was created as a result of conservation of mass-energy. In the β decay process, a neutron in the nucleus decays to a proton and releases an electron and antineutrino. The disintegration energy for β decay is

$$Q = (m_X - m_Y - m_e - m_{\bar{\nu}})c^2 \tag{11.2.12}$$

where m_X is the mass of the parent nucleus, m_Y is the mass of the daughter nucleus, and m_e is the mass of the β particle (an electron). The mass of the antineutrino $m_{\bar{\nu}}$ has traditionally been ignored because people believed neutrinos and antineutrinos were massless. Equation (11.2.12) allows for the possibility that neutrinos and antineutrinos have mass. An example of β decay that is useful in radioactive dating is the decay of carbon-14 to produce nitrogen-14:

$$^{14}_{6}C_8 \rightarrow {}^{14}_{7}N_7 + e^- + \bar{\nu} \tag{11.2.13}$$

An inverse process called electron capture is also possible. In this process, a proton in the nucleus captures an electron from its own orbital atomic electrons and forms a neutron with a change in the nuclear species, thus

$$^{A}_{Z}X_N + e^- \rightarrow {}^{A}_{Z-1}Y_{N+1} + \nu \tag{11.2.14}$$

A neutrino is created as a result of mass-energy conservation. The mass number A remains constant in both β decay and electron capture. An example of electron capture is the production of lithium-7 from beryllium-7 in the process

$$^{7}_{4}Be_3 + e^- \rightarrow {}^{7}_{3}Li_4 + \nu \tag{11.2.15}$$

The γ decay process is the emission of a γ particle (a high energy photon) in the general process

$$^{A}_{Z}X^*_N \rightarrow {}^{A}_{Z}X_N + \gamma \tag{11.2.16}$$

where the asterisk $*$ in $^{A}_{Z}X^*_N$ denotes an excited state of the nucleus $^{A}_{Z}X_N$. An excited state is a state with higher energy than the ground state (lowest energy state) of a quantized system. In the γ decay process, the excited nuclear state relaxes to a lower-energy state and releases a quantum of energy that appears as a γ ray. Light nuclei such as carbon-12 and heavier nuclei such as barium-137 can participate in the γ decay process.

The nuclear decay modes for α decay, β decay, and γ decay result in the emission of relatively small fragments and the formation of a daughter fragment that is comparable in mass to the parent nucleus. In our discussion of binding energy and nuclear stability, we mentioned a process called spontaneous fission that resulted in the production of two daughter nuclei of comparable mass from a parent nucleus that had approximately twice the mass of the daughter nuclei. An example of spontaneous fission is the process

$$^{238}_{92}\text{U}_{146} \rightarrow {}^{145}_{57}\text{La}_{88} + {}^{90}_{35}\text{Br}_{55} + 3\text{n} \qquad (11.2.17)$$

The parent nucleus is uranium-238, the daughter nuclei are lanthanum-145 and bromine-90, and three neutrons are produced in the process. The disintegration energy for Equation (11.2.17) is

$$Q = (m_{\text{U}} - m_{\text{La}} - m_{\text{Pu}} - 3m_{\text{n}})c^2 \qquad (11.2.18)$$

where m_{U} is the mass of uranium-238, m_{La} is the mass of lanthanum-145, m_{Br} is the mass of bromine-55, and m_{n} is the mass of the neutron. The production of neutrons in the fission process is an important by-product that can influence another decay mode: induced fission.

Induced fission is the process in which a nucleus splits into smaller fragments after capturing a slow neutron. A slow, or thermal, neutron is a low-energy neutron with a kinetic energy that depends on the temperature of its environment. The capture of free, slow neutrons can induce fission in some nuclides. These nuclides are called *fissile material* and are nuclei that split into fission fragments after capturing a slow neutron. An example of induced fission that is used in nuclear fission reactors is the uranium-238 decay series [Williams, 1991, page 126]

$$^{238}_{92}\text{U}_{146} + n \rightarrow {}^{239}_{92}\text{U}_{147} + \gamma$$
$$^{239}_{92}\text{U}_{147} \rightarrow {}^{239}_{93}\text{Np}_{146} + e^- + \bar{\nu} \qquad (11.2.19)$$
$$^{239}_{93}\text{Np}_{146} \rightarrow {}^{239}_{94}\text{Pu}_{145} + e^- + \bar{\nu}$$

The mean lives of uranium-238, neptunium-239, and plutonium-239 are 34 minutes, 81 days, and 35,000 years respectively. The long lifetime of plutonium-239 creates a waste management problem for the nuclear fission industry.

Neutrons formed by spontaneous emission of uranium-238 can induce further fission if enough neutrons are present. Therefore, a critical mass of uranium-238 is needed to produce neutrons by spontaneous fission. If enough neutrons are present, a chain reaction will begin. If too many neutrons are present, the chain reaction can proceed too fast and get out of control. Commercial applications of nuclear fission require the ability to control the chain reaction, and this means being able to control the number of neutrons present in a uranium-rich environment. Moderators are used in fission reactors to absorb neutrons and control the nuclear fission reaction. Nuclear fission reactors are discussed in more detail in the following section.

11.3 APPLICATIONS OF NUCLEAR ENERGY

Nuclear reactors are designed for several purposes. The primary commercial purpose is to generate electric power. Nuclear reactors also provide power for ships such as submarines and aircraft carriers, and they serve as facilities for training and research. Our focus here is on the use of nuclear reactors to generate electricity.[2] We then describe the effects of nuclear weapons.

NUCLEAR FISSION REACTORS

The neutrons produced in fission reactions typically have energies ranging from 0.1 MeV (1.6×10^{-14} J) to 1 MeV (1.6×10^{-13} J). Neutrons with energies this high are called *fast neutrons*. Fast neutrons can lose kinetic energy in collisions with other materials in the reactor and eventually equilibrate to thermal conditions. Less-energetic neutrons with kinetic energies on the order of the thermal energy of the reactor are called *slow neutrons* or *thermal neutrons*. Some nuclides tend to undergo a fission reaction after they capture a slow neutron. These nuclides are called *fissile materials* and are valuable fuels for nuclear fission reactors. Fission is induced in fissile materials by slow neutrons. Fission reactors that depend on the nuclear capture of fast neutrons exist, but are beyond the scope of this book.

Moderating materials such as light water (H_2O), heavy water (D_2O), graphite, and beryllium control the number of neutrons in the reactor.

Moderating materials slow down fast neutrons by acquiring kinetic energy in collisions. The best moderating materials present a relatively large cross-section to free neutrons. The neutron cross-section σ_n of a material can be measured by projecting a beam of neutrons with initial current density j_0 onto a volume of the material. Current density is the number of neutrons passing through a unit volume in a unit time. In SI units, current density is the number of neutrons per m^3 per second. Current density $j(x)$ at thickness x of the material obeys an exponential attenuation law

$$j(x) = j_0 \exp(-\Sigma x) = j_0 \exp\left(-\frac{x}{\lambda}\right) \qquad (11.3.1)$$

where λ is the mean free path of neutrons in the material and Σ is called the macroscopic cross-section (even though its unit is reciprocal length, or m^{-1} in SI units). The macroscopic cross-section is the product

$$\Sigma = n_{mod}\, \sigma_n \qquad (11.3.2)$$

where n_{mod} is the number density of atoms (or molecules) of moderating material with the SI unit of particles per m^3, and σ_n is the neutron cross-section with the SI unit m^2. Cross-sections that target nuclei present to projectile particles are often on the order of 1.0×10^{-28} m^2. This unit is called a *barn* from the phrase "as big as a barn."

Nuclear fission reactor design and operation depends on the ratio of neutrons in succeeding generations of decay products. The ratio of the number of neutrons in one generation to the number of neutrons in the preceding generation is called the *neutron multiplication factor*. The neutron multiplication factor k is a positive quantity that has the functional dependence [Lilley, 2001, Chapter 10]

$$k = \eta \varepsilon p f (1 - \ell_f)(1 - \ell_s) \qquad (11.3.3)$$

where the variables are defined as follows:

- η is the number of fast neutrons per thermal neutron absorbed by nuclear fuel.
- ε is the fast fission factor. It is the factor $\varepsilon = 1 + f_f$ where f_f is the fraction of fast neutrons that induce fission before they are slowed

down by a moderating material. In dilute reactors, the number of atoms of fuel is much less than the number of atoms or molecules of moderating material and $\varepsilon \approx 1$, that is, very few fast neutrons induce fission.

- p is the resonance escape probability. It is the probability that uranium-238 will not capture a fast neutron in the energy range $10\,\text{eV}$ ($1.6 \times 10^{-18}\,\text{J}$) to $100\,\text{eV}$ ($1.6 \times 10^{-17}\,\text{J}$) where the uranium-238 neutron capture cross-section has large peaks.
- f is the thermal utilization factor. It is the fraction of thermal neutrons that are absorbed back into the nuclear fuel.
- ℓ_f is the fraction of fast neutrons that leak out of real reactors of finite size.
- ℓ_s is the fraction of slow neutrons that leak out of real reactors of finite size.

If the multiplication factor for a reactor is $k < 1$, the number of fission reactions will diminish from one generation to the next and the chain reaction will die out. The reactor level is considered subcritical. A critical mass of fissile material is present in the reaction chamber when there are enough neutrons being produced by nuclear reactions to balance the loss of neutrons due to the factors shown in Equation (11.3.3). When the reaction is at the critical level, $k = 1$ and the chain reaction can be sustained. If the multiplication factor is $k > 1$, the number of neutrons produced by nuclear reactions exceeds the loss of neutrons and the number of fission reactions will increase from one generation to the next. The reactor level is considered supercritical. The chain reaction in a supercritical reactor will accelerate and, if it proceeds too fast, can have explosive consequences. The nuclear fission bomb is a supercritical reactor.

Coolant materials transport the heat of fission away from the reactor core. Coolants include light water, carbon dioxide, helium, and liquid sodium. In some cases, moderating materials can function as coolants.

A pressurized water reactor heats coolant water to a high temperature and then sends it to a heat exchanger to produce steam. A boiling water reactor provides steam directly. The steam from each of the reactors turns a turbine to generate electricity. A pressurized water reactor is sketched in Figure 11-2. A nuclear reactor is housed in a containment building. The containment building, which is typically a dome, provides protection from internal leaks and external dangers, such as an airplane crash.

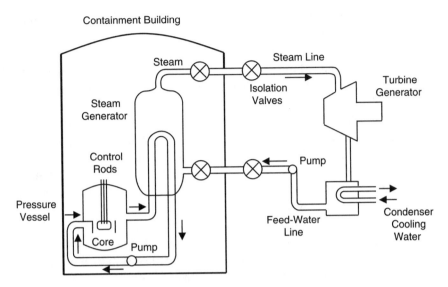

Containment Building

Figure 11-2. Schematic of a pressurized water reactor [after Cassedy and Grossman [1998, page 177]; and *Energy Technologies and the Environment*, U.S. Department of Energy Report No. DOE/EP0026 [June 1981].

NUCLEAR FUSION REACTORS

The idea behind nuclear fusion is quite simple: fuse two molecules together and release large amounts of energy in the process. Examples of fusion reactions include [Lilley, 2001, page 300]

$$p + d \rightarrow {}_{2}^{3}He_1 + \gamma \quad [5.49 \text{ MeV}]$$

$$d + d \rightarrow {}_{2}^{4}He_2 + \gamma \quad [23.85 \text{ MeV}]$$

$$d + d \rightarrow {}_{2}^{3}He_1 + n \quad [3.27 \text{ MeV}] \qquad (11.3.4)$$

$$d + d \rightarrow t + p \quad [4.03 \text{ MeV}]$$

$$d + t \rightarrow {}_{2}^{4}He_2 + n \quad [17.59 \text{ MeV}]$$

where d denotes deuteron (deuterium nucleus) and t denotes triton (tritium nucleus). The energy released is shown in brackets beside each of the fusion reactions. Protons are readily available as hydrogen nuclei. Deuterium is also readily available. Ordinary water contains approximately 0.015 mol %

deuterium [Murray, 2001, page 77], thus 1 atom of deuterium is present in ordinary water for every 6700 atoms of hydrogen.

The fusion reaction can occur only when the atoms of the reactants are heated to a temperature high enough to strip away all of the atomic electrons and allow the bare nuclei to fuse. The state of matter containing bare nuclei and free electrons at high temperatures is called *plasma*. Plasma is an ionized gas. The temperatures needed to create plasma and allow nuclear fusion are too high to be contained by conventional building materials. Two methods of confining plasma for nuclear fusion are being considered.

Magnetic confinement is the first confinement method. It relies on magnetic fields to confine the plasma. An example of a charged particle moving in a magnetic field was presented in Section 2.2. The magnetic confinement reactor is called a tokamak reactor. Tokamak reactors are toroidal (donut shaped) magnetic bottles that contain the plasma that is to be used in the fusion reaction. Two magnetic fields confine the plasma in a tokomak: one is provided by cylindrical magnets that create a toroidal magnetic field, and the other is a poloidal magnetic field that is created by the plasma current. Combining these two fields creates a helical field that confines the plasma. Existing tokamak reactors inject deuterium and tritium into the vacuum core of the reactor at very high energies. Inside the reactor, the deuterium and tritium isotopes lose their electrons in the high-energy environment and become plasmas. The plasmas are confined by strong magnetic fields until fusion occurs.

The second confinement method is inertial confinement. Inertial confinement uses pulsed energy sources such as lasers to concentrate energy onto a small pellet of fusible material, such as a frozen mixture of deuterium and tritium. The pulse compresses and heats the pellet to ignition temperatures.

Although many fusion reactions are possible, a commercial fusion reactor has not yet been constructed. Nuclear fusion reactors are still in the research stage and are not expected to provide commercial power for at least a generation.

EFFECTS OF NUCLEAR WEAPONS

We consider here the effects of nuclear detonations[3] from the point of view of an observer at a distance D from a blast with yield Y. *Yield* is the total effective energy released by a nuclear explosion. The observer's distance D from the blast is the slant range and is expressed in kilometers in subsequent calculations. Yield is expressed in megatons. Intercontinental ballistic missiles can have yields on the order of 20 megatons, and submarine-launched ballistic missiles can have yields on the order of one megaton. Yield raised to the two-thirds power gives equivalent megatons (EMT), thus

$$EMT = Y^{2/3} \tag{11.3.5}$$

EMT is proportional to the area of blast damage caused by a nuclear detonation.

Four important effects of a nuclear detonation depend on slant range D and yield Y. They are overpressure, thermal flux, electromagnetic pulse (EMP), and radiation dosage. *Overpressure* is the increase in ambient pressure caused by the detonation. Blast overpressure is given by

$$P = 1.54 \times 10^5 Z^3 + 1.09 \times 10^5 Z^{3/2}, \quad Z = Y^{1/3}/(D/1609) \tag{11.3.6}$$

where P is in Pascals, D is in meters, and Y is in megatons. The overpressure estimate in Equation (11.3.6) is reasonable for groundbursts but underestimates airburst overpressures. It is sufficient for conveying the concepts of shock wave and atmospheric pressure changes such as wind as a result of a nuclear detonation.

Thermal flux is the total thermal energy propagating outward from the detonation point per unit of cross-sectional area transverse to the direction of propagation. Thermal flux depends on the fraction f of explosive energy appearing as thermal energy in the nuclear fireball, and an atmospheric transmission factor T_A that accounts for the absorption of thermal energy in the atmosphere. Thermal flux Q in J/m^2 can be estimated as

$$Q = \frac{1.26 \times 10^8 fY T_A}{(D/1609)^2} \tag{11.3.7}$$

where D is in meters and Y is in megatons. Thermal flux is deadliest when the observer is in direct view of the detonation. Virtually any shelter can effectively shield a person from thermal flux. Equation (11.3.7) does not take into account firestorms resulting from the thermal flux.

The factor T_A depends on the variability of weather conditions and is a fractional value between 0 and 1. It is usually on the order of 0.5. The factor f depends on the height h at which the detonation occurs. Airbursts, such as the detonations over Japan in World War II, can be hundreds or thousands of meters in the air. An airburst is defined as a blast below 30,500 m (100,000 ft) but high enough that the fireball at maximum luminosity does not touch the earth's surface. If the fireball touches the earth, the detonation is a groundburst. As the height of the blast increases, the suction of dust and debris from the ground into the atmosphere is reduced with a corresponding decline in fallout, or radioactive dust. Once the height h is specified, the variable f can be estimated as 0.15 for a groundburst or 0.35 for an airburst.

The *electromagnetic pulse* (EMP) is a very strong but short-duration electrical field generated by relativistic electrons. Electrons in the atmosphere attain relativistic energies as a result of Compton scattering with gamma rays. Gamma rays are emitted when the excited states of fission fragments decay following an airburst. In Compton scattering, a gamma ray photon transfers some of its energy to an electron in the inner orbital of an atom. The transfer of energy from the photon to the atom can cause the ejection of an energetic electron from the atom. The EMP calculation estimates the effective range R_{EMP} of the EMP from the height h of the detonation and the radius R_{Earth} of the earth as

$$R_{EMP} = \sqrt{2R_{Earth}h} \tag{11.3.8}$$

where h, R_{Earth}, R_{EMP} are in meters. Equation (11.3.8) is the maximum distance that the EMP can reach when a nuclear weapon is detonated above the surface of a spherical object. One of the effects of the EMP is its ability to short-circuit electrical equipment, especially solid-state devices, when the equipment is inside the range of the EMP. Electrical equipment damage is greatest near the detonation point and diminishes as the distance from the detonation point increases. Half of the continental United States would be subjected to the EMP from a detonation at an altitude of 250 km.

One important difference between groundbursts and airbursts is the relative fallout generated by each. Groundbursts generate the most fallout. Radiation dosage for a 1-kiloton fission bomb that has delivered about 250 rads of neutrons at a distance of 0.8 km (0.5 mi) is

$$\text{Dosage} = 250 \frac{1000Y}{16\pi (D/1609)^2} \tag{11.3.9}$$

where D is in meters, Y is in megatons, and Dosage is in rems. The radiation dosage estimate assumes 1 rad is approximately equal to 1 rem. A dosage in excess of 1000 rems is usually fatal [Sartori, 1983, Table 3]. The estimate of radiation dosage, particularly biologically damaging radiation dosage, is the most complicated and uncertain of the nuclear detonation estimates presented here. Equation (11.3.9) is a simple estimate and does not include air-secondary gamma radiation or gamma radiation effects from fission fragments. The basic idea—that radiation is released in lethal amounts by a nuclear blast—is adequately represented.

11.4 AVAILABILITY OF NUCLEAR FUEL

The most abundant fuel for nuclear fission is uranium. Uranium exists in the crust of the earth as the mineral uraninite. Uraninite is commonly called pitchblende and is a uranium oxide (U_3O_8). It is found in veins in granites and other igneous rocks. It is possible to find uranium in sedimentary rocks. In this case, scientists believe that uraninite was precipitated in sedimentary rocks after being transported from igneous rocks by the flow of water containing dissolved uraninite. The volume of uranium V_U depends on the bulk volume V_B and porosity ϕ of the deposit. If we assume that bulk volume contains only uranium ore and pore space, we obtain the uranium volume

$$V_U = V_B(1 - \phi) \tag{11.4.1}$$

Bulk volume V_B and porosity ϕ were introduced in Section 4.6 as physical quantities that describe a porous medium.

Uranium is obtained by mining for the mineral uraninite. Mining methods include underground mining, open pit mining, and in situ leaching. *Leaching* is a process of selectively extracting a metal by a chemical reaction that creates a water-soluble molecule that can be transported to a recovery site. The isotope of uranium that undergoes spontaneous fission (uranium-235) is approximately 0.7% of naturally occurring uranium ore. Uranium must be separated from mined ore and then enriched for use in nuclear fission reactors.

Other isotopes that can be used in the fission process include the fission products plutonium-239 and thorium-232. Specialized reactors called breeder reactors are designed to operate with elements other than uranium. A *breeder reactor* is a nuclear fission reactor that produces more fissile material than it consumes. The breeding ratio B measures the effectiveness of a breeder reactor. It is defined as the number of new fissile atoms produced per atom of existing fissile material consumed. The amount of

Table 11-3
Uranium reserves by mining method, 2001 estimate

Mining method	US $30 per pound Uranium oxide (U_3O_8) million pounds (million kg)	US $50 per pound Uranium oxide (U_3O_8) million pounds (million kg)
Underground	138 (62.6)	464 (210.5)
Open pit	29 (13.2)	257 (116.6)
In situ leaching	101 (45.8)	174 (78.9)
TOTAL	268 (121.6)	895 (406)

Source: Table 3, EIA website, 2002.

fissile material is replaced if $B = 1$; it increases if $B > 1$; and it decreases if $B < 1$.

The amount of uranium that can be recovered from the earth is called *uranium reserves*. Estimates of uranium reserves have been made, and have many of the same uncertainties associated with estimates of fossil fuel reserves. Factors that are not well known include the distribution and extent of uranium deposits, and the price people are willing to pay to recover the resource. Uranium is considered a nonrenewable resource because it exists as a finite volume within the earth. Table 11-3 presents uranium reserves estimates provided by the United States Energy Information Administration for the most common mining methods. One of the appealing features of nuclear fusion is the relative abundance of hydrogen and its isotopes compared to fissile materials.

Example 11.4.1: Nuclear Fission Reactor Performance

The value of uranium reserves to society can be put into perspective by considering the power output of a typical nuclear fission reactor. We need to distinguish here between thermal power input and electrical power output. We append the letter "t" to megawatt to denote thermal megawatts as MWt, and append the letter "e" to denote electrical megawatts as MWe. Murray [2001, page 140] pointed out that a pressurized water reactor generates 1 MWt·day of thermal energy from the consumption of about 1.3 g of uranium-235. If we assume 33% efficiency, a reactor that requires an annual thermal power input of 300 MWt will generate an annual electrical power output of 100 MWe while consuming approximately 142 kg of uranium-235.

Table 11-3 presents a uranium reserve estimate of 268 million kg at a price of US $30 per pound. The mass fraction of uranium in U_3O_8 is about 0.848 and there is about 0.7% uranium-235 in the uranium ore. We therefore estimate a uranium-235 reserve of approximately 1.6 million kg. If we note that 142 kg of uranium-235 can fuel a reactor that outputs 100 MWe in a year, there is enough uranium-235 reserve at the lower price to operate 500 equivalent reactors for 22 years.

11.5   ENVIRONMENTAL AND SAFETY ISSUES

Nuclear energy is a long-term source of abundant energy that has environmental advantages. The routine operation of a nuclear power plant does not produce gaseous pollutants or greenhouse gases such as carbon dioxide and methane. Despite its apparent strengths, the growth of the nuclear industry in many countries has been stalled by the public perception of nuclear energy as a dangerous and environmentally undesirable source of energy. This perception began with the use of nuclear energy as a weapon, and has been reinforced by widely publicized accidents at two nuclear power plants, Three Mile Island, Pennsylvania, and Chernobyl, Ukraine. There are significant environmental and safety issues associated with nuclear energy, especially nuclear fission. We consider a few of the issues in the following sections.[3,4]

RADIOACTIVE WASTE

Nuclear power plants generate radioactive wastes that require long-term storage. The issue of the disposal of radioactive by-products created by nuclear reactors and the effects of nuclear waste on the environment has hindered the expansion of nuclear power. The end products of nuclear fission are highly radioactive and have a half-life measured in thousands of years. They must be disposed of in a way that offers long-term security. One solution is to place used uranium rods containing plutonium and other dangerously radioactive compounds in water, which speeds the decay process of these products. The rods are then buried in a remote location that should not be significantly affected by low levels of radiation. Yucca Mountain, Nevada, is the location for long-term nuclear waste storage in the United States.

A nuclear power plant can contaminate air, water, the ground, and the biosphere. Air can be contaminated by the release of radioactive vapors and gases through water vapor from the cooling towers, gas and steam from the air ejectors, ventilation exhausts, and gases removed from systems having radioactive fluids and gases. The radiation released into the air can return to the earth as radiated rain, which is the analog to acid rain generated by the burning of fossil fuels. Water may be contaminated when radioactive materials leak into coolant water. The contaminated water can damage the environment if it is released into nearby bodies of water such as streams or the ocean. Water and soil contamination can occur when radioactive waste leaks from storage containers and seeps into underground aquifers. The biosphere (people, plants, and animals) is affected by exposure to radioactive materials in the environment. The effect of exposure is cumulative and can cause the immune system of an organism to degrade.

CONTAINMENT FAILURE

Radioactive materials can be released from nuclear reactors if there is a failure of the containment system. Containment can be achieved by the reactor vessel and by the containment dome enclosing the reactor vessel. The two most publicized nuclear power plant incidents were containment failures. They occurred at Three Mile Island, Pennsylvania, in 1979, and at Chernobyl, Ukraine, in 1986.

The Three Mile Island power plant was a pressurized water reactor that went into operation in 1978 and produced approximately a gigawatt of energy. The containment failure at Three Mile Island occurred on March 28, 1979. It began when coolant feedwater pumps stopped and temperature in the reactor vessel began to rise. An increase in pressure accompanied the increasing temperature and caused a pressure relief valve to open. The reactor shut down automatically. Steam from the reactor flowed through the open relief valve into the containment dome. The valve failed to shut at a prespecified pressure and vaporized coolant water continued to flow out of the reactor vessel through the open valve. The water–steam mixture flowing through the coolant pumps caused the pumps to shake violently. Plant operators did not realize they were losing coolant and decided to shut off the shaking pumps. A large volume of steam formed in the reactor vessel and the overheated nuclear fuel melted the metal tubes holding the nuclear fuel pellets. The exposed pellets reacted with water and steam to form a hydrogen gas bubble. Some of the hydrogen escaped into the containment dome. The containment dome did not fail; it contained the hydrogen gas bubble and pressure fluctuations. The operators were eventually able to disperse the hydrogen bubble and regain control of the reactor.

The Chernobyl containment failure occurred in a boiling water reactor that produced a gigawatt of power and had a graphite moderator. Operators at the plant were using one of the reactors, Unit 4, to conduct an experiment. They were testing the ability of the plant to provide electrical power as the reactor was shut down. To obtain measurements, the plant operators turned off some safety systems, such as the emergency cooling system, in violation of safety rules. The operators then withdrew the reactor control rods and shut off the generator that provided power to the cooling water pumps. Without coolant, the reactor overheated. Steam explosions exposed the reactor core and fires started. The Chernobyl reactors were not encased in massive containment structures that are common elsewhere in the world. When the explosions exposed the core, radioactive materials were released into the environment and a pool of radioactive lava burned through the reactor floor. The Chernobyl accident was attributed to design flaws and human error.

Except for the Chernobyl incident, no deaths have been attributed to the operation of commercial nuclear reactors. Carbon [1997, Chapter 5] reported that the known death toll at Chernobyl was less than fifty people. Ristinen and Kraushaar [1999, Section 6.9] reported an estimate that approximately 47,000 people in Europe and Asia will die prematurely from cancer because they were exposed to radioactivity from Chernobyl. The disparity in the death toll associated with the Chernobyl incident illustrates the range of conclusions that can be drawn by different people with different perspectives.

NUCLEAR FALLOUT

Nuclear fallout is the deposition of radioactive dust and debris carried into the atmosphere by the detonation of a nuclear weapon. Nuclear fallout can also be generated by the emission of nuclear material from a nuclear reactor that has been exposed to the atmosphere. The point of deposition of the fallout depends on climatic conditions such as ambient air temperature and pressure, and wind conditions. We can display important factors associated with nuclear fallout by considering a simple model that shows the space–time distribution of fallout.[3]

Let us assume that the distribution of fallout can be described by a distribution for a random walk or Brownian motion [Fanchi, 1986]. Although simplified, the random walk model is a reasonable method for estimating the time it would take for fallout to reach a particular location.

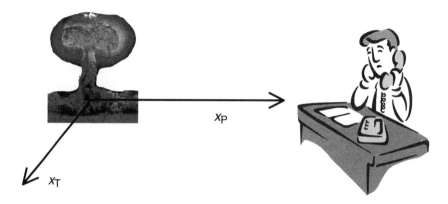

Figure 11-3. Coordinate system for fallout distribution.

We use a Gaussian distribution as the solution of the equation representing the random walk problem. In addition, let us assume that gravity minimizes vertical movement of radioactive particulates so that fallout is distributed primarily in the horizontal plane. With these assumptions, and separating the wind velocity into two components aligned parallel and transverse to the observer's line of sight (Figure 11-3), we obtain the fallout distribution

$$U(x_P, x_T, t) = U_0 \frac{x_P' x_T'}{\sqrt{D_P D_T t^2}} \exp\left[-\frac{(x_P - v_P t)^2}{4 D_P t} - \frac{(x_T - v_T t)^2}{4 D_T t} \right]$$
(11.5.1)

where x, v, D refer to position, velocity, and dispersion of the radioactive particulates, respectively. The subscripts P and T refer to parallel and transverse orientation relative to the observer's line of sight. The velocity typically refers to wind velocity. The constant U_0 is the maximum value of the fallout at the observer's location. A value of the unspecified constant $x_P' x_T'$ is fixed by the condition

$$U\left(x_P^{\text{obs}}, x_T^{\text{obs}}, t_{\text{max}}\right) = U_0$$
(11.5.2)

The coordinates $x_P^{\text{obs}}, x_T^{\text{obs}}$ are the observer's coordinates $x_P^{\text{obs}} = D$, $x_T^{\text{obs}} = 0$ and t_{max} is the time when the fallout $U\left(x_P^{\text{obs}}, x_T^{\text{obs}}, t_{\text{max}}\right)$ is a maximum at the observer's location. Solving the extremum condition $\partial U/\partial t = 0$ for time t_{max} yields

$$t_{max} = \frac{1}{2\gamma}\left\{ 1 + \sqrt{1 + 4\left(\frac{D^2}{4D_P}\right)\gamma} \right\} \tag{11.5.3}$$

where

$$\gamma = \frac{v_P^2}{4D_P} + \frac{v_T^2}{4D_T} \tag{11.5.4}$$

Collecting the results gives the expression for the fallout distribution as a function of time at the observer's location:

$$U(D, 0, t) = U_0 \frac{t_{max}}{t} \exp\left[-\frac{(D - v_P t_{max})^2}{4D_P t_{max}} - \frac{(D - v_T t_{max})^2}{4D_P t} \right.$$

$$\left. + \frac{(v_T^2 t_{max} - v_T^2 t)}{4D_T} \right] \tag{11.5.5}$$

Equation (11.5.5) is an approximation of the fallout distribution at the observer's location at any given time. Fallout effects on biological material are cumulative. An estimate of radiation dosage due to fallout at the observer's location can be estimated by integrating Equation (11.5.5) with respect to time.

A major source of uncertainty in using this or any other fallout model is the determination of the constant U_0. This constant depends on the number and type of nuclear detonations. A superposition principle can be applied for multiple blasts because the equation for the distribution is a linear partial differential equation.

NUCLEAR WINTER

One of the more controversial issues associated with the climatic effects of nuclear fallout is the concept of nuclear winter.[3] *Nuclear winter* is the significant climatic temperature decline resulting from an increase of atmospheric particulates following the detonation of many nuclear weapons in the atmosphere. The temperature decline would generate wintry conditions,

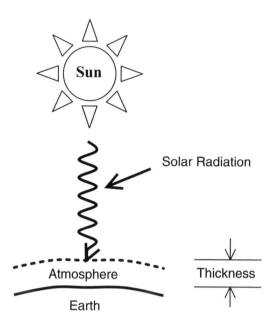

Figure 11-4. Simple atmosphere model.

hence the phrase "nuclear winter." The essence of the ideas on which these predictions are based can be exhibited using the simple climatic temperature model depicted in Figure 11-4.

Figure 11-4 is a sketch of an incident electromagnetic energy source (the sun), a single absorbing layer of material (the earth's atmosphere), and a black body (the earth). The amount of energy reaching the surface of the earth is the amount of solar energy transmitted by the atmosphere. Energy that is absorbed by the atmosphere or reflected into space does not reach the surface of the earth. The transmitted energy is assumed to be responsible for the temperature of the earth's surface. Geothermal sources of energy are neglected in this simple model. Given these assumptions, we find the energy balance at the earth's surface using Stefan's law and an exponential absorption law. The result is

$$I_0 \varepsilon e^{-\delta z} = \sigma T^4 \tag{11.5.6}$$

where I_0 is incident solar radiation, ε is emissivity (or absorptivity of Kirchhoff's radiation law), δ is optical depth, z is atmospheric thickness,

σ is Stefan-Boltzmann's constant, and T is absolute temperature at the earth's surface.

We can estimate the temperature of the earth's surface by solving the energy balance equation, Equation (11.5.6), for T:

$$T = \left(\frac{I_0\varepsilon}{\sigma}\right)^{1/4} e^{-\delta z/4} \tag{11.5.7}$$

Incident solar radiation I_0 can be written as a function of the solar constant I_S and albedo A, which is the fraction of solar radiation reflected by the atmosphere. The relationship is

$$I_0 = I_S(1 - A) \tag{11.5.8}$$

Optical depth depends on weather conditions. An estimate of the exponential factor is obtained by substituting Equation (11.5.8) into Equation (11.5.6) and rearranging to find

$$e^{-\delta z} = \frac{\sigma T^4}{I_S(1 - A)\varepsilon} \tag{11.5.9}$$

Emissivity is 1 for a black body, and a reasonable value for albedo is 0.39. A typical range of the exponential factor is 0.55 to 0.65. Surface temperature estimates are obtained by substituting the exponential decay values back into Equation (11.5.7).

In the nuclear winter scenario, the factor $(1 - A)\varepsilon$ is presumed to get smaller because of increased reflection of solar radiation resulting from additional particulates in the atmosphere. The same type of phenomenon is thought to have occurred during extinction events associated with meteors striking the earth. Opponents of the nuclear winter scenario argue that emissivity is a function of electromagnetic wavelength. Consequently, the greenhouse effect may tend to increase emissivity and the surface temperature of the earth would not change significantly. It is now believed that the temperature decline would not be as severe as originally estimated.

ENDNOTES

1. The principles of nuclear physics are discussed in several sources. See Garrod [1984, Chapter 8], Williams [1991], Serway, et al [1997,

Chapter 13], Bernstein, et al [2000, Chapter 15], and Lilley [2001, especially Chapters 1 and 10].

2. Primary references for the discussion of nuclear fission reactors are Murray [2001, especially Chapters 4 and 11], and Lilley [2001, especially Chapter 10].

3. The effects of nuclear weapons are based on material published by Glasstone and Dolan [1977], Sartori [1983], Hafemeister [1983], and Fanchi [1986]. Additional references are provided in these sources.

4. Environmental and safety issues associated with nuclear energy are discussed extensively in the literature. Primary references for our discussion of environmental and safety issues are Rees [1994], Carbon [1997], Ristinen and Kraushaar [1999, Chapter 6], Garwin and Charpak [2001, especially Chapter 7], and Murray [2001, Part III].

EXERCISES

11-1. A. Plot the number of neutrons $N = A[1-(Z/A)]$ in a nucleus with minimized energy versus the number of protons Z and the condition $N = Z$ versus Z for the range $1 \leq Z \leq 120$.

B. Use the nuclear stability curve to predict the stability of the nickel nuclide $^{60}_{28}Ni_{32}$. Is $^{60}_{28}Ni_{32}$ stable or unstable?

11-2. A. Slow neutrons with a kinetic energy of 1 eV are scattered by hydrogen in water molecules. Suppose the number density of hydrogen atoms is 6.68×10^{22} cm^{-3} and the neutron cross-section is 20 barns. Calculate the macroscopic cross-section Σ.

B. What is the mean free path of a slow neutron in water? Assume the slow neutron is a neutron from Part A.

C. If a beam of slow neutrons from Part A impinges on a tank of water, how deep will the beam go before the current density of slow neutrons is halved?

11-3. A. Suppose a nuclear fission reactor has the following specifications for the parameters in Equation (11.3.3): $\eta = 1.654$, $\varepsilon = 1$, $p = 0.749$, and $f = 0.834$. Assume the reactor is infinite in size so we can neglect leakage. Calculate the multiplication factor k.

B. Use the results of Part A to estimate the percentage of neutrons that could be lost by leakage in a real, finite reactor.

C. Estimate the value of k for a 1% leakage of fast neutrons and a 1% leakage of slow neutrons.

11-4. Approximately one atom of uranium-235 is present for every 140 atoms of uranium in pitchblende. Suppose 1 kg of pitchblende contains 84.8% uranium by mass. How many grams of uranium-235 are in the pitchblende?

11-5. A. Suppose the amount of uranium ore reserves is 75 million tons and has a grade percent of 0.179% uranium oxide (U_3O_8). How many pounds of U_3O_8 are in the ore?
B. Estimate the mass fraction of uranium in U_3O_8. Assume the average mass number of uranium is 238 and the average mass number of oxygen is 16.
C. Assume there is one atom of uranium-235 for every 139 atoms of uranium found in an ore deposit. Estimate the pounds of uranium-235 in the 75 million tons of ore stated in Part A.
D. Convert your answer in Part C to kilograms of uranium-235.

11-6. Estimate the effective range of an EMP from a detonation at an altitude of 250 km.

11-7. Derive Equation (11.5.3) from (11.5.1).

11-8. Calculate the exponential factor $e^{-\delta z}$ in Equation (11.5.9) for a sunny day with a temperature of 100° F and a cloudy day with a temperature of 80° F.

11-9. Normal water contains approximately 99.985 mol % H_2O and 0.015 mol % D_2O (heavy water). How many grams of normal water are needed to provide 1 kg of heavy water? Hint: Complete the following table:

Component	Mass (g)	Molecular wt. (g/mole)	# of moles	Mole fraction
H_2O				0.99985
D_2O	1000			0.00015
Total				1.00000

CHAPTER TWELVE

Alternative Energy: Wind and Water

The energy forms considered so far—fossil energy, nuclear energy, and solar energy—provide most of the energy used in the world today and are expected to contribute to the global energy mix for decades to come. Our purpose here is to introduce additional alternative energy forms, including a range of nonsolar renewable energy sources,[1] and synthetic fuels.

We have defined renewable energy as energy obtained from sources at a rate that is less than or equal to the rate at which the source is replenished. These sources include a variety of options based on wind, water, and biology. Wind and water energy sources rely on gradients in physical properties such as atmospheric pressure and ocean temperature to generate electrical power. The kinetic energy of wind and flowing water are indirect forms of solar energy, and are therefore considered renewable. Wind turbines harness wind energy, and hydroelectric energy is generated by the flow of water through a turbine. Both convert the mechanical energy of a rotating blade into electrical energy in a generator. We begin our discussion with an introduction to fluid flow equations, and then consider energy sources that depend on moving air and water, that is, sources that depend on wind, waves, and tides. We end with an introduction to fractals and their application to the calculation of the length of coastlines.

12.1 FLUIDS IN MOTION

Wind and flowing water are fluids in motion. Techniques for transforming the kinetic energy of wind and flowing water into electrical energy are discussed below. First, we introduce some of the equations that are commonly encountered in the description of fluid flow.

CONTINUITY EQUATION

The flow of a fluid such as air or water through a volume can be described mathematically by the continuity equation [Fanchi, 2002, Chapter 9]. Consider the flow illustrated in Figure 12-1. Flux J is the rate of flow of mass per unit cross-sectional area normal to the direction of flow. The term *mass* refers to the mass of the fluid. Fluid flows into the block at x with fluid flux J_x and out of the block at $x + \Delta x$ with fluid flux $J_{x+\Delta x}$. Applying the principle of conservation of mass, we have the mass balance:

mass in − mass out = mass accumulation   (12.1.1)

The block in Figure 12-1 has length Δx, width Δy, and depth Δz. The bulk volume of the block is $V_B = \Delta x \Delta y \Delta z$.

The mass entering the block in time interval Δt is

$$\left[(J_x)_x \, \Delta y \Delta z + \left(J_y\right)_y \Delta x \Delta z + (J_z)_z \, \Delta x \Delta y \right] \Delta t = \text{mass in} \qquad (12.1.2)$$

Equation (12.1.2) includes terms for flux in the x, y, z directions. The notation $(J_x)_x$ denotes x direction flux at location x and the cross-sectional

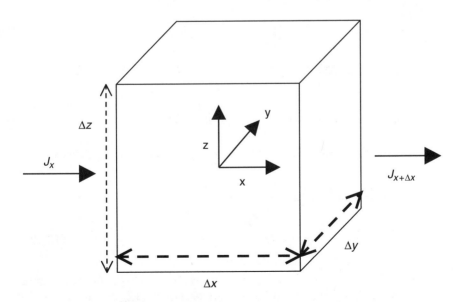

Figure 12-1. Coordinate system for continuity equation.

area that is perpendicular to the direction of fluid flux is $\Delta y \Delta z$. The terms involving $(J_y)_y$, $(J_z)_z$ have analogous meanings.

The mass leaving the block in time interval Δt is

$$
\begin{aligned}
&\left[(J_x)_{x+\Delta x} \Delta y \Delta z + (J_y)_{y+\Delta y} \Delta x \Delta z \right. \\
&\left. + (J_z)_{z+\Delta z} \Delta x \Delta y \right] \Delta t + q \Delta x \Delta y \Delta z \Delta t = \text{mass out}
\end{aligned}
\tag{12.1.3}
$$

The rate term q represents flow of mass directly into or out of the block through an injection or production source. Flow out of the block is represented by $q > 0$, and flow into the block is represented by $q < 0$. There is only convective transport of fluid if $q = 0$.

Mass accumulation in the block is the change in concentration C of the mass in the block during the time interval Δt, where concentration C is defined as the total mass in the block divided by the block volume. The mass accumulation term is

$$
[(C)_{t+\Delta t} - (C)_t] \Delta x \Delta y \Delta z = \text{Mass accumulation}
\tag{12.1.4}
$$

where concentration is evaluated at times $t, t + \Delta t$.

Equations (12.1.2) through (12.1.4) are now substituted into Equation (12.1.1), the mass balance equation, to obtain

$$
\begin{aligned}
&\left[(J_x)_x \, \Delta y \Delta z + (J_y)_y \, \Delta x \Delta z + (J_z)_z \, \Delta x \Delta y \right] \Delta t \\
&- \left[(J_x)_{x+x\Delta} \, \Delta y \Delta z + (J_y)_{y+\Delta y} \, \Delta x \Delta z + (J_z)_{z+z\Delta} \, \Delta x \Delta y \right] \Delta t \\
&- q \Delta x \Delta y \Delta z \Delta t = \left[(C)_{t+\Delta t} - (C)_t \right] \Delta x \Delta y \Delta z
\end{aligned}
\tag{12.1.5}
$$

Dividing Equation (12.1.5) by $\Delta x \Delta y \Delta z \Delta t$ and rearranging gives

$$
\begin{aligned}
&- \frac{(J_x)_{x+\Delta x} - (J_x)_x}{\Delta x} - \frac{(J_y)_{y+\Delta y} - (J_y)_y}{\Delta y} \\
&- \frac{(J_z)_{z+\Delta z} - (J_z)_z}{\Delta z} - q = \frac{(C)_{t+\Delta t} - (C)_t}{\Delta t}
\end{aligned}
\tag{12.1.6}
$$

In the limits $\Delta x \to 0$, $\Delta y \to 0$, $\Delta z \to 0$, $\Delta t \to 0$, the differences are replaced by partial derivatives and Equation (12.1.6) becomes the continuity equation

$$-\frac{\partial J_x}{\partial x} - \frac{\partial J_y}{\partial y} - \frac{\partial J_z}{\partial z} - q = \frac{\partial C}{\partial t} \qquad (12.1.7)$$

If we combine the components of flux in a flux vector $\vec{J} = \{J_x, J_y, J_z\}$, Equation (12.1.7) can be written in vector notation as

$$-\nabla \cdot \vec{J} - q = \frac{\partial C}{\partial t} \qquad (12.1.8)$$

CONVECTION–DISPERSION EQUATION

The continuity equation can be used to describe the mixing of one substance (solute) into another (solvent) by writing flux $\vec{J}$ in the form

$$\vec{J} = C\vec{v} - D\nabla C \qquad (12.1.9)$$

The concentration C is the concentration of the solute in the solvent. The vector $\vec{v}$ is the velocity of the solute, and D is the dispersion of the solute into the solvent. The flux of the solute into the solvent depends on the velocity of the solute, the magnitude of the dispersion, and the gradient of the concentration. Substituting Equation (12.1.9) into Equation (12.1.8) gives

$$-\nabla \cdot C\vec{v} + \nabla \cdot D\nabla C - q = \frac{\partial C}{\partial t} \qquad (12.1.10)$$

If we assume there are no sources or sinks so that $q = 0$ and assume that $\vec{v}, D$ are constant, we can simplify Equation (12.1.10) to the form

$$D\nabla^2 C - \vec{v} \cdot \nabla C = \frac{\partial C}{\partial t} \qquad (12.1.11)$$

Equation (12.1.11) is the convection–dispersion (C-D) equation. The Laplacian operator ∇^2 in Cartesian coordinates has the form

$$\nabla^2 = \frac{\partial^2}{\partial x^2} + \frac{\partial^2}{\partial y^2} + \frac{\partial^2}{\partial z^2} \qquad (12.1.12)$$

The term $D\nabla^2 C$ is the dispersion term, and the term $-\vec{v} \cdot \nabla C$ is the convection term. If the dispersion term is much larger than the convection term, the solution of Equation (12.1.11) can be approximated by the solution of the equation

$$D\nabla^2 C = \frac{\partial C}{\partial t} \qquad (12.1.13)$$

Equation (12.1.13) is a parabolic partial differential equation and behaves mathematically like a heat conduction equation. If the convection term is much larger than the dispersion term, the solution of Equation (12.1.11) can be approximated by the solution of the equation

$$-\vec{v} \cdot \nabla C = \frac{\partial C}{\partial t} \qquad (12.1.14)$$

Equation (12.1.14) is a first-order hyperbolic partial differential equation.

A relatively simple solution of the C-D equation can be obtained in one dimension. In one dimension, we can write Equation (12.1.11) as

$$D \frac{\partial^2 C}{\partial x^2} - v \frac{\partial C}{\partial x} = \frac{\partial C}{\partial t} \qquad (12.1.15)$$

where we have assumed the solute is moving in the x-direction with constant speed v. The concentration $C(x, t)$ is a function of space and time. The dispersion term is $D \partial^2 C / \partial x^2$ and the convection term is $-v \partial C / \partial x$. We must specify two boundary conditions and an initial condition for the concentration $C(x, t)$ to solve the C-D equation. We impose the boundary conditions $C(0, t) = 1$, $C(\infty, t) = 0$ for all time $t > 0$, and the initial condition $C(x, 0) = 0$ for all values of $x > 0$. The boundary condition $C(0, t) = 1$ says that we are injecting 100% solute at $x = 0$ and the boundary condition $C(\infty, t) = 0$ says that the solute never reaches the end of the flow path at $x = \infty$. The initial condition $C(x, 0) = 0$ says that there is no solute in the solvent at the initial time $t = 0$. The solution of the C-D equation is

$$C(x, t) = \frac{1}{2} \left\{ \mathrm{erfc} \left[\frac{x - vt}{2\sqrt{Dt}} \right] + e^{(vx/D)} \mathrm{erfc} \left[\frac{x - vt}{2\sqrt{Dt}} \right] \right\} \qquad (12.1.16)$$

where the complementary error function erfc(y) is defined as [Kreyszig, 1999]

$$\text{erfc}(y) = 1 - \frac{2}{\sqrt{\pi}} \int_0^y e^{-z^2} dz = 1 - \frac{2}{\sqrt{\pi}} \left[y - \frac{y^3}{(1!)\,3} + \frac{y^5}{(2!)\,5} \right.$$

$$\left. - \frac{y^7}{(3!)\,7} + \cdots \right] \tag{12.1.17}$$

The integral in Equation (12.1.17) can be solved using the series expansion in Equation (12.1.17) or a numerical algorithm [Abramowitz and Stegun, 1972].

INCOMPRESSIBLE FLOW

The flow of a gas such as air is compressible flow because gas is highly compressible when compared to other fluids such as water. If the density of a fluid is constant, or nearly so, the description of fluid flow is greatly simplified. To see this, we consider the continuity equation for the flow of a fluid with density ρ, velocity $\vec{v}$, and no source or sink terms, thus

$$\frac{\partial \rho}{\partial t} + \nabla \cdot (\rho \vec{v}) = 0 \tag{12.1.18}$$

A more suitable form of the continuity equation for describing incompressible fluid flow is obtained by writing Equation (12.1.18) as

$$\frac{D\rho}{Dt} + \rho \nabla \cdot \vec{v} = 0 \tag{12.1.19}$$

where we have introduced the differential operator

$$\frac{D}{Dt} = \frac{\partial}{\partial t} + \vec{v} \cdot \nabla \tag{12.1.20}$$

In the case of incompressible fluid flow, the density is constant and Equation (12.1.19) simplifies to the form

$$\nabla \cdot \vec{v} = 0 \tag{12.1.21}$$

Equation (12.1.21) says that the divergence of the velocity of a flowing, incompressible fluid is equal to zero.

NAVIER-STOKES EQUATION

Another useful equation for describing fluid flow is the Navier-Stokes equation. Consider a fluid flowing with velocity $\vec{v}$ in the presence of a pressure distribution P and acceleration of gravity $\vec{g}$. By applying the conservation of momentum, we obtain the Navier-Stokes equation [Sahimi, 1995, Chapter 1; Mathieu and Scott, 2000, Chapter 4]

$$\rho \frac{D\vec{v}}{Dt} = \rho \left(\frac{\partial \vec{v}}{\partial t} + \vec{v} \cdot \nabla \vec{v} \right) = \mu \nabla^2 \vec{v} - \nabla P + \rho \vec{g} \qquad (12.1.22)$$

where we have assumed the fluid has constant density ρ and constant viscosity μ.

The viscosity in Equation (12.1.22) is called dynamic viscosity. *Dynamic viscosity* is a measure of resistance to fluid flow. It is the ratio of shear stress to the rate of change of shear strain. The SI unit of dynamic viscosity is Pascal·second. A widely used unit for dynamic viscosity μ is centipoise. The viscosity of water is approximately 1 cp (10^{-3} Pa·s) at room temperature and pressure. Gas viscosity is much less than water viscosity at ambient conditions.

The relationship between viscosity and flow rate defines the rheology of the fluid. A fluid is a non-Newtonian fluid if its viscosity depends on flow rate, and the fluid is a Newtonian fluid if its viscosity does not depend on flow rate.

If we divide Equation (12.1.22) by ρ, we obtain

$$\frac{D\vec{v}}{Dt} = \left(\frac{\partial \vec{v}}{\partial t} + \vec{v} \cdot \nabla \vec{v} \right) = \eta \nabla^2 \vec{v} - \frac{1}{\rho} \nabla P + \vec{g} \qquad (12.1.23)$$

where we have introduced a second viscosity η called *kinematic viscosity*. Kinematic viscosity η is related to dynamic viscosity μ by the equation $\eta = \mu/\rho$. If fluid density ρ is expressed as g/cc, then kinematic viscosity η has the unit of centistoke, that is, 1 centistoke equals 1 centipoise divided by 1 g/cc.

Two special fluid flow equations are obtained by neglecting specific terms in the Navier-Stokes equation. In the first case, we neglect viscous effects. The viscosity term $\eta\nabla^2\vec{v}$ is then set to zero and Equation (12.1.22) becomes Euler's equation:

$$\rho\frac{D\vec{v}}{Dt} = -\nabla P + \rho\vec{g} \qquad (12.1.24)$$

In the second case, we neglect inertial effects. The term $\rho\,D\vec{v}/Dt$ is set to zero and Equation (12.1.22) becomes Stoke's equation:

$$\eta\nabla^2\vec{v} - \nabla P + \rho\vec{g} = 0 \qquad (12.1.25)$$

REYNOLDS NUMBER

In our discussion of heat exchangers, we saw that the Reynolds number N_{Re} can be used to characterize fluid flow. The Reynolds number expresses the ratio of inertial (or momentum) forces to viscous forces. For fluid flow in a conduit, the Reynolds number is

$$N_{Re} = \frac{\rho v D}{\mu} \qquad (12.1.26)$$

where ρ is fluid density, v is bulk flow velocity, D is tube diameter for flow in a tube, and μ is the dynamic viscosity of the fluid. The choice of units must yield a dimensionless Reynolds number. In SI units, a dimensionless Reynolds number is obtained if fluid density is in kg/m^3, flow velocity is in m/s, tube diameter is in m, and dynamic viscosity is in Pa·s. Note that 1 cp = 1 mPa·s = 10^{-3} Pa·s.

Fluid flow can range from laminar to turbulent flow. There is no fluid motion transverse to the direction of bulk flow in laminar fluid flow. The velocity components of fluid flow fluctuate in all directions relative to the direction of bulk flow when fluid flow is turbulent. For a fluid with a given density and dynamic viscosity flowing in a tube of fixed diameter, the flow regime is laminar at low flow velocities and turbulent at high flow velocities.

SINGLE-PHASE FLOW IN PIPES

Fluid flow in pipes is encountered in many energy technologies. We can begin to appreciate the factors that influence fluid flow in pipe by considering the relatively simple case of single-phase flow in circular pipes.[2]

Laminar flow along the longitudinal axis of a circular pipe is transverse to the cross-sectional area of the pipe. The cross-sectional area A of a circular pipe with internal radius r and internal diameter D is

$$A = \pi r^2 = \pi \left(\frac{D}{2}\right)^2 \qquad (12.1.27)$$

The bulk flow velocity v of a single-phase fluid flowing in the circular pipe is related to volumetric flow rate q by

$$v = \frac{q}{A} = \frac{4q}{\pi D^2} \qquad (12.1.28)$$

Reynolds number for flow in a circular pipe can be written in terms of volumetric flow rate by substituting Equation (12.1.28) into (12.1.26) to give

$$N_{Re} = \frac{\rho v D}{\mu} = \frac{4\rho q}{\pi \mu D} \qquad (12.1.29)$$

where ρ is fluid density and μ is the dynamic viscosity of the fluid. Fluid flow in circular pipes is laminar if $N_{Re} < 2000$, and becomes turbulent when $N_{Re} > 6000$. In SI units, the dimensionless Reynolds number is obtained when fluid density is expressed in kg/m^3, volumetric flow rate is in m^3/s, tube diameter is in m, and dynamic viscosity is in Pa·s.

The relationship between fluid flow velocity and pressure change along the longitudinal axis of the circular pipe is obtained by performing an energy balance calculation. The geometry of an inclined circular pipe with length L along the longitudinal axis and angle of inclination θ is shown in Figure 12-2. The single-phase fluid has density ρ and dynamic viscosity μ. It is flowing in a gravity field with acceleration g.

We make two simplifying assumptions in our analysis that allow us to minimize external factors and consider only mechanical energy terms. We assume no heat energy is added to the fluid, and we assume no work is done on the system by its surroundings—that is, no mechanical devices such as pumps or compressors are adding energy to the system. An energy balance with these assumptions yields the pressure gradient equation

$$\frac{dP}{dL} = \left[\frac{dP}{dL}\right]_{PE} + \left[\frac{dP}{dL}\right]_{KE} + \left[\frac{dP}{dL}\right]_{fric} \qquad (12.1.30)$$

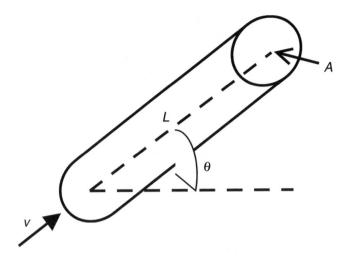

Figure 12-2. Flow in an inclined circular pipe.

where P is pressure. We have written the pressure gradient along the longitudinal axis of the pipe as the sum of a potential energy term

$$\left[\frac{dP}{dL}\right]_{PE} = \rho g \sin \theta \qquad (12.1.31)$$

a kinetic energy term

$$\left[\frac{dP}{dL}\right]_{KE} = \rho \, v \frac{dv}{dL} \qquad (12.1.32)$$

and a friction term

$$\left[\frac{dP}{dL}\right]_{fric} = f \frac{\rho v^2}{2D} \qquad (12.1.33)$$

that depends on a dimensionless friction factor f. If the flow velocity of the fluid does not change appreciably in the pipe, the kinetic energy term can be neglected and the pressure gradient equation reduces to the simpler form

$$\frac{dP}{dL} \approx \rho g \, \sin \theta + f \frac{\rho v^2}{2D} \qquad (12.1.34)$$

Equation (12.1.34) is valid for single-phase, incompressible fluid flow. If we further assume that the right-hand side is constant over the length L of the pipe, Equation (12.1.34) can be integrated to give the pressure change

$$\Delta P \approx \rho g L \sin \theta + f \frac{\rho v^2}{2D} L \qquad (12.1.35)$$

The friction factor f depends on flow regime. For laminar flow with Reynolds number $N_{Re} < 2100$, the friction factor is inversely proportional to Reynolds number:

$$f = \frac{16}{N_{Re}} \qquad (12.1.36)$$

For turbulent flow with Reynolds number $N_{Re} > 2100$, the friction factor depends on Reynolds number and pipe roughness. Pipe roughness can be quantified in terms of relative roughness ζ. Relative roughness is a fraction and is defined relative to the inner diameter of the pipe as

$$\zeta = \frac{\ell_p}{D} < 1 \qquad (12.1.37)$$

The length ℓ_p is the length of a protrusion from the pipe wall. Typical values of pipe relative roughness ζ range from 0.0001 (smooth) to 0.05 (rough). The length of protrusions inside the pipe may change during the period that the pipe is in service. For example, build-up of scale or pipe wall corrosion can change the relative roughness of the pipe. An estimate of friction factor for turbulent flow is [Beggs, 1991, page 61]

$$\frac{1}{\sqrt{f}} = 1.14 - 2 \log \left[\zeta + \frac{21.25}{N_{Re}^{0.9}} \right] \qquad (12.1.38)$$

12.2   WIND

Wind has been used as an energy source for thousands of years. Historical applications include sailing and driving windmills. Windmills have been used for grinding grain and pumping water. Wind is still used today as a source of power for sailing vessels and parasailing. The use of wind as a source of energy for generating electrical power is a relatively new application and is the primary focus of this section. We can understand some of the essential concepts associated with wind-generated electrical power by estimating the power output of a wind turbine.

WIND TURBINE

Modern wind turbines are classified as either horizontal axis turbines or vertical axis turbines. A vertical axis turbine has blades that rotate around a vertical axis, and its visual appearance has been likened to an eggbeater. A horizontal axis turbine has blades that rotate around a horizontal axis (see Figure 12-3). Horizontal axis turbines are the most common turbines in use today. A typical horizontal axis turbine consists of a rotor with two or more blades attached to a machine cabin set atop a post that is mounted on a foundation block. The machine cabin contains a generator attached to the wind turbine. The rotor blades can rotate in the vertical plane and the machine cabin can rotate in the horizontal plane. The ability of the turbine-generator assembly to rotate around a vertical axis is called the *yaw* effect. The angle between the rotor blade and the plane of rotation of the rotor blade is the pitch angle. The pitch angle of the rotor blade can be used to control the rotation rate of the rotor blade.

Figure 12-3 illustrates a cylinder of air approaching a rotating horizontal axis wind turbine. The cylinder of air shown in Figure 12-3 has the volume

$$\Delta V = A\,\Delta L \tag{12.2.1}$$

where A is the cross-sectional area and ΔL is the length of the cylinder of air. Let us assume that the density of air ρ_{air} is approximately constant. The mass of the cylinder of air is

$$\Delta m_{\text{air}} = \rho_{\text{air}}\,\Delta V = \rho_{\text{air}} A\,\Delta L \tag{12.2.2}$$

Suppose the cylinder of air in Figure 12-3 is moving with speed v_{air} directly at the turbine. The air speed v_{air} is the speed of the wind. The kinetic energy of the moving air is

$$\Delta \text{KE}_{\text{air}} = \frac{1}{2}\Delta m_{\text{air}} v_{\text{air}}^2 = \frac{1}{2}\rho_{\text{air}} A\,\Delta L v_{\text{air}}^2 \tag{12.2.3}$$

The length ΔL of the cylinder of air that reaches the wind turbine in a time interval Δt is

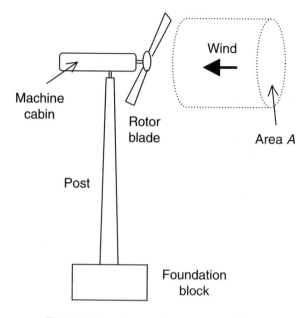

Figure 12-3. Schematic of a wind turbine.

$$\Delta L = v_{\text{air}} \Delta t \tag{12.2.4}$$

Substituting Equation (12.2.4) into Equation (12.2.3) gives

$$\Delta \text{KE}_{\text{air}} = \frac{1}{2} \rho_{\text{air}} A v_{\text{air}}^3 \Delta t \tag{12.2.5}$$

The rate of arrival of air is the wind power, or

$$P_{\text{wind}} = \frac{\Delta \text{KE}_{\text{air}}}{\Delta t} = \frac{1}{2} \rho_{\text{air}} A v_{\text{air}}^3 \tag{12.2.6}$$

Wind power is proportional to the cube of wind speed.

The area A is the surface area of the circle formed by the rotating tip of the rotor blade. If the rotor blade has radius R, the area is

$$A = \pi R^2 \tag{12.2.7}$$

We can use Equation (12.2.7) to write wind power in the form

$$P_{\text{wind}} = \frac{\pi}{2}\rho_{\text{air}}R^2v_{\text{air}}^3 \qquad (12.2.8)$$

Equation (12.2.8) shows that wind power is proportional to the square of the radius of the fan created by the rotating rotor blade.

So far in our analysis, we have assumed that the wind direction is perpendicular to the plane of rotation of the rotor blade, and that the wind speed is constant. Wind power is at maximum if the wind direction is perpendicular to the plane of rotation of the rotor blade, otherwise the wind power will be less than maximum. If the wind direction is parallel to the plane of rotation of the rotor blade for an infinitesimally thin rotor blade, the wind turbine will not provide any wind power. A change in wind direction can put stress on wind turbines because of the gyroscope effect. A *gyroscope* is a symmetrical rigid body that is free to turn about a fixed point and is subjected to an external torque. The wind turbine is a symmetrical rigid body that is free to turn about the fixed post. The change in wind direction subjects the wind turbine to a torque that causes the wind turbine to behave like a gyroscope and precess.[3] In addition, wind speed is seldom constant; it can vary from still to tornado or hurricane speed. The speed of rotation of the tip of the wind turbine is

$$v_{\text{tip}} = R\omega \qquad (12.2.9)$$

where ω is the angular frequency of the turbine. If v_{tip} is sufficiently large, it can be lethal to animals entering the fan area of the rotor blade. This creates an environmental hazard that must be considered when selecting locations for wind turbines.

Electrical power output from a wind turbine is a product of the efficiency η_{wind} times the input wind power. The optimum power output is approximately

$$P_{\text{out}} = \eta_{\text{wind}}P_{\text{wind}} = \frac{\pi}{2}\eta_{\text{wind}}\rho_{\text{air}}R^2v_{\text{air}}^3 \qquad (12.2.10)$$

The efficiency η_{wind} depends on several factors. One factor that affects the efficiency of a wind turbine is the efficiency of converting mechanical energy of the rotor blade into electrical energy. Another factor is the reliability of the wind turbine. The rate of rotation of the rotor blade depends on wind speed. If the wind speed is too large, the rotor blade can turn too fast and damage the system. To avoid this problem, wind turbines may have to be taken off-line in high wind conditions.

POWER COEFFICIENT AND BETZ' LIMIT

The analysis of wind power presented in the preceding section assumes that all of the kinetic energy of the wind incident on the turbine is converted to rotational energy of the rotor blades. In reality, the wind speed v_{upwind} upwind of the turbine is reduced to a wind speed $v_{downwind}$ downwind of the turbine. Following the analysis presented by Shepherd and Shepherd [1998, Chapter 10], we can estimate wind power for a more realistic system.

We again consider the cylinder of air in Figure 12-3. The kinetic energy of the moving air that is extracted for power production is

$$\Delta KE_{air} = \frac{1}{2} \Delta m_{air} \left(v_{upwind}^2 - v_{downwind}^2 \right) \qquad (12.2.11)$$

The mass of air that is needed to move the rotor blade in a time interval Δt is

$$\Delta m_{air} = \rho_{air} A v_{actuate} \Delta t \qquad (12.2.12)$$

where $v_{actuate}$ is the wind velocity that actuates the rotor and A is the cross-sectional area shown in Figure 12-3. If we assume that the reduction in kinetic energy in the upwind air stream is transferred to the wind turbine, conservation of energy and the continuity equation can be used [Sørenson, 2000, Section 4.1.4] to show that $v_{actuate}$ is given by

$$v_{actuate} = \frac{v_{upwind} + v_{downwind}}{2} \qquad (12.2.13)$$

Substituting Equations (12.2.12) and (12.2.13) into Equation (12.2.11) gives

$$\Delta KE_{air} = \frac{1}{2} \rho_{air} A \frac{v_{upwind} + v_{downwind}}{2} \left(v_{upwind}^2 - v_{downwind}^2 \right) \Delta t$$
$$(12.2.14)$$

The extracted wind power is

$$P_{extracted} = \frac{\Delta KE_{air}}{\Delta t} = \frac{1}{2}\rho_{air}A\frac{v_{upwind} + v_{downwind}}{2}\left(v_{upwind}^2 - v_{downwind}^2\right)$$
(12.2.15)

Equation (12.2.15) can be written in the simplified form

$$P_{extracted} = \frac{1}{4}\rho_{air}Av_{upwind}^3\left[1 + \beta_{wind} - \beta_{wind}^2 - \beta_{wind}^3\right]$$

$$= C_p \cdot \frac{1}{2}\rho_{air}Av_{upwind}^3$$
(12.2.16)

where $\beta_{wind} = v_{downwind}/v_{upwind}$ is the ratio of downwind velocity to upwind velocity, and C_p is the dimensionless power coefficient

$$C_p = \frac{1}{2}\left[1 + \beta_{wind} - \beta_{wind}^2 - \beta_{wind}^3\right]$$
(12.2.17)

The power coefficient is typically in the range $0 \le C_p \le 0.4$ for actual wind turbines. If we compare Equation (12.2.16) to Equation (12.2.10), we see that the power coefficient C_p serves as an efficiency factor for converting input wind power to output rotor blade power.

A theoretical maximum power coefficient is obtained by solving the extremum condition

$$\frac{dP_{extracted}}{d\beta_{wind}} = 0$$
(12.2.18)

for β_{wind}. The physically meaningful solution to Equation (12.2.18) is

$$(\beta_{wind})_{max} = \frac{1}{3}$$
(12.2.19)

Substituting Equation (12.2.19) into Equation (12.2.17) gives the power coefficient

$$(C_p)_{max} = \frac{16}{27} \approx 0.593$$
(12.2.20)

Equation (12.2.20) is the theoretical maximum power coefficient and is called Betz' limit, after Albert Betz, the person who first made the calculation in 1928 [Taylor, 1996, page 289]. Betz' limit says that approximately 59.3% of the power in the wind is the maximum percentage of wind power that can be extracted. It is obtained when the downwind velocity is one third of the upwind velocity, as shown in Equation (12.2.19).

WIND FARMS

A wind farm or wind park is a collection of wind turbines. The areal extent of the wind farm depends on the radius R of the rotor blades (Figure 12-4). A wind turbine must have enough space around the post to allow the fan of the rotor blade to face in any direction. If we neglect the finite extent of the machine cabin and assume the rotor blade with radius R is horizontal, the blade will sweep out a surface area πR^2 if the rotor blade is directly above the post of the wind turbine and the machine cabin is rotated through 360°. If we include the finite extent of the machine cabin, the surface area will be $\pi R_{\text{eff}}^2 \geq \pi R^2$. The minimum spacing between the posts of two equivalent wind turbines must be $2R_{\text{eff}}$ to avoid collisions between rotor blades. If we consider the aerodynamics of wind flow, which is the factor that controls turbine spacing, the turbine spacing in a wind farm increases to at least 5 to 10 times rotor diameter $2R$ [Sørensen, 2000, page 435] behind the plane of the rotor blade. The additional distance between posts is designed to minimize turbulence between wind turbines and enable the restoration of the wind stream to its original undisturbed state after it passes by one turbine on its way to the next turbine. Wind turbine spacing is an important factor in determining the surface area, or footprint, needed by a wind farm.

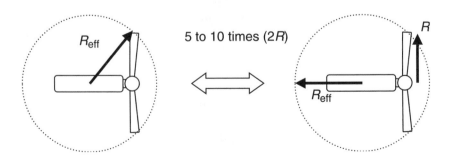

Figure 12-4. Wind turbine spacing.

ENVIRONMENTAL IMPACT

Wind energy is a renewable energy that is considered a clean energy because it has a minimal impact on the environment compared to other forms of energy. Wind turbines provide electrical energy without emitting greenhouse gases. On the other hand, we have already observed that the harvesting of wind energy by wind turbines can have environmental consequences.

Rotating wind turbines can kill birds and interfere with migration patterns. Wind turbines with slowly rotating, large-diameter blades and judicious placement of wind turbines away from migration patterns can reduce the risk to birds. Wind farms can have a significant visual impact that may be distasteful to some people. Wind turbines produce some noise when they operate. In the past, wind turbines with metal blades could interfere with television and radio signals. Today, turbine blades are made out of composite materials that do not interfere with electromagnetic transmissions.

12.3 HYDROPOWER

Water can generate power when it moves from a high potential energy state to a low potential energy state. We can see how this happens by considering a mass of water Δm_{water} falling a height h through a gravitational field with constant gravitational acceleration g. The change in potential energy ΔPE_{water} is

$$\Delta PE_{water} = (\Delta m_{water})\, gh \tag{12.3.1}$$

The height h is called the elevation or hydraulic head, and $g \approx 9.8$ m/s^2 on the surface of the earth.

One of the most common examples of hydropower, or power generated by the change in potential energy of water moving through a gravitational field, is the hydroelectric power plant, discussed in Chapter 2. Dams with turbines and generators are used to convert the change in potential energy in Equation (12.3.1) into mechanical kinetic energy. The water behind a dam falls through an elevation h. If the density of water ρ_{water} is considered a constant, the mass of falling water can be written as

$$\Delta m_{water} = \rho_{water}\, \Delta V_{water} \tag{12.3.2}$$

where ΔV_{water} is the volume of falling water. Substituting Equation (12.3.2) into (12.3.1) gives

$$\Delta PE_{\text{water}} = \rho_{\text{water}}\, gh \Delta V_{\text{water}} \qquad (12.3.3)$$

The power generated by the falling water in a time interval Δt is the hydropower

$$P_{\text{hydro}} = \frac{\Delta PE_{\text{water}}}{\Delta t} = \rho_{\text{water}}\, gh \frac{\Delta V_{\text{water}}}{\Delta t} \qquad (12.3.4)$$

The term $\Delta V_{\text{water}}/\Delta t$ is the volumetric water flow rate q_{water}, or flow rate in volume of water per unit time. In terms of q_{water}, Equation (12.3.4) can be written in the form

$$P_{\text{hydro}} = \rho_{\text{water}}\, gh\, q_{\text{water}} \qquad (12.3.5)$$

The output power depends on the efficiency η_{hydro} of the hydropower system, thus

$$P_{\text{out}} = \eta_{\text{hydro}}\, P_{\text{hydro}} = \eta_{\text{hydro}}\, \rho_{\text{water}}\, gh\, q_{\text{water}} \qquad (12.3.6)$$

We see from Equation (12.3.6) that power output depends on efficiency η_{hydro}, elevation h, and volumetric flow rate of water q_{water}.

The height h, or head, in the preceding equations should be replaced by the effective head h_{eff} for realistic systems. The effective head is less than the actual head h because water flowing through a conduit such as a pipe will lose energy to friction and turbulence. In this case, Equations (12.3.5) and (12.3.6) become

$$P_{\text{hydro}} = \rho_{\text{water}}\, gh_{\text{eff}}\, q_{\text{water}} \qquad (12.3.7)$$

and

$$P_{\text{out}} = \eta_{\text{hydro}}\, P_{\text{hydro}} = \eta_{\text{hydro}}\, \rho_{\text{water}}\, gh_{\text{eff}}\, q_{\text{water}} \qquad (12.3.8)$$

The rate q_{water} that water falls through the effective head h_{eff} depends on the volume of the penstock shown in Figure 12-5. If the penstock volume

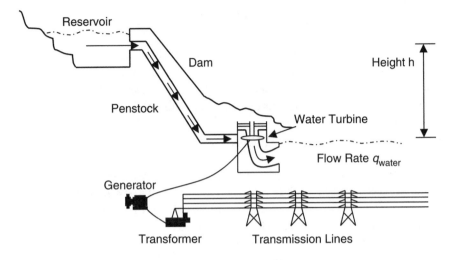

Figure 12-5. Hydropower.

is too small, the output power will be less than optimum because the flow rate q_{water} could have been larger. On the other hand, the penstock volume cannot be arbitrarily large because the flow rate q_{water} through the penstock depends on the rate that water fills the reservoir behind the dam.

The volume of water in the reservoir, and corresponding height h, depends on the water flow rate into the reservoir. During drought conditions, the elevation can decline because there is less water in the reservoir. During rainy seasons, the elevation can increase as more water drains into the streams and rivers that fill the reservoir behind the dam. Hydropower facilities must be designed to balance the flow of water through the electric power generator with the water that fills the reservoir through such natural sources as rainfall, snowfall, and drainage. Typical hydropower plant sizes are shown in Table 12-1.

Table 12-1
Hydropower plant sizes

Size	Electrical energy generating capacity (MW)
Micro	< 0.1
Small	0.1 to 30
Large	> 30

Source: DoE hydropower, 2002.

12.4   THE OCEAN

The oceans are another solar-powered source of energy. The mechanical energy associated with waves and tides, and thermal energy associated with temperature gradients in the ocean, can be used to drive electric generators. We consider each of these energy sources in this section.

WAVES AND TIDES

In our discussion of hydropower, we saw that water can generate power when it moves from a high potential energy state to a low potential energy state. The change in potential energy for a volume of water ΔV_{water} with density ρ_{water} falling through a height h in a gravitational field with constant acceleration g is

$$\Delta PE_{water} = \rho_{water}\, gh\Delta V_{water} \tag{12.4.1}$$

If the moving water is an ocean wave, the elevation h varies sinusoidally with time, thus

$$h = h_0 \sin \alpha t \tag{12.4.2}$$

where h_0 is the amplitude of the wave and α is the frequency of oscillation. The amplitude h_0 will depend on weather conditions; it will be small during calm weather and can be very large during inclement weather, such as hurricanes. The change in the potential energy of wave motion can be transformed into energy for performing useful work.

Suppose we lay a paddle with density ρ_{paddle}, length L_{paddle}, width w_{paddle}, and thickness h_{paddle} on the water. The paddle should be buoyant enough to move up and down with the wave (Figure 12-6). The change in potential energy of the paddle is

$$\Delta PE_{paddle} = \rho_{paddle}\, gh\, L_{paddle}\, w_{paddle}\, h_{paddle} \tag{12.4.3}$$
$$= \rho_{paddle}\, gh_0\, L_{paddle}\, w_{paddle}\, h_{paddle}\, \sin \alpha t$$

where we have used Equation (12.4.2). The magnitude of ΔPE_{paddle} depends on the amplitude h_0 of the water wave, the size of the paddle, and the density of the paddle.

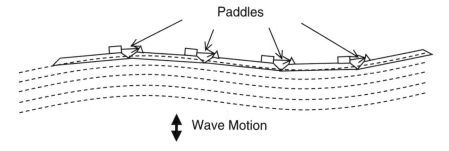

Paddles

Wave Motion

Figure 12-6. Capturing wave energy.

The energy density of waves breaking on a coastline in favorable locations can average 65 MW/mi (40 MW/km) of coastline [DoE Ocean, 2002]. The motion of the wave can be converted to mechanical energy and used to drive an electricity generator. The wave energy can be captured by floats or pitching devices like the paddles presented in Figure 12-6. Another approach to capturing wave energy is to install an oscillating water column. The rise and fall of water in a cylindrical shaft drives air in and out of the top of the shaft. The motion of the air provides power to an air-driven turbine. The output power depends on the efficiency of converting wave motion to mechanical energy, and then converting mechanical energy to electrical energy. A third approach to capturing wave energy is the wave surge or focusing technique. A tapered channel installed on a shoreline concentrates the water waves and channels them into an elevated reservoir. Water flow out of the reservoir is combined with hydropower technology to generate electricity.

The ebb and flow of tides produces tidal energy that can be captured to produce electricity. Figure 12-7 illustrates a tidal energy station. A dam with a sluice is erected across the opening to a tidal basin to capture tidal energy. The sluice is opened to let the tide flow into the basin, and then closed when sea level drops. Hydropower technology can be used to generate electricity from the elevated water in the basin.

The amount of hydropower obtained from tidal motion depends on the potential energy of the water that flows in and out of the tidal basin. The center of gravity of the water in the flooded area A of the tidal basin rises and falls through a height $h/2$. We can estimate the potential energy using Equation (12.4.1) with the volume of water given by

$$\Delta V_{\text{water}} = A\frac{h}{2} \qquad (12.4.4)$$

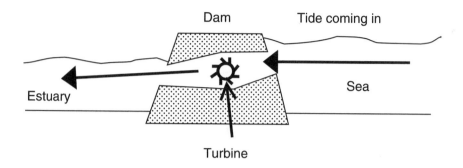

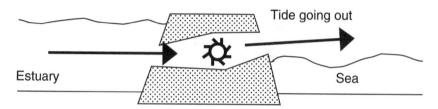

Figure 12-7. Capturing tidal energy [after Shepherd and Shepherd [1998, page 212]; and "Renewable Energy—A Resource for Key Stages 3 and 4 of the UK National Curriculum," Renewable Energy Enquines Bureau, 1995, ETSU, Harwell, Oxfordshire, United Kingdom].

so that the potential energy is

$$\Delta PE_{water} = \frac{1}{2}\rho_{water}\, gh^2 A \tag{12.4.5}$$

The water can flow through the gate four times per day because the tide rises and falls twice a day. The maximum daily power available from tidal motion is

$$P_{max} = 4\frac{\Delta PE_{water}}{\Delta t} = 2\frac{\rho_{water}\, gh^2 A}{\Delta t} \tag{12.4.6}$$

where Δt is one day.

Tides can rise as high as 15 m on the Rance River in France. The Rance tidal energy station has the potential of generating 240 MW of power [DoE Ocean, 2002; Serway and Faughn, 1985, page 114]. Tidal energy stations can have an environmental impact on the ecology of the tidal basin because of reduced tidal flow and silt buildup.

OCEAN THERMAL

Temperature gradients in the ocean exist between warm surface water and cooler water below the surface. Near ocean-bottom geothermal vents and underwater volcanoes, temperature gradients exist between hot water near the heat source and cooler waters away from the heat source. If the temperature gradients are large enough, they can be used to generate power using ocean thermal energy conversion (OTEC) power plants.

Three types of OTEC systems are recognized [DoE Ocean, 2002]: closed-cycle plants, open-cycle plants, and hybrid plants. A closed-cycle plant circulates a heat transfer fluid through a closed system. The fluid is heated with warm seawater until it is flashed to the vapor phase. The vapor is routed through a turbine and then condensed back to the liquid phase with cooler seawater. Pressure changes in an open-cycle plant make it possible to flash warm seawater to steam and then use the steam to turn a turbine. A hybrid plant is a combination of an open-cycle plant with a heat transfer fluid. Warm seawater is flashed to steam in a hybrid plant and then used to vaporize the heat transfer fluid. The vaporized heat transfer fluid circulates in a closed system as in the closed-cycle plant.

An ocean thermal energy conversion system can be built on land near a coast, installed near the shore on a continental shelf, or mounted on floating structures for use offshore. In some parts of the world, a desirable by-product of an OTEC system is the production of desalinated water. Salt from the desalination process must be disposed of in a manner that minimizes its environmental impact.

12.5 FRACTALS AND GEOGRAPHICAL LENGTHS

Benoit Mandelbrot [1967] introduced the concept of fractional dimension, or fractal, to describe the complexities of geographical curves. One of the motivating factors behind this work was an attempt to determine the length of coastlines. This problem is discussed here as an introduction to fractals and closely follows the discussion in Fanchi [2000b, Section 1.5]. The length of coastlines is an important parameter for estimating tidal energy.

We can express the length of a coastline L_c by writing

$$L_c = N\varepsilon \tag{12.5.1}$$

Australia

Figure 12-8. An irregular boundary: the coastline of Australia.

where N is the number of measurement intervals with fixed length ε. For example, ε could be the length of a meter stick. Geographers have long been aware that the length of a coastline depends on its regularity and the unit of length ε. They found that L_c increases for an irregular boundary, such as the coastline of Australia (Figure 12-8), as the scale of measurement ε gets shorter. This behavior is caused by the ability of the smaller measurement interval to more accurately include the lengths of irregularities in the measurement of L_c.

Richardson [see Mandelbrot, 1983] suggested that N has a power law dependence on ε such that

$$N \propto \varepsilon^{-D} \tag{12.5.2}$$

or

$$N = F\varepsilon^{-D} \tag{12.5.3}$$

where F is a proportionality constant and D is an empirical parameter. Substituting Equation (12.5.3) into Equation (12.5.1) gives

$$L_c(\varepsilon) = F\varepsilon^{1-D} \tag{12.5.4}$$

Table 12-2
Examples of fractals

Example	Fractional dimension D
West coast of Britain	1.25
German land-frontier, 1900 A.D.	1.15
Australian coast	1.13
South African coast	1.02
Straight line	1.00

Taking the log of Equation (12.5.4) yields

$$\log[L_c(\varepsilon)] = \log(F) + (1 - D)\log(\varepsilon) \tag{12.5.5}$$

If we define variables $x = \log(\varepsilon)$ and $y = \log(L_c(\varepsilon))$, Equation (12.5.5) becomes an equation for a straight line $y = mx + b$, where m is the slope of the line and b is its intercept with the y-axis. A plot of $\log(L_c)$ versus $\log(\varepsilon)$ should yield a straight line with slope $1 - D$.

Richardson applied Equation (12.5.4) to several geographic examples and found that the power law did hold [Mandelbrot, 1983, page 33]. Values of D are given in Table 12-2. Notice that the value of D is equal to 1 for a straight line. Mandelbrot [1967] reinterpreted Richardson's empirical parameter D as a fractional dimension, or fractal, and pointed out that Equation (12.5.4) applies to a measurement of the length of a coastline because a coastline is self-similar, that is, its shape does not depend on the size of the measurement interval.[4]

ENDNOTES

1. The primary references for nonsolar renewable energy are Sørensen [2000], Shepherd and Shepherd [1998], and Kraushaar and Ristinen [1993]. General, supplemental references are Fay and Golomb [2002], Young and Freedman [2000], Bernstein, et al. [2000], Serway, et al. [1997], and Boyle [1996].
2. The primary references for single-phase flow in pipes are Beggs [1991, Chapter 3], Economides, et al. [1994, Chapter 7], and Brill and Mukherjee [1999, Chapter 2].
3. Young and Freedman [2000, Section 10-8] and Fowles [1970, Section 9.9] present an analysis of gyroscopic precession.

4. For more discussion of self-similarity and fractals, see a book such as Devaney [1992], or Strogatz [1994].

EXERCISES

12-1. Plot the concentration $C(x, t)$ in Equation (12.1.16) given the following physical parameters: $D = 0.1$ ft^2/day, $v = 1$ ft/day. The plot should present $C(x, t)$ in the range $0 \leq x \leq 1$ for the three times $t = 0.1$ day, 0.2 day, 0.3 day. This data represents fluid movement in porous media, such as water draining through sand on a beach.

12-2. A. Suppose water is flowing through a circular pipe with volumetric flow rate $q = 1000$ barrels/day. The water density is $\rho = 1$ g/cc $= 1000$ kg/m^3 and the dynamic viscosity of water is $\mu = 1$ cp $= 1000$ Pa·s. The pipe length is 8000 ft and has a 5-inch inner diameter. The relative roughness of the pipe wall is 0.000144. What is the flow regime of the flowing water? Hint: Calculate Reynolds number for flow.
B. What is the friction factor?
C. Plot pressure gradient dP/dL versus inclination angle θ. Use 10° increments for the inclination angle in the range $-90° \leq \theta \leq 90°$. Express dP/dL in SI units and θ in degrees.
D. What is the pressure gradient dP/dL at $\theta = 90°$? Express your answer in psi/ft.

12-3. Energy intensity may be defined as average power per unit area. Fill in the following table to compare the energy intensity of a hydropower dam and a solar collector.

Facility	Power (MW)	Area (km^2)	Energy intensity (W/m^2)
Dam	600	400,000	
Solar collector	10	0.5	

12-4. A. The reservoir behind a dam covers an area of 20,000 hectares and has an average depth of 100 m. What is the volume of water in the reservoir (in m^3)?
B. The electric power output capacity of the dam is 2000 MWe and the dam produced 5 billion kWh of energy. What is the average power output for the year?

C. What is the capacity factor of the dam? Capacity factor is defined as the average power output divided by the power output capacity.

12-5. A. A wind turbine has a diameter of 44 m and spins at a rate of 28 revolutions per minute. What is the angular speed (in m/s) of the tip of the turbine blade?

B. Express the angular speed in kilometers per hour and miles per hour.

12-6. Suppose the average wave power along a coast is 30 MW per kilometer of coastline. If the efficiency of power production from wave power is 25%, estimate the length of coastline that would have to be dedicated to power production to produce 1000 MW of power.

12-7. The power output of a wind turbine $P_{turbine}$ has the functional dependence $P_{turbine} \propto \eta D^2 v_{wind}^3$ where D is the diameter of the turbine, v_{wind} is the wind speed, and η is efficiency. A wind turbine with $D = 20$ m and $\eta = 0.5$ produces 2.5 MW in a wind speed of 25 m/s. What is the power output of a wind turbine with diameter of 44 m and efficiency of 50% in a wind with wind speed of 20 m/s?

12-8. A. A wind turbine with a 20 m long rotor blade produces 1.2 MW power in a 45 mph (20 m/s) wind. How many wind turbines are needed to produce 1000 MW power?

B. Suppose each turbine occupies an area of 1600 m². What is the area occupied by a wind farm that produces 1000 MW power in a 45 mph (20 m/s) wind? Express your answer in m² and km².

12-9. A. Derive Equation (12.2.16) from Equation (12.2.15).

B. Solve Equation (12.2.18) for β_{wind} by first differentiating Equation (12.2.16) with respect to β_{wind}. What are the two possible values of β_{wind}?

C. What is the value of the power coefficient for the physically meaningful value of β_{wind}?

12-10. A. Plot the power coefficient for a wind turbine given by Equation (12.2.17) as a function of β_{wind} for the range $0 \leq \beta_{wind} \leq 1$.

B. What value of β_{wind} gives the maximum value of the power coefficient?

12-11. Suppose the flooded area of a tidal basin is 20 mi², and the variation in height between high and low tides is 6 feet. What is the maximum available power that can be obtained from a tidal power station each day? Convert your units to SI units and express your answer in MW.

12-12. A river has a flow rate of 6000 m³/s. It flows through a penstock with an effective head of 100 m. How much power is generated by a hydropower plant that is operating with a plant efficiency of 78%? Assume the density of water is 1.0 g/cc. Express your answer in gigawatts.

CHAPTER THIRTEEN

Alternative Energy: Biomass and Synfuels

The alternative energy forms discussed in the previous chapter rely on wind and water. In this chapter, we consider biomass and synthetic fuels. We begin with an introduction to the underlying science of life.[1]

13.1 THE SYNTHETIC THEORY OF EVOLUTION

The year 1859 marked the beginning of the modern theory of biological evolution.[2] It is the year English naturalist Charles Robert Darwin published his theory of evolution. Darwin's theory, like all theories of evolution, sought to explain how new species arise, and how a species is able to survive within a particular environment.[3]

Darwin accumulated observational evidence for his theory while serving as naturalist on board the H.M.S. *Beagle* during its survey of the South American coast. Although his theory was essentially complete by 1838, it was not published until 20 years later when Darwin learned that a fellow Englishman, naturalist Alfred Russel Wallace (1823–1913), had independently developed the same theory and was ready to present his work.[4] In 1858 the work of both men was presented to the Linnaen Society of England. A year later Darwin published *The Origin of Species*.

The younger Wallace arrived at his theory by a stroke of creative genius. His concepts were essentially the same as Darwin's, but not supported by as much observational data. Darwin's achievement rested not only on the development of a plausible theory, but also on the presentation of substantial supporting evidence.

Darwin and Wallace viewed evolution as a struggle for survival. In this view, Darwin benefited greatly from a familiarity with the social philosophy of T.R. Malthus (1766–1834). Malthus hypothesized that human societies were in competition for scarce resources. Philosopher Herbert Spencer (1820–1903) embraced the concept of competition for scarce resources and coined the phrase "survival of the fittest." Darwin applied the concept of competition for scarce resources to explain the evolution of species. Those species with an unusual characteristic, such as great speed or strength or intellect, would survive and reproduce. Other species, being at a disadvantage, would be unable to compete for essential life-sustaining resources and would eventually become extinct. Thus small herbivores that were swift of foot or clever had a better chance of surviving than comparably sized herbivores that could not outrun or outsmart predators.

Another concept stemming from Darwin's theory was the notion that species evolve gradually and continuously. Darwinian evolution disputed the theory of catastrophism pioneered by the French biologist Baron Georges Cuvier (1769–1832). Cuvier was a founder of the science of paleontology, the study of fossils. He demonstrated that the fossil record could be used to yield meaningful information about extinct animals. He believed that the fossils contained in a particular rock stratum could be used as an identifying characteristic of that stratum. Cuvier explained the abrupt differences in fossil content from one stratum to the next in terms of widespread environmental catastrophes. These catastrophes were supposed to have destroyed all animals in a given geologic stratum. Life in subsequent geologic strata resulted from new creations. In this way Cuvier could also explain the biblical Noah's flood.

Cuvier's catastrophism seemed to work well for the observations he had, but his detailed observations were limited to the Paris vicinity. The English geologist William Smith made detailed observations throughout England.[5] Analysis of Smith's work showed that evidence for Cuvier's catastrophes in the fossil record could be found in many strata, but not everywhere. It became clear that Cuvier's catastrophes were not worldwide, nor even continent wide. Something was wrong with Cuvier's catastrophism.

Scottish geologist Charles Lyell (1797–1875) provided an alternative explanation. Lyell believed that geologic processes at work in his time— wind, rain, erosion, volcanoes, and so on—were responsible for the shape of the earth. He argued that only contemporary processes should be used to understand the development of the earth over geologic time. These processes worked slowly but continuously over long periods of time. His ideas, now identified by the term *uniformitarianism*, greatly influenced the young Darwin on his voyage of discovery.

Darwin applied Lyell's geological concepts to the biological observations he accumulated during his tenure as naturalist on the H.M.S. *Beagle*. To explain the great variety of species he observed, Darwin postulated that the characteristics of species were always changing.

One possible cause of the changes Darwin noted was attributed to the external environment. This idea was first advocated by the French naturalist Georges Buffon (1707–1788), and developed further by Buffon's countryman Jean Baptiste Lamarck (1744–1829). According to Lamarck, parents in a given species must either adapt to the changing environment or eventually become extinct. In those species that adapt, their new characteristics are passed on to their offspring by heredity. Lamarck's ideas have been superseded by mutation theory, which is based on the genetic theory of Gregor Johann Mendel (1822–1884).

Mendel studied the humble edible pea in a monastery garden at Brunn, Austria. By crossing pea plants, he was able to produce obvious differences in characteristics, such as changes in the size of the plant and the color of its flowers. Mendel realized that it was not the characteristics that were inherited, but the factors determining the characteristics. Today it is known that hereditary characteristics are conveyed from one generation to another through genes in the cells. These genes are not altered by structural adaptations of the parent.

Lamarck and Mendel's work grew in significance after Darwin's work appeared. Darwin did not actually have a mechanism to explain the variations he postulated. In this sense Darwinian evolution is incomplete. A synthesis of Mendelian genetics and Darwinian evolution occurred in the 1940s. According to Ridley [1996, page 14], this synthesis is known by many names: the synthetic theory of evolution, neo-Darwinism, or the modern synthesis. Variations of the synthetic theory of evolution are considered next.

13.2   EVOLUTION: GRADUALISM OR PUNCTUATED EQUILIBRIUM?

Charles Darwin and Alfred Russel Wallace pioneered the theory that is now known as Darwinian evolution. In this theory, the variety of life we see on Earth today descended from a common ancestor. The changes in the lineage of a species from one generation to the next were attributed to natural selection, or survival of the fittest. One of the main scientific difficulties with Darwinian evolution was the ambiguity associated with the concept of natural selection. What is the mechanism that is responsible for modifications to a species?

Gregor Mendel proposed an answer. In his experiments with peas, he observed that the traits of the offspring depended on the traits of the parents, and even earlier generations. Mendel's work led to the development of the concept of genes as the carriers of information about the characteristics of an organism. Genes are contained in the cells of organisms. A cell is a biochemical factory that is encased in a porous membrane and includes genetic material called nucleic acid.

Changes in genes, called *mutations*, provide a mechanism for explaining how variations within a population could occur. The combination of Darwin and Wallace's descent with modification from a common ancestor and Mendel's genetics became the basis for the synthetic theory of evolution. In the synthetic theory of evolution, life progresses from simple to complex and evolution by natural selection is driven by successful mutations. Species diversify as time moves from the past to the present.

The discovery of more fossils in geologic formations, such as the Burgess Shale in British Columbia, Canada, led scientists to question the hypothesis that the evolution of life is a progression.[6] Many life forms that were found in the Burgess Shale have not survived and do not have any surviving descendants. The unsuccessful lineage was decimated during the course of geologic history. The differences between diversification and decimation are illustrated in Figure 13-1. Furthermore, there is no obvious trend towards greater complexity and diversity. Instead it appears that evolution is driven by random variation and away from a simple beginning. A new view of evolution was needed.

In the synthetic theory of evolution, genetic variation occurs within a population as a result of random mutation or recombination of genes. Populations evolve when the frequency of occurrence of a gene changes. The changes can be due to random genetic drift or natural selection. The rate of genetic variation in the original version of the synthetic theory was gradual and accounted for the observed diversification of species. The process was called *speciation*. Speciation led to the gradual evolution of reproductive isolation among populations. Today, researchers are considering the validity of the hypothesis that evolution was gradual.

Proponents of the synthetic theory of evolution subscribe to a few basic evolutionary mechanisms: organisms vary within a population; the variation can be inherited; more offspring are produced each generation than can survive; and the organisms best suited to their environment tend to produce more viable offspring. Evolution is viewed as a gradual, adaptive process called *anagenesis*. According to this view, transitional forms linking one major taxonomic group (or taxa) with another must have existed. Why, then, are there very few examples of transitional forms in the fossil record? This scarcity is a legitimate scientific problem for the synthetic theory.

A. Diversification

Present

Past

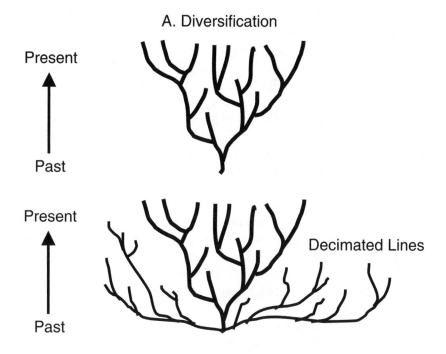

Present

Decimated Lines

Past

B. Diversification and Decimation

Figure 13-1. Diversification and decimation.

It has motivated alternative approaches to evolution. One of the most promising alternatives is punctuated equilibrium.

There are two schools of thought regarding the rate of evolution. The original school of thought is referred to as *gradualism*. Gradualists view evolution as a gradual, continuous process. The new school of thought is called *punctuated equilibrium*. Proponents of punctuated equilibrium believe that evolution has occurred rapidly over geologic time. The periods of rapid change were followed by periods of stasis, or relative calm. Thus, periods of evolutionary equilibrium were disrupted, or punctuated, by periods of rapid change. Two events that are used as evidence to support the hypothesis of punctuated equilibrium are the Permo-Triassic extinction and the Cretaceous-Tertiary extinction. The Permo-Triassic extinction is characterized by the disappearance of trilobites from the fossil record at the end of the Permian Period, and the Cretaceous-Tertiary extinction is characterized by the disappearance of dinosaurs from the fossil record at the end of the Cretaceous Period.

As early as the 1930s, some biologists suggested that evolution might proceed at a rapid, but sporadic, rate. Their views conflict with the neo-Darwinian belief in a slow and steady evolutionary process. Within the past two decades, the concept of an evolutionary process in which long periods of stasis, or species stability, is interrupted by short episodes of rapid change has gained a substantial following. The view known as punctuated equilibrium was introduced by Niles Eldridge and Stephen Jay Gould in 1972.

Proponents of punctuated equilibrium argue that the fossil record supports an evolutionary process in which stasis is the norm and morphological change is the exception. When morphological change does occur, it is rapid and usually results in speciation, the creation of new species by the branching or fission of lineages. The lineages are organized into clades. A *clade* is a set of species that are descended from a common ancestor. The process of branching is known as *cladogenesis*. Many punctuationalists believe cladogenesis is instigated when species are forced to adapt to a new environment. In addition to the lack of transitional forms in the fossil record, they point to examples of geographical isolation, such as Darwin's Galapagos Island finches, as observational evidence.

The views of proponents of the synthetic theory of evolution, or neo-Darwinian evolutionists, clash with the views of punctuationalists in many areas, including the tempo and mode of evolution, and the taxonomic classification of species. Neo-Darwinian evolutionists rely on an evolutionary taxonomy that employs theoretical evolutionary linkages and concepts (such as anagenesis) to classify lineages. Punctuationalists prefer a taxonomy that is as independent as possible of any evolutionary theory. They tend to support a taxonomy first introduced by Willi Hennig in 1966, and now known as cladistic taxonomy or cladistics. Cladistics depends on the fact that speciation creates new taxonomic characteristics. A taxon (plural taxa) is any formally recognized taxonomic group such as the genus *Homo* or the species *Homo sapiens*. Taxa are classified according to the degree to which they share these new characteristics with their relatively recent ancestry. Taxonomists are divided on whether to retain evolutionary taxonomy, or replace it with cladistic taxonomy. Both taxonomies presume that organisms have changed over long periods of time.

13.3   EVOLUTION OF HUMANS

For centuries the most popular explanations of the existence of life in its myriad forms required divine intervention[7]. In 1859, the British naturalist

Charles Darwin published a theory of evolution that removed the need for divine intervention. Darwin argued that all forms of life have evolved from a common ancestor. His evidence was extensive and he introduced a plausible mechanism to explain evolutionary change. Today, evidence supporting the theory of evolution is voluminous.[8] It is drawn from the study of fossils, from comparing the anatomy of different organisms, from similarities in the biochemical composition of organisms, from the geographical distribution of species, and from many other sources.

Comparison of the embryos of different organisms in the early stages of development can show remarkable similarities in their embryonic structures. As each embryo develops, differences become more pronounced until, by the latter stages of development, distinguishing features of individual species are prominent. The study of embryonic development led Ernst Haeckel to formulate his biogenetic law: the embryonic development of an individual organism retraces the evolutionary steps of its ancestral group. New species may arise by altering the timing of embryonic development. For example, the slowing or retardation of embryonic development, a process known as *neoteny*, may be exemplified by human evolution from an apelike ancestor. Human and ape embryos pass through a stage in which their craniums are bulbous. Unlike humans, who retain the bulbous cranium, apes have low foreheads and narrow skulls. Human embryonic cranial development does not proceed beyond the point when the cranium is bulbous.

The synthetic theory of evolution provides a rational explanation for the observed developmental sequence of embryos. According to the synthetic theory, the embryonic development of an organism resembles to some extent the embryonic development of its ancestors. The more closely related two species are, such as human and ape, the greater the similarities. This idea is supported by so much experimental data that it has been given a name: *recapitulation*.

Recapitulation is one example of the volumes of evidence that can be cited to support the synthetic theory of evolution. Though much evidence exists, the theory still contains gaps. Modern researchers are trying to fill the gaps. If we accept the basic tenets of Darwinian evolution, what does it say about human beings in relation to other species?

A concise tabulation of the steps in the biological evolution of people is presented in Table 13-1. The times in the table are based on radioactive dating of fossils and rocks. The timetable is based on the presumption that the naturally occurring processes at work today were also present in the early history of the earth. This presumption is known as the principle

Table 13-1
History and sequence of biological evolution

Era	Event	Years ago
Archeozoic	Prebiotic soup: formation of amino acids	4 billion
	Formation of molecular attributes	
	Prototypes of DNA and RNA enclosed in membranes	
	First bacteria form	
Protozoic	Simplest algae and fungi form	2 billion
Paleozoic	First fishes and invertebrates	600 million
	First terrestrial plants	
Mesozoic	First mammals and birds	200 million
Cenozoic	First primates	60 million
	Java and Peking man	500 thousand
	Neanderthal man	
	Cro-Magnon man	35 thousand

of uniformitarianism. In contrast to the principle of uniformitarianism, the principle of catastrophism holds that changes on the earth occurred abruptly, not gradually, because of external influences. The external influences of catastrophism were once thought to be divine in origin.[9] The biblical flood of Noah is an example of a divinely inspired catastrophic event. Divine influences have given way in our more modern, rational time, to rare celestial encounters. For example, modern catastrophists explain the extinction of the dinosaurs at the end of the Mesozoic Era as a consequence of a comet or asteroid smashing into the earth. Such collisions would have created huge clouds of dust and debris that could have cooled the earth and made it uninhabitable for large land animals. Uniformitarian proponents argue that the same effect could have been achieved by global volcanic activity.

Some four billion years ago, so our scenario goes, amino acids, sugars, and other essential biochemicals formed a prebiotic soup. Molecules within this rich mixture coalesced to form molecular aggregates. Among these polymers were proteins and the nucleic acids. These molecules, as we have seen, were capable of replicating themselves. Eventually they became encased in chemical membranes to form the earliest prokaryotes. Among the earliest prokaryotes were bacteria and cyanobacteria (blue-green algae). Cyanobacteria were the first oxygen producers. An important energy source for the prokaryotes was solar ultraviolet light. These organisms, particularly cyanobacteria, were responsible for changing the atmosphere of the earth from its primitive hydrogen-bearing state to its modern oxygen-bearing state. The appearance of oxygen in the atmosphere, and the accompanying development of an ozone layer, created an absorptive barrier to block solar ultraviolet light. A new energy source was needed for life to evolve further. The new energy source was photosynthesis, and the next billion years saw the appearance of eukaryotes (algae and fungi) in the oceans of the primitive Earth.

Eukaryotes may have begun as a colony of prokaryotic organisms. As a colony, the prokaryotes would have found advantages in specialization. Some could have focused on the development of chloroplasts to enhance the efficiency of the colony's fledgling photosynthetic activities. Others could have specialized in the development of mitochondrion for production of adenosine triphosphate, an essential biochemical for energy conversion within an organism. The symbiotic relationship was found to be so successful in competing for natural resources that the colony became a permanent assemblage in the form of the earliest eukaryotes. Direct evidence of the history of life on Earth during the Precambrian is limited, but the discovery of fossilized microorganisms three and a half billion years old by paleobiologist William Schopf has created opportunities for gathering more information about life on the primordial Earth.[10]

The processes of natural selection, including mutation, continued to function as time passed. Simple unicellular organisms evolved into simple multicellular organisms such as the syncytial flatworm. Both vertebrates and invertebrates have an ancestral lineage traceable to syncytial flatworms. Among the first invertebrates were sponges, jellyfish, and extinct trilobites. Precambrian fossil evidence for these organisms has been found in the Ediacara Hills of South Australia (ca. 650 million years ago) and the Burgess Shale in the Rockies of British Columbia (ca. 530 million years ago). Both sites contain fossils of life forms that lived on mud flats and were originally trapped in mud slides. No organisms with major hard parts such as skeletons, shells, or backbones were found at either site.

Jawless fish without fins and with bodies covered by bony plates were the first vertebrates. Once fish appeared, their evolution proceeded rapidly, particularly during the Devonian Period some four hundred million years ago. Among the diversity of fish evolving in this period were fish with fins, like the shark, and amphibious fish with lungs and the capacity to breathe air. The earth's air had acquired a significant amount of free oxygen by the beginning of the Devonian Period: roughly half of what it has today. It was becoming possible for life to move onto land.

Several problems had to be overcome before life could exist on land. Organisms had to prevent dehydration; they had to develop structural strength to support themselves against gravity, and they had to develop a gas exchange mechanism for using atmospheric oxygen and expelling carbon dioxide. Plants solved the first two problems by developing a waterproof protective coating. Arthropods such as insects were the first land animals. They developed an exoskeleton to prevent dehydration and provide structural support. Gas exchange was handled by the evolution of respiratory systems. Given these capabilities, some organisms found life on land to be beneficial in their struggle for survival.

Terrestrial plants evolved from algaelike seaweed into ferns and primitive trees. The earliest land plants reproduced both sexually and asexually. Seed-bearing plants with unprotected seeds called gymnosperms were successors to the earliest terrestrial plants. They included extinct seed ferns and extant conifers. Gymnosperms were the predecessors of angiosperms, which include modern flowering plants with protected seeds. The seeds of angiosperms consist of an unfertilized ovule enclosed in a protective environment called an ovary. The ovule is fertilized by pollen that must tunnel through the ovary.

Amphibious fish found breathing air to be a distinct advantage in some survival situations, such as avoiding water-bound predators. Amphibians were the first true land vertebrates. They were not completely weaned from the sea, however. They required freestanding water to reproduce. Eventually an organism acquired the capability to lay water-bearing amniotic eggs: reptiles had evolved.

Reptiles thrived during the Mesozoic Era. Many types of reptiles were able to survive in the climate of the Mesozoic Era. These types included huge predatory dinosaurs such as Tyrannosaurus rex, and reptiles such as the flying archeopteryx, the predecessor of modern birds. Our own reptilian ancestor, says the theory of evolution, was a small, four-legged creature with a relatively long, pointed tail, and a mouthful of fearsome teeth. This creature is called a therapsid.

Evolving from the therapsid around sixty-five million years ago were the first mammals. The extinction of dinosaurs, and many other species of plants and animals, at the end of the Cretaceous period made it possible for mammals to become the dominant species. The early mammals walked on four legs and had tails. Like their reptilian ancestors, one group of mammals continued to lay eggs. Another group of mammals—marsupials such as the kangaroo—retained their young in the mother's reproductive tract. Yet a third group—our group—gave birth to the young only after they were relatively well developed. Mammals in this group are called *placental mammals* because their young are nourished within the uterus of the mother through the placenta. The placenta is an organ that develops in the mother during pregnancy. It is the interface between the mother and her developing child.

Good vision and an ability to grasp objects with an appendage proved to be valuable survival tools for the order of placental mammals known as primates. Some primates learned to stand on their hind legs for greater range of vision and easier access to tree leaves and fruits. These primates include apes, the apelike gibbon, and humans. Fossil evidence indicates that the first species of human evolved two to three million years ago.

13.4   MODERN TAXONOMY

For centuries biologists have sought to develop schemes for classifying living organisms of all kinds. The first major classification scheme, or taxonomy, was developed by Carl Linnaeus and published as *Systema naturae* in the eighteenth century.[11] His scheme, like most taxonomies, used physical characteristics to distinguish one organism from another. It categorized organisms into one of two kingdoms: Plant and Animal. Linnaeus' two kingdoms were the standard until the 1960s. Advances in our knowledge of evolution and the variety of organisms, especially microorganisms, led researchers to suggest new classification schemes.

Whittaker replaced the two-kingdom approach with five kingdoms. His five kingdoms are the Monera, Protista (or Protoctista), Fungi, Plantae, and Animalia. Monera are unicellular prokaryotic organisms such as bacteria and cyanobacteria. Prokaryotes, you may recall, have a single strand of DNA, no nucleus, and are asexual. Protista are unicellular eukaryotic organisms such as protozoa, single-celled fungi, and algae. Unlike prokaryotes, eukaryotes have nuclei; their DNA exists as a double helix with two interwoven strands; and reproduction is sexual. Sexual reproduction allows the mixing of genetic information from two parents. Whittaker's kingdom Fungi are the multicellular fungi. Kingdoms Plantae and Animalia represent multicellular plants and animals respectively. They include the organisms known to Linnaeus and his contemporaries.

Carl Woese replaced Linnaeus' two kingdoms with three domains: the Archaea, the Bacteria, and the Eucarya. The three domains encompass Whittaker's five kingdoms. By examining the taxonomic classification for humans, we gain another perspective on our niche in nature.

Our genetic material, our chromosomes, is contained in a distinct nucleus. This makes us members of the Eucarya domain. We are part of the kingdom Animalia. Our embryonic development includes a stage during which we have gill pouches on our throats. Though the gill pouches disappear before birth, their presence in our development, together with our skeletal system and nervous system, are used to classify us as members of the Phylum Chordata. Our two arms and two legs place us in the Tetrapoda (four appendage) superclass. The existence of body hair and mammary glands in the skin associate us with the class Mammalia. As mentioned previously, a grasping ability with our hands, and good vision place us in the order Primates. An erect posture, relatively long arms, and no tail are the additional features exhibited by our family, the Hominidae family. We share this family with the apes. Distinguishing us from the apes is an enlarged cerebral cortex, and a unique social behavior. Human intelligence is needed for survival within a complex social community, and provides the basis for a uniquely human capability: self-awareness. These are the hallmarks of our genus: *Homo*.

Homo first appeared two to three million years ago[12] in the form of *Homo habilis*, or "handy man." The name was proposed because of an apparent connection to stone tools. The genus *Homo* is comprised of the Java and Peking men in Asia, and the Neanderthal man in Europe. Many anthropologists believe our species, *sapiens*, arose in Africa about one hundred thousand years ago. *Sapiens* is a term meaning "wise." It is attached to our species because our cranial development is more pronounced than other species. Many other traits are needed besides brain size to classify a fossil as a modern human. They include shape of the skull, size of teeth and nose, and general body shape. Our place in nature is categorized in Table 13-2.

Table 13-2
Human taxonomy

Domain	Eucarya
Kingdom	Animalia
Phylum	Chordata
Class	Mammalia
Order	Primates
Family	Hominidae
Genus	Homo
Species	Sapiens

13.5 POPULATION MODELS

Biological populations play an important role in evolutionary biology and the study of ecosystems.[13] From our perspective, the demand for energy depends in part on the size of the population that needs energy. The behavior of a biological population may be treated as a mass balance problem in which the mass is biomass. The mass balance equation for biomass is

$$\begin{pmatrix} \text{Rate of} \\ \text{mass accumulation} \end{pmatrix} = (\text{mass flux in}) - (\text{mass flux out})$$
$$+ \begin{pmatrix} \text{net rate} \\ \text{of reaction} \end{pmatrix} \tag{13.5.1}$$

Equation (13.5.1) can be quantified as

$$V\frac{dX}{dt} = (QX)_{\text{in}} - (QX)_{\text{out}} + R_{\text{Kin}} \tag{13.5.2}$$

where X is biomass (mass per volume), V is volume, Q is flow rate (volume per time), t is time, and R_{Kin} is the reaction term (mass per time) that represents kinetic processes that describe the growth or decay of the population. If we consider an isolated population with a reaction term that is proportional to X, the flux vanishes and Equation (13.5.2) becomes

$$V\frac{dX}{dt} = R_{\text{Kin}} = VkX \tag{13.5.3}$$

The proportionality constant k has the unit 1/time and describes the kinetic reaction. Dividing by the volume gives the equation

$$\frac{dX}{dt} = kX \tag{13.5.4}$$

which has the exponential solution

$$X(t) = X_0 \exp(kt) \tag{13.5.5}$$

where X_0 is the initial value of X. Two important observations are worth noting here. First, the isolated population will grow at an exponential rate if $k > 0$. In the case of growth, k is called the specific growth-rate coefficient. Second, the isolated population will decay at an exponential rate if $k < 0$. The exponential growth model represented by Equation (13.5.4) is a simple model that does not account for feedback mechanisms that affect population growth or decay. For example, we do not expect an isolated population to grow indefinitely because it will eventually outgrow its food supply. A more sophisticated model of population behavior is described in the following section.

THE LOGISTIC GROWTH MODEL

The logistic growth model takes into account the dependence of population size on the ability of the environment to support the population. This ability is called the *carrying capacity* of the environment. The exponential growth model is modified by a term that depends on population size and a parameter called the carrying capacity K. Carrying capacity is a positive, non-zero number that has the same unit as the population X. The logistic growth model is

$$\frac{dX}{dt} = k_{max} \left(1 - \frac{X}{K} \right) X \tag{13.5.6}$$

where k_{max} is called the maximum specific growth-rate coefficient and has the unit 1/time. Equation (13.5.6) has the solution

$$X(t) = \frac{K}{1 + \left[\left(\frac{K - X_0}{X_0} \right) \exp\left(-k_{max} t \right) \right]} \tag{13.5.7}$$

and X_0 is the initial size of the population. The rate of change of the population dX/dt vanishes as the size of the population approaches the carrying capacity of the environment, that is, as $X \to K$. This happens when the denominator in Equation (13.5.7) approaches 1.

THE MAY EQUATION

Biologist Robert May [1976] recognized that a nonlinear difference equation of the form

$$x_{n+1} = \lambda x_n (1 - x_n) \tag{13.5.8}$$

could be used to describe the evolution of a biological population. Equation (13.5.8) is now known as the May equation. May viewed

Equation (13.5.8) as a biological model representing the evolution of a population x_n. A study of the behavior of the May equation provides insight into the behavior of biological populations and lets us introduce concepts in nonlinear dynamics and chaos theory.

The May equation has the quadratic form

$$x_{n+1} = F(x_n) = \lambda x_n - \lambda x_n^2 \qquad (13.5.9)$$

The function $F(x_n)$ is called the logistic function in nonlinear dynamics. The linear term λx_n represents linear growth ($\lambda > 1$) or death ($\lambda < 1$). The nonlinear term $-\lambda x_n^2$ represents a nonlinear death rate that dominates when the population becomes sufficiently large. Once we specify an initial state x_0, the evolution of the system is completely determined by the relationship $x_{n+1} = F(x_n)$. The May equation is an example of a dynamical system.

Each dynamical system consists of two parts: (1) a state and (2) a dynamic. The dynamical system is characterized by the state, and the dynamic is the relationship that describes the evolution of the state. The function $F(x_n)$ in Equation (13.5.9) is the dynamic that maps the state x_n to the state x_{n+1}. The quantity λ is the parameter of the map F. Applying the map F for N iterations generates a sequence of states $\{x_0, x_1, x_2, \ldots, x_N\}$. The sequence $\{x_0, x_1, x_2, \ldots, x_N\}$ is known as the orbit of the iterative mapping as N gets very large. The behavior of the dynamical system approaches steady-state behavior as $N \to \infty$. For large N, the steady-state x_N must be bounded.

Several plots can be made to help visualize the behavior of a dynamical system. Three are summarized in Table 13-3 [Fanchi, 2000b, Section 10.4].

May's logistic function $F(x_n)$ in Equation (13.5.9) has a linear term and a nonlinear term. Let us normalize the population so that $0 \leq x_n \leq 1$. If the original population x_0 is much less than 1, then the nonlinear term is negligible initially and the population at the end of the first iteration is

Table 13-3
Helpful nonlinear dynamical plots

Assume a discrete map $x_{i+1} = f_\lambda(x_i)$ with parameter λ has been given.

Type	Plot	Comments
Time evolution	i versus (x_i or x_{i+1})	Vary initial value x_0 and λ to study sensitivity to initial conditions
Return map	x_i versus x_{i+1}	Vary initial value x_0 and λ to look for instabilities
Logistic map	x_{i+1} versus λ	Search for order (patterns)

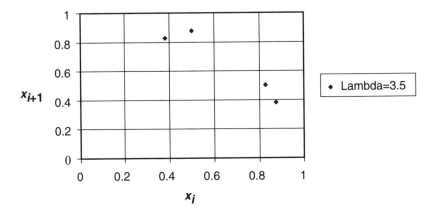

Figure 13-2. May equation for $\lambda = 3.5$.

proportional to x_0. The parameter λ is the proportionality constant. The population will increase if $\lambda > 1$, and it will decrease if $\lambda < 1$. If $\lambda > 1$, the population will eventually grow until the nonlinear term dominates, at which time the population will begin to decline because the nonlinear term is negative.

The logistic map is the equation of an inverted parabola. When $\lambda < 1$, all populations will eventually become extinct, that is, x_{n+1} will go to zero. If the parameter λ is a value between 1 and 3, almost all values of the state x_{n+1} approach, or are attracted to, a fixed point. If λ is larger than 3, the number of fixed points begins to increase until, for sufficiently large λ, the values of x_{n+1} do not converge to any fixed point and the system becomes chaotic. To see the onset of chaos, we plot return maps. Figures 13-2 and 13-3 show return maps for an initial condition $x_0 = 0.5$ and the parameter λ equals 3.5 and 3.9. There are only four fixed points when $\lambda = 3.5$. The number of fixed points increases dramatically when $\lambda = 3.9$.

13.6   POPULATIONS AND CHAOS

Chaotic behavior of a dynamical system may be viewed graphically as two trajectories that begin with nearby initial conditions (Figure 13-4). If the trajectories are sensitive to initial conditions, so much so that they diverge at an exponential rate, then the dynamical system is exhibiting chaotic behavior.[14] A quantitative characterization of this behavior is expressed in terms of the Lyapunov exponent.

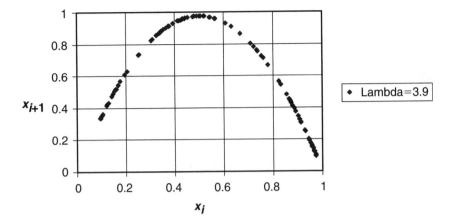

Figure 13-3. May equation for $\lambda = 3.9$.

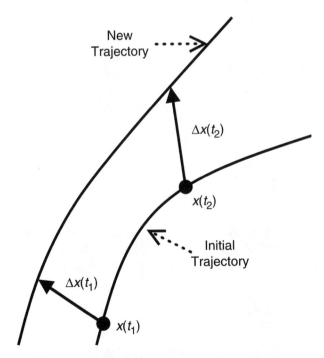

Figure 13-4. Chaos and trajectories [after Fanchi, 2000b, Section 10.4].

We can motivate the definition of chaos as a dynamical system with positive Lyapunov exponents [Jensen, 1987] by deriving its one-dimensional form from the first-order differential equation

$$\frac{dx_0}{dt} = V(x_0) \tag{13.6.1}$$

The population models in Section 13.5 have the form shown in Equation (13.6.1). A solution of Equation (13.6.1) for a given initial condition is called a *trajectory*, and the set of all trajectories is the flow of Equation (13.6.1). The mapping V is said to generate the flow.

The separation Δx between two trajectories is calculated by slightly displacing the original trajectory. The difference between the original (x_0) and displaced $(x_0 + \Delta x)$ trajectories is

$$\frac{d(x_0 + \Delta x)}{dt} - \frac{dx_0}{dt} = V(x_0 + \Delta x) - V(x_0) \tag{13.6.2}$$

Substituting a first-order Taylor series expansion of $V(x_0 + \Delta x)$ in Equation (13.6.2) gives

$$\frac{d(\Delta x)}{dt} = \left[\frac{dV(x)}{dx} \right]_{x_0} \Delta x \tag{13.6.3}$$

where the derivative $dV(x)/dx$ is evaluated at x_0. Equation (13.6.3) has the solution

$$\Delta x = \Delta x_0 \exp \left\{ \int_0^t \left[\frac{dV(x)}{dx} \right]_{x_0} dt' \right\} \tag{13.6.4}$$

where t' is a dummy integration variable and the separation Δx_0 is the value of Δx at $t = 0$. A measure of the trajectory separation is obtained by calculating the magnitude of Δx. The trajectory separation $d(x_0, t)$ is

$$d(x_0, t) = \left| \sqrt{(\Delta x)^2} \right| = |\Delta x| \tag{13.6.5}$$

Following Lichtenberg and Lieberman [1983], a chaotic system is a system that exhibits an exponential rate of divergence of two trajectories that were initially close together. We quantify this concept by assuming the exponential relationship

$$d(x_0, t) = d(x_0, 0) \exp\left[\sigma(x_0, t)t\right] \tag{13.6.6}$$

where $d(x_0, 0)$ is the trajectory separation at $t = 0$ and σ is a factor that we have not yet determined. A positive value of σ in Equation (13.6.6) implies a divergence of trajectories, whereas a negative value implies convergence. Solving Equations (13.6.4) through (13.6.6) for σ and recognizing that $\sigma(x_0, t)t$ is positive for diverging trajectories lets us write

$$\sigma(x_0, t) = \frac{1}{t} \int_0^t \left[\frac{dV}{dx}\right]_{x_0} dt' \tag{13.6.7}$$

where Equations (13.6.4) and (13.6.5) have been used. The Lyapunov exponent σ_L is the value of $\sigma(x_0, t)$ in the limit as t goes to infinity, thus

$$\sigma_L = \lim_{t \to \infty} \sigma(x_0, t) = \lim_{t \to \infty} \left\{ \frac{1}{t} \int_0^t \left[\frac{dV}{dx}\right]_{x_0} dt' \right\} \tag{13.6.8}$$

Equation (13.6.8) is an effective definition of chaos if the derivative dV/dx is known at x_0, and both the integral and limit can be evaluated.

13.7 BIOMASS

People have used biomass for fuel ever since we learned to burn wood.[15,16] *Biomass* is matter that was recently formed as a result of photosynthesis. Biomass includes wood and other plant or animal matter that can be burned directly or can be converted into fuels. In addition, products derived from biological material are considered biomass. Methanol, or wood alcohol, is a volatile fuel that has been used in racing cars for years. Another alcohol, clean-burning ethanol, which can be produced from sugarcane, can be blended with gasoline to form a blended fuel (gasohol) and used in conventional automobile engines, or used as the sole fuel source for modified engines.

Table 13-4
Energy density of common materials

Material	MJ kg^{-1}	MJ m^{-3}
Crude oil[a]	42	37,000
Coal[a]	32	42,000
Dry wood or sawmill scrap[b]	12.5	10,000
Methanol[a]	21	17,000
Ethanol[a]	28	22,000

[a] *Sørensen [2000, page 552].*
[b] *Sørensen [2000, page 473].*

Availability is one advantage biomass has relative to other forms of renewable energy. Energy is stored in biomass until it is needed. The availability of other renewable energy forms, such as wind and solar energy, depend on environmental conditions that can vary considerably.

Biomass can be used as a fuel in the solid, liquid, or gaseous state. Technologies exist to convert plants, garbage, and animal dung into natural gas. An example of a biomass project is the production of gas from a landfill. A landfill is a pit filled with garbage. When the pit is full, we can cover it with dirt, and drill a well through the dirt into the pit. The well provides a conduit for the natural gas that is generated from the decay of biological waste in garbage. The landfill gas is filtered, compressed, and routed to the main gas line for delivery to consumers.

The amount of energy that can be produced from biomass depends on the heat content of the material when it is dry. Table 13-4 shows the energy densities for several common materials. A kilogram of dry wood has a smaller energy density than crude oil and coal. The relative value of the energy density of biomass to fossil fuel is further diminished because oil and natural gas burners are more efficient than boilers used with wood or straw [Sørensen, 2000, page 473].

Some types of biomass used for energy include forest debris, agricultural waste, wood waste, animal manure, and the nonhazardous, organic portion of municipal solid waste. Developing countries are among the leading consumers of biomass because of their rapid economic growth and increasing demand for electricity in municipalities and rural areas.

WOOD

Wood has historically been a source of fuel. We have already learned that wood was a primary energy source through much of history [Nef, 1977], but deforestation became such a significant problem in sixteenth-century England that the English sought, and found, an alternative source of fuel: coal. Today, many people rely on wood as a fuel source in underdeveloped parts of the world. Economic sources of wood fuels include wood residue from manufacturers, discarded wood products, and nonhazardous wood debris.

An increased reliance on wood as a fuel has environmental consequences, such as an increased rate of deforestation and an increase in the production of by-products from wood burning. One way to mitigate the environmental impact is to implement reforestation programs. Reforestation programs are designed to replenish the forests. Research is underway to genetically engineer trees that grow quickly, are drought resistant, and easy to harvest. Fast-growing trees are an example of an *energy crop*, that is, a crop that is genetically designed to become a competitively priced fuel.

The environmental impact of increased emission of combustion byproducts must also be considered. When a carbon-based fuel burns, the carbon can react with oxygen to form carbon dioxide or carbon monoxide in the basic combustion reactions

$$C + O_2 \rightarrow CO_2 \tag{13.7.1}$$

and

$$2C + O_2 \rightarrow 2CO \tag{13.7.2}$$

The heat of combustion of the reaction in Equation (13.7.1) is 14,600 BTU/lbm of carbon (3.4×10^7 J/kg), and the heat of combustion of the reaction in Equation (13.7.2) is 4,430 BTU/lbm (1.0×10^7 J/kg) of carbon. If hydrogen is present, as it would be when a hydrocarbon is burned, hydrogen reacts with oxygen to form water in the reaction

$$2H_2 + O_2 \rightarrow 2H_2O \tag{13.7.3}$$

The heat of combustion of the reaction in Equation (13.7.3) is 62,000 BTU/lbm of hydrogen (1.4×10^8 J/kg). These are exothermic reactions, that is, reactions that have a net release of energy.

An exothermic reaction between two reactants A, B has the form

$$A + B \rightarrow \text{products} + \text{energy} \tag{13.7.4}$$

A + B → Products + Energy

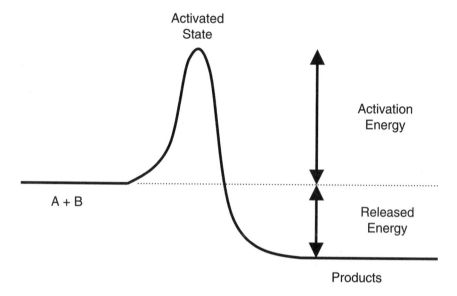

Figure 13-5. Activation energy for exothermic reaction.

Although the exothermic reaction releases energy, it may actually require energy to initiate the reaction. The energy that must be added to a system to initiate a reaction is called *activation energy*, and is illustrated in the energy diagram in Figure 13-5.

Activation energy is added to reactants to create an activated state, which we write as $(A + B)^*$. The asterisk * denotes the addition of energy to the reactants. The activated state is usually a short-lived, intermediate state. In terms of the activated state, a more complete description of a reaction may be written in the form

$$A + B + \varepsilon_1 \rightarrow (A + B)^* \rightarrow \text{products} + \varepsilon_2 \qquad (13.7.5)$$

where ε_1, ε_2 are energies. The energy ε_1 is added to the reactants and the energy ε_2 is released by the reaction. The reaction is exothermic if $\varepsilon_2 > 0$, and endothermic if $\varepsilon_2 < 0$. If enough energy is added to the system to form the activated state, the reaction can proceed. An exothermic reaction will form products with a net release of energy.

We know from observation and the chemistry of wood combustion that burning wood consumes carbon, hydrogen, and oxygen. The combustion reactions produce water, carbon monoxide, and carbon dioxide, a greenhouse gas. Efforts to reduce greenhouse gas emissions, such as sequestering greenhouse gases in geologic formations, are as important to biomass consumption as they are to fossil fuel consumption.

ETHANOL

Ethanol is an alcohol that can be added to gasoline to increase the volume of fuel available for use in internal combustion engines. Ethanol is made from a biomass feedstock. A *feedstock* is the raw material supplied to an industrial processor. Residues that are the organic by-products of food, fiber, and forest production are economical biomass feedstock. Examples of residue include sawdust, rice husks, corn stover, and bagasse. *Bagasse* is the residue that is left after juice has been extracted from sugar cane. *Corn stover* is used to produce ethanol. It is a combination of corn stalks, leaves, cobs, and husks.

Ethanol production proceeds in several steps. First, the feedstock must be delivered to the feed handling area. The biomass is then conveyed to a pretreatment area where it is soaked for a short period of time in a dilute sulfuric acid catalyst at a high temperature. The pretreatment process liberates certain types of sugars and other compounds. The liquid portion of the mixture is separated from the solid sludge and washed with a base such as lime to neutralize the acid and remove compounds that may be toxic to fermenting organisms. A beer containing ethanol is produced after several days of anaerobic fermentation. The beer is distilled to separate ethanol from water and residual solids.

The production of ethanol from corn requires distillation. During distillation, a mixture is boiled and the ethanol-rich vapor is allowed to condense to form a sufficiently pure alcohol. Distillation is an energy-consuming process. The use of ethanol in petrofuels to increase the volume of fuel would yield a favorable energy balance if the energy required to produce the ethanol were less than the energy provided by the ethanol during the combustion process. Researchers are attempting to determine if ethanol use serves as a source of energy or as a net consumer of energy.

BIOPOWER

Biomass has historically been used to provide heat for cooking and comfort.[4] Today, biomass fuels can be used to generate electricity and produce natural gas. The power obtained from biomass, called *biopower*, can be provided in scales ranging from single-family homes to small cities. Biopower is typically supplied by one of four classes of systems: direct-fired, cofired, gasification, and modular systems.

A direct-fired system is similar to a fossil fuel–fired power plant. High-pressure steam for driving a turbine in a generator is obtained by burning biomass fuel in a boiler. Biopower boilers presently have a smaller capacity than coal-fired plants. Biomass boilers provide energy in the range of 20 to 50 MW, and coal-fired plants provide energy in the range of 100 to 1500 MW. The technology exists for generating steam from biomass with an efficiency of over 40%, but actual plant efficiencies tend to be on the order of 20% to 25%.

Biomass may be combined with coal and burned in an existing coal-fired power plant. The process is called cofiring, and is considered desirable because the combustion of the biomass–coal mixture is cleaner than burning only coal. Biomass combustion emits smaller amounts of pollutants, such as sulphur dioxide and nitrogen oxides. The efficiency of the cofiring system is comparable to the coal-fired power plant efficiency when the boiler is adjusted for optimum performance. Existing coal-fired power plants can be converted to cofiring plants without major modifications. The efficiency of transforming biomass energy to electricity is comparable to the efficiency of a coal-fired power plant, and is in the range of 33% to 37%. The use of biomass that is less expensive than coal can reduce the cost of operating a coal-fired power plant.

Biomass gasification converts solid biomass to flammable gas. An example conversion process is described in more detail in Section 13.8. Cleaning and filtering can remove undesirable compounds in the resulting biogas. The gas produced by gasification can be burned in combined-cycle power generation systems. The combined-cycle system combines gas turbines and steam turbines to produce electricity with efficiency as high as 60%.

Modular systems use the technologies described in the preceding paragraphs, but on a scale that is suitable for applications that demand less energy. A modular system can be used to supply electricity to villages, rural sites, and small industry.

13.8   SYNFUELS

Synthetic fuels, or synfuels, are fossil fuel substitutes created by chemical reactions using such basic resources as coal or biomass. Society can extend the availability of fossil fuels such as natural gas and oil by substituting synthetic fuels for conventional fossil fuels. Here we consider coal gasification, biomass conversion, and gas-to-liquids conversion.

COAL GASIFICATION

Large coal molecules are converted to gaseous fuels in coal gasification.[17] Coal is thermally decomposed at temperatures on the order of 600° C to 800° C. The products of decomposition are methane and a carbon-rich char. If steam is present, the char will react with the steam in the reaction

$$C(s) + H_2O(g) \rightarrow CO(g) + H_2(g) \qquad (13.8.1)$$

where $C(s)$ denotes carbon in the solid phase, and (g) denotes the gas phase. The carbon monoxide-hydrogen mixture is called *synthesis gas*, or *syngas*. Carbon monoxide can react with steam in the reaction

$$CO(g) + H_2O(g) \rightarrow CO_2(g) + H_2(g) \qquad (13.8.2)$$

If the carbon dioxide is removed, the hydrogen-enriched mixture will participate in the reaction

$$CO(g) + 3H_2(g) \rightarrow CH_4(g) + H_2O(g) \qquad (13.8.3)$$

to form methane. The coal gasification process can be used to synthesize methane from coal.

BIOMASS CONVERSION

There are several ways to convert biomass into synthetic fuels. Oils produced by plants such as rapeseed (canola), sunflowers, and soybeans can be extracted and refined into a synthetic diesel fuel that can be burned in diesel engines. Thermal pyrolysis and a series of catalytic reactions can convert the hydrocarbons in wood and municipal wastes into a synthetic gasoline. Fermentation is a process that uses microorganisms to convert fresh biological material into simple hydrocarbons or hydrogen. We illustrate fermentation processes by describing the anaerobic digestion process, a process that is well suited for producing methane from biomass.[18]

The anaerobic digestion process proceeds in three stages. In the first stage, the complex biomass is decomposed by the first set of microorganisms. The decomposition of cellulosic material $(C_6H_{10}O_5)_n$ into the simple sugar glucose $(C_6H_{12}O_6)$ occurs in the presence of enzymes provided by the microorganisms. The overall reaction may be written as

$$(C_6H_{10}O_5)_n + nH_2O \xrightarrow{\text{enzymes}} nC_6H_{12}O_6 \qquad (13.8.4)$$

Stage one does not require an anaerobic (oxygen-free) environment. In the second stage, hydrogen atoms are removed in a dehydrogenation process that requires acidophilic (acid-forming) bacteria. The net reaction for stage two can be written as

$$nC_6H_{12}O_6 \xrightarrow{\text{bacteria}} 3nCH_3COOH \qquad (13.8.5)$$

The molecule CH_3COOH is acetic acid, which is commonly found in vinegar. In the third stage, a mixture of carbon dioxide and methane called *biogas* is produced from the acetic acid produced in stage two. The third stage requires the presence of anaerobic bacteria known as methanogenic bacteria in an oxygen-free environment. The reaction may be written as

$$3nCH_3COOH \xrightarrow{\text{anaerobic bacteria}} 3nCO_2 + 3nCH_4 \qquad (13.8.6)$$

Biomass, particularly biological waste, appears to be a plentiful source of methane. A simple methane digester that converts biological feed such as dung to methane is shown in Figure 13-6. The screw agitator mixes the liquid slurry containing the feed. The mixing action facilitates the release of methane from the decaying biomass. Methane trapped by the metal dome is recovered through the gas outlet.

Biomass conversion turns a waste product into a useful commodity. One difficulty with the exploitation of biomass fuels is the potential impact on the ecology of the region. For example, excessive use of dung and crop residues for fuel instead of fertilizer can deprive the soil of essential nutrients that are needed for future crops.

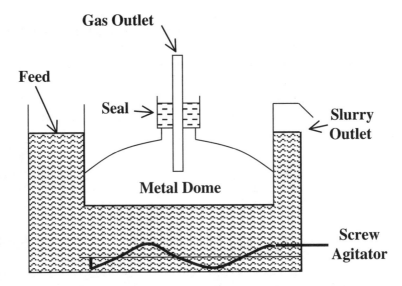

Figure 13-6. Methane digester. [after Cassedy and Grossman [1998, page 298]; P.D. Dunn, *Renewable Energies: Sources, Conversion and Application*, 1986, P. Peregrinius, Ltd., London, United Kingdom.]

GAS-TO-LIQUID CONVERSION

Synthetic liquid hydrocarbon fuels can be produced from natural gas by a gas-to-liquids (GTL) conversion process that consists of the following three steps[19]:

- Natural gas is partially oxidized with air to produce synthetic gas (syngas).
- The synthetic gas is reacted in a Fischer-Tropsch (F-T) reactor to polymerize it into liquid hydrocarbons of various carbon-chain lengths. An example of an F-T reactor is a fluidized bed with a catalyst such as nickel or cobalt. The F-T process produces a hydrocarbon mixture with a range of molecular weight components by reacting hydrogen and carbon monoxide in the presence of a catalyst.
- The high molecular weight products are separated and cracked into lower-molecular-weight transportation fuels. *Cracking* is the process of breaking down a high-molecular-weight substance into lower-molecular-weight products. Cracking is achieved using catalysts, heat, or hydrocracking (cracking with catalysts in the presence of hydrogen).

The primary product of the GTL process is a low-sulfur, low-aromatic, high-cetane diesel fuel. Aromatic molecules have multiple bonds between some of the carbon atoms and one or more carbon rings. Benzene is an example of an aromatic molecule. The cetane number of diesel fuel is a measure of the ignition delay of diesel. *Ignition delay* is the time interval between the start of injection and the start of combustion (ignition) of the fuel. Higher cetane fuels in a particular diesel engine will have shorter ignition delay periods than lower cetane fuels.

ENDNOTES

1. A discussion of basic biology can be found in a variety of sources, such as Kimball [1968], Attenborough [1979], Arms and Camp [1982], Wallace [1990], and Purves, et al. [2001].
2. Moore [1971], Folsom [1979], Gould [1993, 2002], Ridley [1996], and Purves, et al. [2001] are basic references for much of the material on evolution.
3. Several authors discuss the role of climate and atmosphere in the origin of life, such as Kasting, et al. [1988], Horowitz [1986], Gould [1993], Lahav [1999], and Ward and Brownlee [2000].
4. In addition to the references in Endnote 2, see Brackman [1980], Gjertsen [1984, Chapter 13], and Horvitz [2002, Chapter 10].
5. See Winchester [2001].
6. See Gould [1989].
7. See Cowen [2000, pg. 7], and Smith and Szathmáry [1999] for more detailed discussions of the conditions that are needed for life to begin.
8. Gould [1993], Ridley [1996], and Cowen [2000] are modern references for the discussion of evolution. For a discussion of the relationship between progress and evolution, see Gould [1993, 1996]. Gould [1989] discusses the history and significance of the Burgess Shale.
9. An example of catastrophism is the view of dinosaur extinction given by Russell [1982] and Alvarez [1987]. According to this view, a celestial body struck the earth and changed the climate enough to cause the extinction of many species of plants and animals. This view relies on the observation of an impact crater of a comet or asteroid on the Yucatan peninsula. Gould [1993], and Ridley [1996] review alternative theories for explaining mass extinctions.

10. See Schopf [1999].
11. Gjertsen [1984] discusses Linnaeus' work. Whittaker [1969], and Tudge [2000] discuss classification schemes that have replaced the two-kingdom approach.
12. The taxonomy of *Homo sapiens* is still evolving; for example, see Gould [1993], Leakey [1994], and Tudge [2000]. For a fictional, anthropologically supported reconstruction of the daily life of Lucy, one of the earliest ancestors of *Homo sapiens*, see Johanson and O'Farrell [1990]. The geographic distribution of humans is discussed in several sources, including Campbell [1985], Gamble [1993], and Tudge [2000].
13. For more detailed discussion of ecosystems, see Mihelcic [1999, Section 5.2.2].
14. Further terminology and discussion of nonlinear dynamics and chaos may be found in the literature; for example, see Hirsch and Smale [1974], Jensen [1987], Devaney [1992], and Strogatz [1994].
15. The primary references for nonsolar renewable energy are Sørensen [2000], Shepherd and Shepherd [1998], and Kraushaar and Ristinen [1993].
16. Biomass energy resources include the references in Endnote 15 and the United States Department of Energy [DoE Biomass, 2002].
17. Silberberg [1996, page 245] describes the coal gasification process.
18. Sørensen [2000, pages 481–484], and Kraushaar and Ristinen [1993, pages 277–278] describe the anaerobic digestion (or anaerobic fermentation) process.
19. Couvaras [1999] describes the gas-to-liquids conversion process.

EXERCISES

13-1. A way to see the onset of chaos is to plot the number of fixed points versus the parameter λ. This plot is called the logistic map. Plot the logistic map for the May equation in the range $3.5 \le \lambda \le 4.0$ and initial condition $x_0 = 0.5$.

13-2. Use chaos theory to calculate the trajectory separation Δx and Lyapunov exponent σ_L for a map $V(x) = \lambda x$. The equation $dx/dt = \lambda x$ obtained from the map $V(x) = \lambda x$ is the differential analog of the May equation with the linear growth or decay term, and without the quadratic death rate term.

13-3. Show that Equation (13.5.7) is a solution of (13.5.6).

13-4. A. Calculate the population size $X(t)$ for long times, that is, as $t \to \infty$.

B. Calculate the population size $X(t)$ when $K = X_0$.

13-5. The growth in world population from 1980 to 2000 is given by the following data from the Energy Information Agency website in 2002.

Year	Worldwide population (millions)	Year	Worldwide population (millions)
1980	4433.313	1991	5332.173
1981	4513.113	1992	5413.369
1982	4593.499	1993	5497.501
1983	4674.402	1994	5580.870
1984	4757.237	1995	5663.354
1985	4838.936	1996	5745.035
1986	4931.791	1997	5825.217
1987	5019.45	1998	5910.747
1988	5089.189	1999	5992.127
1989	5168.745	2000	6075.076
1990	5252.611		

A. Fit an exponential growth curve $X(t) = X_0 \exp(kt)$ to the data and plot the results.

B. Express the values of X_0, k in millions of people and year^{-1} respectively.

C. If this growth continues, what will the population be in the years {2010, 2020, 2100}?

D. Do you believe these numbers? Justify your answer.

13-6. A. Using the year 2000 as our initial year, we can estimate the initial population size for all people on the earth as $X_0 = 6075$ million people. Assume the population has a specific growth-rate $k = 0.0157$ years^{-1}. Estimate the population size in years {2010, 2050, 2100} using the logistic growth model with carrying capacities $K = 10,000$ million people and $K = 50,000$ million people.

B. How do these results compare to a population size forecast of 29.4 billion people for the year 2100 obtained using an exponential growth model?

13-7. A. Suppose an acre of land is exposed to 7×10^9 kcal of sunlight each year. What is the solar intensity (in W/m^2)?
B. Assume 0.5% of the incident sunlight is fixed as organic matter, that is, 0.5% of the incident sunlight is effectively used to produce an organic molecule. What is the effective solar intensity (in W/m^2) given the data in Part A?

13-8. A. A cord of wood is a pile of wood logs that is 4 ft long and stacked in a pile that is 8 ft wide and 4 ft high. What is the volume (in m^3) occupied by the stack of wood and air?
B. Suppose the heat content of a cord of wood is 3×10^{10} J and we can obtain half a cord of firewood from an acre of forest. If we use the thermal energy of firewood from an acre of forest during a year, what is the average annual power output (in W) of the firewood per square meter of forest?
C. What is the average annual power output of the firewood in Part B per acre of forest?

13-9. A. Suppose a cord of wood occupies a volume of 3.6 m^3, has a density of 1.6 g/cc, and a heat content of 3×10^{10} J/cord. Calculate the heat content of a kilogram of wood and express it in J/kg.
B. How much wood energy does a camper consume each day if the camper burns 3 kg wood per day?

13-10. A. Ethanol (EtOH) has a heat value of 76,000 BTU/gallon. Approximately 2.5 gallons of EtOH can be produced from one bushel of corn, and one acre of farm land can yield about 120 bushels of corn per year. Estimate the amount of EtOH energy (in BTU) that can be produced from one acre of corn each year.
B. Express the rate of EtOH energy production per acre of land in W/m^2.

13-11. A. A hectare of land is exposed to 1000 $kWh/m^2/yr$ solar radiation. Approximately 20% of the light is reflected. The growing season is approximately one third of a year. How much light energy is available for photosynthesis during the growing season? Express your answer in GJ.
B. Suppose one fifth of the available light energy in Part A reaches the growing leaves. How much light energy can be used for photosynthesis if half of the light reaching the growing leaves has the correct frequency range for photosynthesis?

CHAPTER FOURTEEN

Energy, Economics, and Environment

A central theme of our study is the recognition that society's dependence on fossil energy in the twenty-first century is in a state of transition to a more varied energy mix. We have discussed many technical factors in previous chapters, but the energy mix of the future will depend on additional factors, such as the economics of energy options and their environmental impact. In this chapter, we discuss several nontechnical factors that will influence the twenty-first century energy mix.

14.1   ENERGY CONSERVATION AND COGENERATION

Energy can be conserved using a number of very simple techniques. Some simple energy conservation methods that each of us can adopt include walking more and driving less; carpooling; planning a route that will let you drive to all of your destinations in the shortest distance; and using passive solar energy to dry your clothes or heat your home. Other energy conservation methods are more complicated; they are designed to reduce energy losses by improving the energy conversion efficiency of energy consuming devices. For example, improving gas mileage by reducing the weight of a vehicle or increasing the efficiency of internal combustion engines can reduce energy loss. We can decrease the demand for energy consuming activities, such as air conditioning in the summer or heating in the winter, by developing and using more effective insulating materials.

The social acceptability of energy conservation varies widely around the world. In some countries, such as Germany, energy conservationists and environmentalists are a political force (the Green Party). In other countries, people may espouse conservation measures but be unwilling to partic- ipate in or pay for energy conservation practices, such as recycling or driving energy efficient vehicles. Some governments, especially in energy importing nations, are encouraging or requiring the development of energy conserving technologies. The technical basis for energy conservation and an example of an energy conserving technology are discussed here.

CONSERVATION

The conservation of energy is a law of nonrelativistic physics. Energy conservation requires that there be a balance between the energy input E_{input} to a process and the sum of energy output E_{output} by the process plus energy loss E_{loss}. A general expression for the energy balance is

$$E_{input} = E_{output} + E_{loss} \qquad (14.1.1)$$

The efficiency η of converting energy from one form to another is the ratio of energy output to energy input, thus

$$\eta = \frac{E_{output}}{E_{input}} = 1 - \frac{E_{loss}}{E_{input}} \qquad (14.1.2)$$

Energy losses often occur as heat lost by conduction E_{cond}, convection E_{conv}, and radiation E_{rad}. A general expression for energy lost by these mechanisms is

$$E_{loss} = E_{cond} + E_{conv} + E_{rad} \qquad (14.1.3)$$

If we compare Equation (14.1.3) with (14.1.2), we see that an increase in energy loss lowers the efficiency of energy conversion. Conversely, if we can decrease energy loss, we can increase energy conversion efficiency. One of the goals of energy conservation is to increase energy conversion efficiency by reducing energy loss. In the next section, we discuss cogeneration as an example of an energy conserving technology.

COGENERATION

One method for improving energy conversion efficiency is to find a way to use energy that would otherwise be lost as heat. *Cogeneration* is the simultaneous production and application of two or more sources of energy. The most common example of cogeneration is the simultaneous generation of electricity and useful heat. In this case, a fuel such as natural gas can be burned in a boiler to produce steam. The steam drives an electric generator and is recaptured for such purposes as heating or manufacturing. Cogeneration is most effective when the cogeneration facility is near the site where excess heat can be used. The primary objective of cogeneration is to reduce the loss of energy in Equation (14.1.2) by converting part of the energy loss to an energy output.

14.2   ENERGY AND THE ENVIRONMENT

We have discussed many energy sources in previous chapters, and we have considered the feasibility of using each energy source as a fuel. The energy densities shown in Table 14-1 are among the most important factors considered in selecting a fuel source. In addition to energy density, such factors as cost, reliability, and social acceptability must be considered.

The fuel used to drive a power plant is called the *primary fuel*. Examples of primary fuels include oil, coal, natural gas, and uranium. Primary fuels are the source of *primary energy*, which is energy that has not been obtained by anthropogenic conversion or transformation. At present, fossil fuels are the primary fuel for the majority of power plants in the world. Fossil fuel resources are finite, however, and power plants that burn fossil fuels emit greenhouse gases and other pollutants. Before judging fossil fuels too harshly, however, it must be realized that every energy source has advantages and disadvantages. Energy density, cost, and reliability are among the advantages that fossil fuels enjoy relative to other energy sources.

Table 14-1
Energy density of common materials*

Material	Energy density	
	MJ kg^{-1}	MJ m^{-3}
Crude oil	42	37,000
Coal	32	42,000
Dry wood	12.5	10,000
Hydrogen, gas	120	10
Hydrogen, liquid	120	8,700
Methanol	21	17,000
Ethanol	28	22,000

Source: Sørensen, 2000, page 552.

A reliable energy source is a resource that is available almost all the time. Some downtime may be necessary for facility maintenance or other operating reasons. The United States Department of Energy has defined a reliable energy source as an energy source that is available at least 95% of the time [DoE Geothermal, 2002].

Society is searching for environmentally compatible, reliable energy sources. One of the characteristics of an environmentally compatible energy source is its cleanliness. A clean energy source emits negligible amounts of greenhouse gases or other pollutants. Even though fossil fuels have serious pollution problems, we have seen that mitigation technologies such as greenhouse gas sequestration can reduce the impact and improve their acceptance as an energy source. In an objective assessment of competing energy sources, we must recognize that clean energy sources such as hydroelectric, solar, and wind energy can have a significant environmental impact that can adversely affect their environmental compatibility. The environmental impact of some clean energy sources is summarized in the following list:

- Hydroelectric facilities, notably dams, can flood vast areas of land. The flooded areas can displace people and wildlife, and affect the ecosystems of adjacent areas with consequences that may be difficult to predict. Dams can change the composition of river water downstream of the dam, and can deprive land areas of a supply of silt for agricultural purposes. A dam on a river can prevent the upstream migration of certain species of fish, such as salmon.
- Geothermal power plants can emit toxic gases such as hydrogen sulfide, or greenhouse gases such as carbon dioxide. The produced water from a geothermal reservoir will contain dissolved solids that can form solid precipitates when the temperature and pressure of the produced water changes.
- Solar power plants are relatively inefficient, and a solar power plant, such as the Solar Electric Generating Station in Southern California, has a large footprint and can be visually offensive.
- Wind farms can interfere with bird migration patterns and can be visually offensive.

The selection of an environmentally compatible primary fuel is not a trivial problem. Two areas of special concern for the twenty-first century energy mix are nuclear fission and nuclear fusion. We consider them in more detail here.

NUCLEAR FISSION

Nuclear fission plants can produce more energy and operate continuously for longer periods of time than can other power plants. Compared to fossil fuel–driven power plants, nuclear fission plants require a relatively small mass of resource to fuel the nuclear plant for an extended period of time. Nuclear fission plants rely on a nonrenewable resource: uranium-235. The earth's inventory of uranium-235 will eventually be exhausted. Breeder reactors use the chain reaction that occurs in the reactor control rods to produce more fissionable material (specifically plutonium-239).

One of the main concerns of nuclear fission technology is to find a socially and environmentally acceptable means of disposing of fuel rods containing highly radioactive waste. The issue of waste is where most of the debate about nuclear energy is focused. The waste generated by nuclear fission plants emits biologically lethal radiation and can contaminate the site where it is stored for thousands of years. On the other hand, environmentally compatible disposal options are being developed. One disposal option is to store spent nuclear fuel in geologically stable environments. Yucca Mountain, Nevada, has been selected as a nuclear waste disposal site in the United States.

NUCLEAR FUSION

The resources needed for the nuclear fusion reaction are abundant; they are the isotopes of hydrogen. The major component of fusion, deuterium, can be extracted from water, which is available around the world. Tritium, another hydrogen isotope, is readily available in lithium deposits that can be found on land and in seawater. Unlike fossil fuel driven power plants, nuclear fusion does not emit air pollution.

Nuclear fusion reactors are considered much safer than nuclear fission reactors. The amounts of deuterium and tritium used in the fusion reaction are so small that the instantaneous release of a large amount of energy during an accident is highly unlikely. The fusion reaction can be shut down in the event of a malfunction with relative ease. A small release of radioactivity in the form of neutrons produced by the fusion reaction may occur, but the danger level is much less than that of a fission reactor. The main problem with nuclear fusion energy is that the technology is still under development; commercially and technically viable nuclear fusion reactor technology does not yet exist. A panel working for the United States Department of Energy has suggested that nuclear fusion could be providing energy to produce electricity by the middle of the twenty-first century if adequate support is provided to develop nuclear fusion technology [Dawson, 2002].

14.3 ECONOMICS

One essential component of energy resource management is economics.[1] An economic analysis is used to weigh various options and decide on an appropriate course of action. The management challenge is to simultaneously optimize technical objectives and the performance of economic measures. We provide a brief introduction to economics in this section, beginning with the concept of time value of money. We then discuss cash flow and measures of economic performance.

TIME VALUE OF MONEY

The value of money can change with time. The future amount A of a sum of money P that is invested at an annual interest rate i_{int} is

$$A = P\left(1 + i_{int}\right)^N \tag{14.3.1}$$

where N is the total number of years into the future. The effect of inflation is to reduce the future amount A by a factor $(1 + i_{inf})$ each year, where i_{inf} is the inflation rate per year. The combined effect of inflation rate and interest rate on the future value of money is given by

$$A = P\left(\frac{1 + i_{int}}{1 + i_{inf}}\right)^N \tag{14.3.2}$$

An effective discount rate i_{eff} can be calculated by writing

$$i_{eff} = \left(\frac{1 + i_{int}}{1 + i_{inf}}\right) - 1 = \frac{i_{int} - i_{inf}}{1 + i_{inf}} \tag{14.3.3}$$

so that Equation (14.3.5) becomes

$$A = P\left(1 + i_{eff}\right)^N \tag{14.3.4}$$

We can rearrange Equation (14.3.1) to determine the present value P of a future amount of money A, thus

$$P = \frac{A}{(1 + i_{\text{dis}})^N} \qquad (14.3.5)$$

The quantity i_{dis} in this case is called the discount rate, and the factor $(1 + i_{\text{dis}})^{-N}$ is called the present worth factor $\text{PWF}(i_{\text{dis}}, N)$. The present worth factor

$$\text{PWF}(i_{\text{dis}}, N) = \frac{1}{(1 + i_{\text{dis}})^N} \qquad (14.3.6)$$

depends on discount rate i_{dis} and the number of years N. Equation (14.3.5) can be written in terms of $\text{PWF}(i_{\text{dis}}, N)$ as

$$P = A \times \text{PWF}(i_{\text{dis}}, N) = \frac{A}{(1 + i_{\text{dis}})^N} \qquad (14.3.7)$$

In terms of effective interest rate i_{eff}, the present value of money becomes

$$P = A \times \text{PWF}(i_{\text{eff}}, N) = \frac{A}{(1 + i_{\text{eff}})^N} \qquad (14.3.8)$$

CASH FLOW AND ECONOMIC INDICATORS

An economic analysis of competing investment options usually requires the generation of cash flow predictions. The cash flow of an investment option is the net cash generated by the investment option or expended on the investment option as a function of time. The time value of money is included in the economic analysis by applying a discount rate to adjust the value of money in future years to the value of money during a base year. The resulting cash flow is called the *discounted cash flow*. The net present value (NPV) of the discounted cash flow is the value of the cash flow at a specified discount rate. We can quantify these ideas as follows.

Net present value NPV is the difference between revenue R and expenses E, thus

$$\text{NPV} = R - E \qquad (14.3.9)$$

Net present value, revenue, and expenses depend on the time value of money. We can account for the time value of money by introducing a discount rate r in the calculation. Revenue can be expressed as

$$R = \sum_{n=1}^{N} \frac{P_n Q_n}{(1 + r)^n} \qquad (14.3.10)$$

where N is the total number of years, P_n is price per unit quantity produced during year n, and Q_n is the quantity produced during year n. The quantity produced can be volume of oil or gas, kilowatt-hours of electricity, tonnes of coal or uranium, or any other appropriate measure of resource production. Expenses include capital expenditures CAPEX_n during year n, operating expenditures OPEX_n during year n, and taxes TAX_n during year n. The resulting expression for expenses is

$$E = \sum_{n=1}^{N} \frac{\text{CAPEX}_n + \text{OPEX}_n + \text{TAX}_n}{(1 + r)^n} \qquad (14.3.11)$$

Combining the expressions for revenue and expenses in the net present value equation gives

$$\text{NPV} = \sum_{n=1}^{N} \frac{P_n Q_n - \text{CAPEX}_n - \text{OPEX}_n - \text{TAX}_n}{(1 + r)^n} \qquad (14.3.12)$$

The time dependence of NPV is illustrated in Figure 14-1. This figure shows that NPV can be negative. A negative NPV says that the investment option is operating at a loss. The loss is usually associated with initial capital investments and operating expenses that are incurred before the investment option begins to generate revenue. For example, mirrors have to be installed in a solar electric generating system and a generator built before electricity can be sold to the transmission grid. The installation of mirrors and a generator are capital expenses that occur before revenue is generated by the sale of electricity. Similarly, large investments in the design and construction of offshore platforms must be made before wells can be drilled and oil produced. The eventual growth in positive NPV occurs when revenue exceeds expenses. The point in time when $\text{NPV} = 0$ is the payout time. In the example shown in Figure 14-1, payout time is

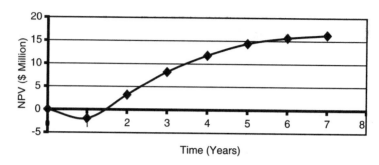

Time (Years)

Figure 14-1. Typical cash flow.

approximately 1.5 years. The concept of payout time can be applied to discounted cash flow or undiscounted cash flow.

Several commonly used indicators of economic performance are listed in Table 14-2. The discount rate at which the maximum value of NPV is zero is called the discounted cash flow return on investment (DCFROI) or internal rate of return (IRR). Payout time, NPV, and DCFROI account for the time value of money. An indicator that does not account for the time value of money is the profit-to-investment (PI) ratio. The PI is the total undiscounted cash flow without capital investment divided by total investment. It is often useful to prepare a variety of plots, such as NPV versus time and NPV versus discount rate, to show how economic indicators perform as functions of time.

Economic analyses using indicators of economic performance provide information about the relative performance of different investment options. The economic viability of an investment option is usually decided after considering a combination of economic indicators. For example, an investment option may be considered economically viable if the DCFROI > 30% and payout time is less than 2 years. It should be remembered, however, that quantitative indicators provide useful information, but not complete information. Economic viability is influenced by both tangible

Table 14-2
Indicators of economic performance

Discount rate	Factor to adjust the value of money to a base year.
Net present value (NPV)	Value of cash flow at a specified discount rate.
DCFROI or IRR	Discount rate at which NPV = 0.
Payout time	Time when NPV = 0.
Profit-to-investment (PI) ratio	Undiscounted cash flow without capital investment divided by total investment.

and intangible factors. The tangible factors, such as building a generator or drilling a well, are relatively easy to quantify. Intangible factors such as environmental and socio-political concerns are relatively difficult to quantify, yet may be more important than tangible factors.

14.4   LIFE CYCLE ANALYSIS

The selection of an energy source depends on such factors as availability, accessibility, environmental acceptability, capital cost, and ongoing operating expenses. The analysis of the costs associated with an energy source should take into account the initial capital expenditures and annual operating expenses for the life of the system. This analysis is called *life cycle analysis*, and the costs are called *life cycle costs*.[2] Life cycle costing requires the analysis of all direct and indirect costs associated with the system for the entire expected life of the system. According to Sørensen [2000, page 762], life cycle analysis includes analyzing the impact of "materials or facilities used to manufacture tools and equipment for the process under study, and it includes final disposal of equipment and materials, whether involving reuse, recycling, or waste disposal." A list of life cycle costs is presented in Table 14-3.

It is important to recognize that the future cost of some energy investment options may change significantly as a result of technological advances. The cost of a finite resource can be expected to increase as the availability of the resource declines, but the cost of an emerging technology will usually decline as the infrastructure for supporting the technology matures. Table 14-4 illustrates the sensitivity of oil producing techniques to the price of oil. The table shows that more-sophisticated technologies can be justified as the price of oil increases. It also includes a price estimate for

Table 14-3
Life cycle costs of an energy system

Capital equipment costs
Acquisition costs
Operating costs for fuels, etc.
Interest charges for borrowed capital
Maintenance, insurance, and miscellaneous charges
Taxes (local, state, federal)
Other recurring or one-time costs associated with the system
Salvage value (usually a cost) or abandonment cost

Source: Goswami, et al., 2000, page 528.

Table 14-4
Sensitivity of oil recovery technology to oil price

Oil recovery technology	Oil price (US$ per barrel in year 2000 US$)
Conventional	10–30
Enhanced oil recovery (EOR)	20–40
Extra heavy oil (e.g. tar sands)	25–45
Alternative energy sources	40+

alternative energy sources, such as wind and solar. In some cases there is overlap between one technology and another. For example, steam flooding is an EOR process that can compete with oil recovery techniques, and chemical flooding is an EOR process that can be as expensive as many alternative energy sources.

The initial costs of one energy system may be relatively low compared with competing systems. If we consider only initial cost in our analysis, we may adopt an energy option that is not optimum. For example, the annual operating expenses for an option we might choose based on initial cost may be significantly larger than those of an alternative option. In addition, projections of future cost may be substantially in error if the cost of one or more of the components contributing to an energy system changes significantly in relation to our original estimate. To avoid making less-than-optimum decisions, we should consider all of the life cycle costs of each investment option, and evaluate the sensitivity of cash flow predictions to plausible changes in cost as a function of time. Inherent in life cycle analysis is an accurate determination of end use efficiency.

End use efficiency is the overall efficiency of converting primary energy to a useful form of energy. It should include an analysis of all factors that affect the application. As a simple example, consider the replacement of a light bulb. The simplest decision is to choose the least expensive light bulb. On a more sophisticated level, we need to recognize that the purpose of the light bulb is to provide light, and some light bulbs can provide light longer than other light bulbs. In this case we need to consider the life of the light bulb in addition to its price. But there are still more factors to consider. If you live in an equatorial region, you might prefer a light bulb that emits light and relatively little heat, so you can reduce air conditioning expenses. On the other hand, if you live in a cooler northern climate in the northern hemisphere, you might desire the extra heat and choose a light bulb that can also serve as a heat source. Once you select a light bulb, you want to use it where it will do the most good. Thus, if you chose a more expensive light bulb that has a long life and generates little heat, you would probably prefer to use the light bulb in a room where the bulb would be used frequently, such as a kitchen or office, rather than a closet where the bulb would be used less frequently. All of these factors should be taken into account in determining the end use efficiency associated with the decision to select a light bulb.

One of the goals of life cycle analysis is to make sure that decision makers in industry, government, and society in general are aware of all of the costs associated with a system. In the context of energy resource management, Sørensen [2000, Section 7.4.3] has identified the following impact areas: economic, environmental, social, security, resilience, development, and political. We have already discussed economic and environmental impacts. The remaining impact areas are designed to raise awareness of issues that are external to the technical energy production process. Some typical questions that must be answered include the following: Does use of the resource have a positive social impact, that is, does resource use provide a product or service without adversely affecting health or work environment? Is the resource secure, or safe, from misuse or terrorist attack? Is the resource resilient—that is, is the resource relatively insensitive to system failure, management errors, or future changes in the way society assesses its impact? Does the resource have a positive or negative effect on the development of a society—that is, does the resource facilitate the goals of a society, such as decentralization of energy generating facilities or satisfying basic human needs? What are the political ramifications associated with the adoption of an energy resource? Is the resource vulnerable to political instability or can the resource be used for political leverage? A thorough life cycle analysis will provide answers to all of these questions. Of course, the validity of the answers will depend on our ability to accurately predict the future.

RISK ANALYSIS AND REAL OPTIONS ANALYSIS

A characteristic of natural resource management is the need to understand the role of uncertainty in decision making. The information we have about a natural resource is usually incomplete. What information we do have may contain errors. Despite the limitations in our knowledge, we must often make important decisions to advance a project. These decisions should be made with the recognition that risk, or uncertainty, is present and can influence investment decisions. Here, risk refers to the possibility that an unexpected event can adversely affect the value of an asset. Uncertainty is not the same as risk. *Uncertainty* is the concept that our limited knowledge and understanding of the future does not allow us to predict the consequences of our decisions with 100% accuracy. Risk analysis is an attempt to quantify the risks associated with investing under uncertainty.

One of the drawbacks of traditional risk analysis is the limited number of options that are considered. The focus in risk analysis is decision making based on current expectations about future events. For example, the net present value analysis discussed in Section 14.3 requires forecasts of revenue and expenses based on today's expectations. Technological advances or political instabilities are examples of events that may significantly alter our expectations. We might overlook or ignore options that would have benefited from the unforeseen events. An option in this context is a set of policies or strategies for making current and future decisions. Real options analysis attempts to incorporate flexibility in the management of investment options that are subject to considerable future uncertainty.

The best way to incorporate options in the decision making process is to identify them during the early stages of analysis. Once a set of options has been identified for a particular project, we can begin to describe the uncertainties and decisions associated with the project. By identifying and considering an array of options, we obtain a more complete picture of what may happen as a consequence of the decisions we make. Real options analysis helps us understand how important components of a project, particularly components with an element of uncertainty, influence the value of the project.

14.5 SUSTAINABLE DEVELOPMENT: A COMPELLING SCENARIO

The concept of sustainable development was introduced in 1987 in a report prepared by the United Nations' World Commission on Environment and Development. The Commission, known as the Brundtland Commission after chairwoman Gro Harlem Brundtland of Norway, said that society should adopt a policy of sustainable development that allows society to meet its present needs while preserving the ability of future generations to meet their own needs [WCED, 1987]. The concept of sustainable development is a road map of how we should prepare for the future, and it is a vision of what the future should be. It is a scenario that has been adopted by the United Nations and helps explain evolving business practices in the energy industry. We can better appreciate the importance of the sustainable development scenario by contrasting the concept of scenario with other forms of prognostication. We then consider the concept of sustainable development in more detail.

STORIES, SCENARIOS, AND MODELS

One of the problems facing society is the need to develop and implement a strategy that will provide energy to meet future global energy needs and satisfy environmental objectives. The development of strategies depends on our view of the future. At best, we can only make educated guesses about what the future will bring. The quality of our educated guesses depends on our understanding and the information we have available. We can distinguish between different levels of guessing by giving different names to our predictions.

Figure 14-2 displays three levels of predicting the future: stories, scenarios, and models. A story can be used to provide a qualitative picture of the future. Stories are relatively unclear because our understanding is limited and the information we have is relatively incomplete. As we gain information and understanding, we can begin to discuss scenarios. Scenarios let us consider different stories about complex situations. They let us incorporate more detail into plausible stories. Unlike forecasts, which let

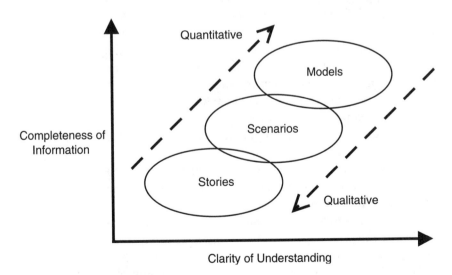

Figure 14-2. Stories, scenarios, and models.

us extrapolate historical behavior to predict the future, scenarios let us consider the effects of discontinuities and sudden changes. Forecasts assume a certain degree of continuity from past to future, although the future may in fact be altered dramatically by an unexpected development. It is somewhat reminiscent of the distinction between gradualism and catastrophism we encountered in our discussion of evolution. In the context of energy, revolutionary developments in nuclear fusion technology or unexpected cost reductions in solar energy technology could lead to abrupt changes in historical trends of energy production and consumption. These changes could invalidate any forecast that was based on a continuous extrapolation of historical trend and lead to a future that would have been considered implausible based on past performance.

We can construct models that allow us to quantify our scenarios as our knowledge and understanding increase. The models can be used to make quantitative predictions. We can compare model predictions with the actual performance of a system as time passes and we acquire additional information. The intermediate position of scenarios between stories and models in Figure 14-2 shows that scenarios can be thought of as tools that allow us to integrate the intuition we express in stories with quantitative modeling.

SUSTAINABLE DEVELOPMENT

An emerging energy mix is needed to meet energy demand in the twenty-first century. The demand for energy is driven by factors such as increasing trends in population and consumption. The ability to meet the demand for energy depends on such factors as price volatility, supply availability, and efficiency of energy use. One measure of how efficiently a country is using its energy is energy intensity. In the context of energy policy, *energy intensity* may be defined on the national level as the total domestic primary energy consumption divided by the gross domestic product. Countries that have low energy consumption and high domestic productivity will have relatively low energy intensities. Countries with high energy consumption and low domestic productivity will have relatively high energy intensities. By considering the change in energy intensity as a function of time, we can see if a country is improving its efficiency of energy consumption relative to its domestic productivity. If one of our goals is to maintain or improve quality of life with improved energy efficiency, we would like to see the energy intensity of a nation decrease as a function of time. This would indicate that the ratio of energy consumption to domestic productivity is decreasing.

The emerging energy mix is expected to rely on clean energy—that is, energy that is generated with minimal environmental impact. The goal is sustainable development: the integration of social and environmental concerns into development activities that optimize economic profitability and value creation as the world undergoes the transition from non-renewable fossil fuels to renewable fuels and a sustainable, secure energy infrastructure. Society desires, and industry is seeking, to achieve sustainable development. One industry response to environmental and social concerns in the context of sustainable development is the "triple bottom line" [Whittaker, 1999].

The three components of sustainable development, and the three goals of the triple bottom line (TBL), are economic prosperity, social equity, and environmental protection. From a business perspective, the focus of TBL is the creation of long-term shareholder value by recognizing that corporations are dependent on licenses provided by society to do business. If business chooses not to comply with sustainable development policies, society can enforce compliance by imposing additional government regulation and control.

Sustainable development is a scenario that is concerned about the rights of future generations. The concept of rights is a legal and philosophical one. It is possible to argue that future generations do not have any legal rights to current natural resources and are not entitled to rights. From this perspective, each generation must do the best it can with available resources. On the other hand, many societies are choosing to adopt the value of preserving natural resources for future generations. National parks are examples of natural resources that are being preserved. An energy trend that appears to be consistent with the sustainable development scenario is described next.

DECARBONIZATION

Energy forecasts rely on projections of historical trends. Table 14-5 is based on historical data presented by the United States Energy Information Administration for the last four decades of the twentieth century. The table shows historical energy consumption in units of quads. The row of data labeled "Geothermal, etc." includes net electricity generation from wood, waste, solar, and wind. We recall here that published statistical data are subject to revision, even if the data are historical data that have been published by a credible source. Data revisions may change specific numbers as new information is received and used to update the database, but it is

Table 14-5
World primary energy production

Primary energy	Primary energy production (quads)			
	1970	1980	1990	2000
Coal	62.96	72.72	94.29	92.51
Natural gas	37.09	54.73	75.91	90.83
Crude oil	97.09	128.12	129.50	145.97
Natural gas plant liquids	3.61	5.10	6.85	9.28
Nuclear electric power	0.90	7.58	20.31	25.51
Hydroelectric power	12.15	18.06	22.55	27.46
Geothermal, etc.	1.59	2.95	3.94	5.36

Source: 1970–2000 [EIA Table 11.1, 2002].

reasonable to expect the data presented in Table 14-5 to show qualitatively correct trends.

The data in Table 14-5 are graphically displayed in Figure 14-3.

The first four energy sources in Figure 14-3—coal, natural gas, crude oil, and natural gas plant liquids—are fossil fuels. The data show the dominance of fossil fuels in the energy mix at the end of the twentieth century.

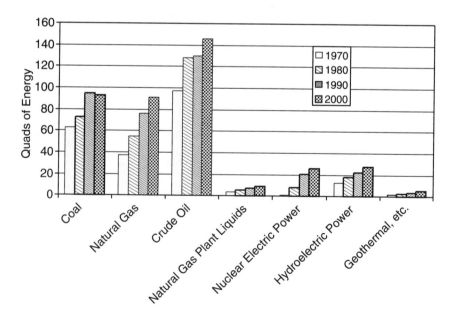

Figure 14-3. Historical energy consumption.

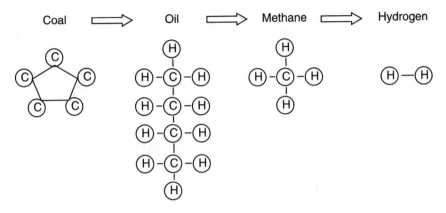

Figure 14-4. Decarbonization.

The data also show that the amount of produced energy from non-fossil fuels is increasing.

The trend in the twentieth century has been a "decarbonization" process, that is, a move away from fuels with many carbon atoms to fuels with few or no carbon atoms. Ausubel [2000, page 18] defines *decarbonization* as "the progressive reduction in the amount of carbon used to produce a given amount of energy." Figure 14-4 illustrates how the carbon to hydrogen ratio (C:H) declines as the fuel changes from carbon-rich coal to carbon-free hydrogen. The use of hydrogen as a fuel is discussed in more detail in the next chapter.

Figure 14-5 shows the historical pattern and postulated future trends presented by Ausubel [2000]. The figure uses fractional market share M

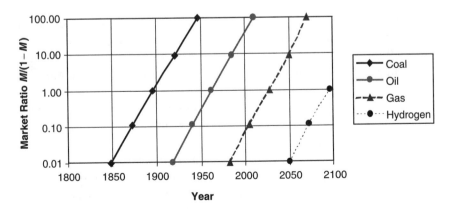

Figure 14-5. Postulated trends in global energy consumption.

to present the market ratio $M/(1 - M)$ of a fuel as a function of time. The historical slopes of the market ratio for coal and oil are assumed to hold true for natural gas and hydrogen during the twenty-first century. The historical trend suggests that the twenty-first century will see a gradual transition from a dominance of fossil fuels in the current energy mix to a more balanced distribution of energy options.

14.6   ENERGY AND ETHICS

One issue that must be considered in the context of sustainable development is the distribution of energy. Should energy be distributed around the world based on need, ability to pay, or some other value? This question is an ethical issue because the answer depends on the values we choose to adopt.

The distribution of energy in the future will depend on whether or not a nation has a large per capita energy base.[3] Should nations with energy resources help those in need? If so, how should they help? Traditional ethics would favor a policy of helping those nations without energy resources, but opinions differ on how to proceed. Two of the more important ethical positions are identified as lifeboat ethics and spaceship ethics. These positions are considered here for two reasons: they are diametrically opposed ethical positions that apply to the global distribution of energy; and they illustrate that people of good will can take opposite positions in a significant debate.

Proponents of lifeboat ethics oppose the transfer of wealth by charitable means. In this view, the more-developed industrial nations are considered rich boats, and the less-developed, overcrowded nations are poor boats. The rich boats should not give the poor boats energy because their help would discourage the poor boats from making difficult choices such as population control and investment in infrastructure. Lifeboat ethics is a "tough love" position; it encourages nations to seek self-sufficiency. On the other hand, it might make some nations desperate and encourage the acquisition of energy resources by military means.

Proponents of spaceship ethics argue that everyone is a passenger on spaceship Earth. In this view, some passengers travel in first class while others are in steerage. A more equitable distribution of energy is needed because it is morally just, and it will prevent revolts and social turmoil. Thus, the wealthy should transfer part of their resources to the poor for both moral and practical reasons. On the other hand, nations that receive charitable donations of energy may be unwilling to make the sacrifices needed to become self-sufficient.

Both ethical positions have advantages and disadvantages. Is there a middle ground? Can the amount of useful energy be increased so that the need for sacrifice is lessened and the cost of energy is decreased? These are questions facing society in general and the energy professional in particular.

14.7   ENERGY AND GEOPOLITICS

Quality of life, energy, and the distribution of energy are important components of global politics.[4] Readily available, reasonably priced energy is a critical contributor to the economic well-being of a nation. We have already seen that deforestation in England motivated the search for a new primary fuel. The need for oil encouraged Japanese expansion throughout Asia in the 1930s and was one of the causes of World War II. The 1973 Arab–Israeli war led to the first oil crisis with a short-term, but significant increase in the price of oil. This oil price shock was followed by another in 1979 after the fall of the Shah of Iran. These oil price increases are considered shocks because they were large enough to cause a significant decline in global economic activity [Verleger, 2000, page 76]. Our ability to correctly forecast energy demand depends on our understanding of technical and socio-political issues. In this section, we give a brief introduction to global politics and then discuss its implications for models of future energy demand.

CLASH OF CIVILIZATIONS

The world has been undergoing a socio-political transition that began with the end of the Cold War and is continuing today. Huntington [1996] provided a view of this transition that helps clarify historical and current events, and provides a foundation for understanding the socio-political issues that will affect energy demand.

Huntington argued that a paradigm shift was occurring in the geopolitical arena. A *paradigm* is a model that is realistic enough to help us make predictions and understand events, but not so realistic that it tends to confuse rather than clarify issues. A paradigm shift is a change in paradigm. A geopolitical model has several purposes. It lets us order events and make general statements about reality. We can use the model to help us understand causal relationships between events and communities. The communities can range in size from organizations to alliances of nations. The geopolitical model lets us anticipate future developments and, in some instances, make predictions. It helps us establish the importance of information in relation to the model, and it shows us paths that might help us reach our goals.

The Cold War between the Soviet Union and the Western alliance led by the United States established a framework that allowed people to better understand the relationships between nations following the end of World War II in 1945. When the Cold War ended with the fall of the Berlin wall and the break up of the Soviet Union in the late 1980s and 1990s, it signaled the end of one paradigm and the need for a new paradigm. Several geopolitical models have been proposed. Huntington considered four possible paradigms for understanding the transition (Table 14-6).

The paradigms in Table 14-6 cover a wide range of possible geopolitical models. The One Unified World paradigm asserts that the end of the Cold War signaled the end of major conflicts and the beginning of a period of relative calm and stability. The Two Worlds paradigm views the world in an "us versus them" framework. The world was no longer divided by political ideology (democracy versus communism); it was divided by some other issue. Possible divisive issues include religion and rich versus poor (generally a North–South geographic division). The world could also be split into zones of peace and zones of turmoil. The third paradigm, Anarchy, views the world in terms of the interests of each nation, and considers the relationships between nations to be unconstrained. According to Huntington, these three paradigms, One Unified World, Two Worlds, and Anarchy, range from too simple (One Unified World) to too complex (Anarchy).

The Chaos paradigm says that post–Cold War nations are losing their relevance as new loyalties emerge. In a world where information flows freely and quickly, people are forming allegiances based on shared traditions and value systems. The value systems are notably cultural and, on a more fundamental level, religious. The new allegiances are in many cases a rebirth of historical loyalties. New alliances are forming from the new allegiances and emerging as a small set of civilizations. The emerging civilizations are characterized by ancestry, language, religion, and way of life.

Table 14-6
Huntington's possible
geopolitical paradigms

1   One Unified World
2   Two Worlds (West versus non-West)
3   Anarchy (184+ Nation-states)
4   Chaos

Table 14-7
Huntington's major contemporary civilizations

Civilization	Comments
Sinic	China and related cultures in Southeast Asia
Japanese	The distinct civilization that emerged from the Chinese civilization between 100 and 400 C.E.
Hindu	The peoples of the Indian subcontinent that share a Hindu heritage
Islamic	A civilization that originated in the Arabian peninsula and now includes subcultures in Arabia, Turkey, Persia, and Malaysia
Western	A civilization centered around the northern Atlantic that has a European heritage and includes peoples in Europe, North America, Australia, and New Zealand
Orthodox	A civilization centered in Russia and distinguished from Western civilization by its cultural heritage, including limited exposure to Western experiences (such as the Renaissance, the Reformation, and the Enlightenment)
Latin America	Peoples with a European and Roman Catholic heritage who have lived in authoritarian cultures in Mexico, Central America, and South America

Huntington considered the fourth paradigm, Chaos, to be the most accurate picture of current events and recent trends. He argued that the politics of the modern world can be best understood in terms of a model that considers relationships between the major contemporary civilizations shown in Table 14-7. The existence of a distinct African civilization has been proposed by some scholars, but is not as widely accepted as the civilizations identified in the table.

Each major civilization has at least one core state [Huntington, 1996, Chapter 7]. France and Germany are core states in the European Union. The United States is a core state in Western civilization. Russia and China are core states, perhaps the only core states, in Orthodox civilization and Sinic civilization respectively. Core states are sources of order within their civilizations. Stable relations between core states can help provide order between civilizations.

Within the context of the multicivilization geopolitical model, the two world wars in the twentieth century began as civil wars in Western civilization, and engulfed other civilizations as the hostilities expanded. The Cold War and the oil crises in the latter half of the twentieth century were conflicts between civilizations. Western civilization has been the most powerful civilization for centuries, where power in this context refers to the ability to control and influence someone else's behavior. The trend in global politics is a decline in the political power of Western civilization as other civilizations develop technologically and economically. Energy is a key factor in this model of global politics.

Forecasts of energy consumption depend on our understanding of the factors that influence the growth of non-Western civilizations. Forecasts of energy production depend on the ability of energy producers to have access to natural resources. Access depends, in turn, on the nature of relationships between civilizations with the technology to develop natural resources and civilizations with territorial jurisdiction over the natural resources. Examples of forecasts are discussed in the next chapter.

ENDNOTES

1. Economics references include Sørensen [2000], Thompson and Wright [1985], and Newendorp and Schuyler [2000]. References on investment decision analysis techniques include Dixit and Pindyck [1994], Newendorp and Schuyler [2000], and Copeland, et al. [2000].
2. See Goswami, et al. [2000, Section 12.2], and Sørensen [2000, Section 7.4] for more discussion about life cycle analysis.
3. For a more in-depth discussion of ethics and energy, see Cassedy and Grossman [1998, Chapter 5].
4. There is a large literature on global politics and energy. References of note here are *The Prize* by Daniel Yergin [1992], and *The Clash of Civilizations* by Samuel P. Huntington [1996].

EXERCISES

14-1. A. Use the information in Table 14-1 to fill in the following table. For comparison, the density of water at room temperature and pressure is approximately 1000 kg/m^3.

Material	Energy density MJ/kg	Energy density MJ/m^3	Density Kg/m^3
Crude oil	42	37,000	
Coal	32	42,000	
Dry wood	12.5	10,000	
Hydrogen, gas	120	10	
Hydrogen, liquid	120	8,700	
Methanol	21	17,000	
Ethanol	28	22,000	

B. Use the information from Part A to calculate the mass and volume of material needed to produce 1 quad of energy. Fill in the following table.

**Mass and volume needed to produce
1 quad of energy**

Material	Mass (kg)	Volume (m³)
Crude oil		
Coal		
Dry wood		
Hydrogen, gas		
Hydrogen, liquid		
Methanol		
Ethanol		

C. Use the information from Part B to determine the relative mass and relative volume needed to produce 1 quad of energy. Define relative mass as the mass of the material divided by the mass of crude oil, and relative volume as the volume of material divided by the volume of crude oil. In this case, crude oil mass and volume are the bases for comparison. Fill in the following table.

**Relative mass and volume needed to produce
1 quad of energy**

Material	Relative mass	Relative volume
Crude oil		
Coal		
Dry wood		
Hydrogen, gas		
Hydrogen, liquid		
Methanol		
Ethanol		

14-2. A. Suppose the efficiency of converting energy from one form to another is 40% for an energy production process. An energy conservation process reduces the energy loss by 10% of the input energy. What is the efficiency of the new process?

B. What is the percent increase in energy output relative to the original energy output?

14-3. A. Use Equation (14.3.1) to find out how many years N are needed to double your principal P for each of the following interest rates: $i_{int} = \{3\%, 6\%, 9\%\}$.
B. Does N depend on principal?

14-4. Plot effective interest rate i_{eff} versus interest rate i_{int} for the following inflation rates: $i_{inf} = \{3\%, 10\%, 20\%\}$. Let the interest rate range be $0\% \leq i_{int} \leq 20\%$.

14-5. A. The following table presents data for an economic analysis. Prices are inflated by 3% and expenses are inflated by 5% . Use the data in the table to plot NPV (Net Present Value) versus time for a discount rate of 15% .

Year (n)	Price	Quantity	CAPEX	OPEX	TAX
0	20.00	0	0	0	0
1	20.60	0	50000000	0	0
2	21.22	1000000	0	100000	5304500
3	21.85	860708	0	86071	4702594
4	22.51	818731	0	81873	4607443
5	23.19	778801	0	77880	4514218
6	23.88	740818	0	74082	4422878
7	24.60	704688	0	70469	4333387
8	25.34	670320	0	67032	4245707
9	26.10	637628	0	63763	4159801
10	26.88	606531	0	60653	4075632
11	27.68	576950	0	57695	3993167
12	28.52	548812	0	54881	3912371
13	29.37	522046	0	52205	3833209
14	30.25	496585	0	49659	3755649
15	31.16	472367	0	47237	3679658
16	32.09	449329	0	44933	3605205
17	33.06	427415	0	42741	3532259
18	34.05	406570	0	40657	3460788
19	35.07	386741	0	38674	3390764
20	36.12	367879	0	36788	3322156

B. When does payout occur?
C. What is the DCFROI (Discounted Cash Flow Return On Investment)?

14-6. A. Find the total world primary energy production using data presented in Table 14-5.
B. Plot world primary energy production by energy type versus year. Express energy in quads, and present your data as a line plot.
C. Is the trend in energy production increasing or decreasing for each energy type?
D. Will the trend continue?

14-7. Plot the primary energy production data presented in Table 14-5 as a bar chart with energy production displayed in quads along the vertical axis. You should display all four periods (1970, 1980, 1990, 2000) on one bar chart.

14-8. A. Fill in the following table using the primary energy production data presented in Table 14-9. The column labeled "HC" should include all energy types that are predominantly hydrocarbon based. The column labeled "Other" should include all energy types that are not fossil fuels.

Year	Coal	HC	Other	Total
1970				
1980				
1990				
2000				

B. Use the above information to fill in the following table. The column labeled "Fossil fuel" should include all energy types that are fossil fuels.

Year	Fossil fuel	Other	Total
1970			
1980			
1990			
2000			

C. Is the relative amount of fossil fuel increasing or decreasing in the total energy mix?

14-9. A. Suppose 100,000 gallons of oil is spilled at sea. How many barrels of oil were spilled?
B. If the oil has a specific gravity of 0.9, determine the mass of oil spilled. Express your answer in kg and lbm.

C. Estimate the area covered by the spill if the thickness of the spill is 1 mm. Express your answer in m^2, km^2, mi^2, and acres.

14-10. A. Complete the following table.

Fuel	Price	Energy density	$ per MJ
Coal	$50 per tonne	42 MJ/kg	
Oil	$20 per barrel	42000 MJ/m^3	
Methane	$70 per 1000 m^3	38 MJ/ m^3	

B. Which of the fuels in the table is the least expensive per MJ for the prices quoted in the table?

14-11. A. The following table shows the amount of CO_2 that is emitted by the combustion of four types of fuel. Suppose each fuel is used as the primary energy source for a 1000 MW power plant. If the power plant operates for 20 years, how much energy will be provided during the period of operation? Express your answer in GJ.

Fuel	CO_2 emission* [kg per GJ heat output]
Coal	120
Oil	75
Methane	50
Wood	77

* Ramage and Scurlock [1996, Table 4.1, page 143]

B. How many kg of CO_2 will be emitted by each fuel?
C. How many moles of CO_2 will be emitted by each fuel?
D. Estimate the bulk volume V_B of a reservoir that would be needed to sequester the produced CO_2 for each fuel in part C. Assume reservoir porosity ϕ is 18%, average reservoir pressure P after CO_2 injection is 2000 psia, average reservoir temperature T after CO_2 injection is 150° F, and the water saturation S_w after CO_2 injection is 30%. Assume the volume of sequestered CO_2 can be approximated using the ideal gas law. Express bulk volume in m^3.
E. Express bulk volume in Part D in acre-ft.

14-12. A. The following table presents the gross domestic product (GDP in billions of 1995 US dollars) and primary energy consumption (in quadrillion BTUs) for a few countries for years 1990, 1995, and 2000. Calculate the energy intensity (in barrels of oil equivalent per US$ 1000). Data are from the U.S. Energy Information Administration website and were recorded in 2002.

Country	Year	GDP	Primary energy consumption	Energy intensity
China	1990	398	27.0	
	1995	701	35.2	
	2000	1042	36.7	
France	1990	1478	8.81	
	1995	1555	9.54	
	2000	1764	10.4	
Japan	1990	4925	17.9	
	1995	5292	20.8	
	2000	5342	21.8	
Saudi Arabia	1990	114	3.15	
	1995	128	3.85	
	2000	139	4.57	
United Kingdom	1990	1041	9.29	
	1995	1127	9.60	
	2000	1295	9.88	
United States	1990	6580	84.4	
	1995	7400	91.0	
	2000	9049	98.8	

B. In which countries is the energy intensity decreasing?
C. In which countries is the energy intensity relatively constant?
D. In which countries is the energy intensity increasing?

CHAPTER FIFTEEN

The Twenty-First Century Energy Mix

Many scenarios of the twenty-first century energy mix expect society to use several different energy sources. They view hydrogen as a major carrier of energy, but not as an energy source. Even though the assumptions, methods, and results presented in these scenarios are debatable, they all show an energy infrastructure in transition. We discuss a set of energy forecasts later in this chapter. First, we consider the role hydrogen may play as a major component of the twenty-first century energy mix.

15.1   HYDROGEN AND FUEL CELLS

Hydrogen is found almost everywhere on the surface of the earth as a component of water. It has many commercial uses, including ammonia (NH_3) production for use in fertilizers, methanol (CH_3OH) production, hydrochloric acid (HCl) production, and use as a rocket fuel. Liquid hydrogen is used in cryogenics and superconductivity. Hydrogen is important to us because it can be used as a fuel. We first describe the properties of hydrogen before discussing the production of hydrogen and its use as the primary fuel in fuel cells.

PROPERTIES OF HYDROGEN

Hydrogen is the first element in the Periodic Table. The nucleus of hydrogen is the proton, and hydrogen has only one electron. At ambient conditions on Earth, hydrogen is a colorless, odorless, tasteless, and nontoxic gas of diatomic molecules (H_2). Selected physical properties of hydrogen, methane, and gasoline are shown in Table 15-1. A kilogram of hydrogen, in either the gas or liquid state, has a greater energy density than the most widely used fuels today, such as oil and coal (Appendix D-1). The heating value of a fuel is the heat released when the fuel is completely

Table 15-1
Selected physical properties of hydrogen, methane, and gasoline[a]

Property	Hydrogen (gas)	Methane (gas)	Gasoline (liquid)
Molecular weight (g/mol)[a]	2.016	16.04	~110
Mass density (kg/m^3)[a,b]	0.09	0.72	720–780
Energy density (MJ/kg)	120[a]	53[c,d]	46[a,c]
Volumetric energy density (MJ/m^3)[a]	11[a]	38[c,d]	35,000[a,c]
Higher heating value (MJ/kg)[a]	142.0	55.5	47.3
Lower heating value (MJ/kg)[a]	120.0	50.0	44.0

[a] *Ogden [2002, Box 2, page 71].*
[b] *at 1 atm and 0° C.*
[c] *Hayden [2001, page 183].*
[d] *Ramage and Scurlock [1996, Box 4.8, page 152].*

burned in air at room temperature and the combustion products are cooled to room temperature [Çengel and Boles, 2002, page 254]. The *lower heating value* is the heating value obtained when the water produced by combustion is allowed to leave as a vapor. The *higher heating value* is the heating value obtained when the water produced by combustion is completely condensed. In this case, the heat of vaporization is included in the higher heat of combustion of the fuel. The energy conversion efficiencies of cars and jet engines are normally based on lower heating values, and the energy conversion efficiencies of furnaces are based on higher heating values. Lower heating value and higher heating value are sometimes abbreviated as LHV and HHV respectively.

Hydrogen is considered a clean, reliable fuel once it is produced because the combustion of hydrogen produces water vapor. Hydrogen can react with oxygen to form water in the exothermic combustion reaction $2H_2 + O_2 \rightarrow 2H_2O$. The heat of combustion of the reaction is 62,000 BTU/lbm of hydrogen (1.4×10^8 J/kg). Hydrogen combustion does not emit toxic greenhouse gases such as carbon monoxide or carbon dioxide. When hydrogen is burned in air, it does produce traces of nitrogen oxides. Hydrogen is considered a carrier of energy because it can be used as a fuel to provide energy. It is not considered an energy source because energy is required to produce molecular hydrogen H_2 from water.

HYDROGEN PRODUCTION

Hydrogen can be produced by electrolysis.[1] Electrolysis is the non-spontaneous splitting of a substance into its constituent parts by supplying electrical energy. In the case of water, electrolysis would decompose the water molecule into its constituent elements by the addition of electrical energy, thus

$$H_2O + \text{electrical energy} \rightarrow H_2 + \frac{1}{2}O_2 \qquad (15.1.1)$$

It is difficult to electrolyze very pure water because there are few ions present to flow as an electrical current. Electrolyzing an aqueous salt solution enhances the production of hydrogen by the electrolysis of water. An aqueous salt solution is a mixture of ions and water. The addition of a small amount of a nonreacting salt such as sodium sulfate (Na_2SO_4) accelerates the electrolytic process. The salt is called an *electrolyte* and provides ions that can flow as a current.

We illustrate the electrolysis process in Figure 15-1 using the electrolysis of molten table salt (sodium chloride or NaCl). Electrolysis is an oxidation-reduction (redox) reaction. Recall that oxidation involves the loss of electrons, and reduction involves the gain of electrons. The redox reaction takes place in an electrolytic cell. The electrolytic cell is a cell with two electrodes: the anode, and the cathode. Oxidation occurs at the anode, and reduction occurs at the cathode. The voltage source in Figure 15-1

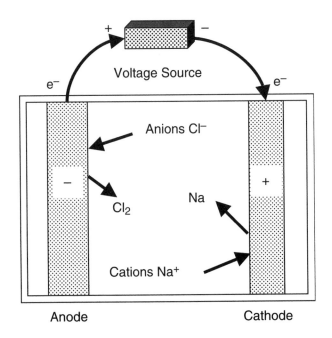

Figure 15-1. Electrolysis of molten salt.

provides the potential difference needed to initiate and support the redox reaction. In the figure, a negatively charged chlorine ion with one extra electron loses its excess electron at the anode and combines with another chlorine atom to form the chlorine molecule (Cl_2). The electrons flow from the anode to the cathode, where a positively charged sodium ion gains an electron and becomes an electrically neutral atom of sodium.

The redox reaction can be treated as two half-reactions. The first half-reaction is oxidation at the anode. Oxidation occurs when electrons are released by the reducing agent in the oxidation half-reaction. In the case of electrolysis of a water–table salt solution, two oxidation half-reactions occur at the anode:

anode − oxidation :

$$2H_2O(l) \rightarrow 4H^+(aq) + O_2(g) + 4e^- \qquad (15.1.2)$$
$$2Cl^-(l) \rightarrow Cl_2(g) + 2e^-$$

The parenthetic symbols denote liquid phase (ℓ), aqueous phase (aq), and gas phase (g). A reducing agent is a substance that donates electrons in a reaction. The hydrogen atom from a water molecule is the reducing agent at the anode. Electrolysis of a water–table salt solution is illustrated in Figure 15-2.

The second half-reaction is reduction at the cathode. Reduction occurs when the oxidizing agent acquires electrons in the reduction half-reaction. In the case of electrolysis of a water–table salt solution, two reduction half-reactions occur at the cathode:

cathode − reduction :

$$4H_2O(l) + 4e^- \rightarrow 2H_2(g) + 4OH^-(aq) \qquad (15.1.3)$$
$$Na^+(aq) + e^- \rightarrow Na(s)$$

The parenthetic symbol (s) denotes solid phase. An oxidizing agent is a substance that accepts electrons in a reaction. The hydroxyl radical OH from a water molecule is the oxidizing agent at the cathode for the upper reaction, and the sodium cation is the oxidizing agent for the lower reaction.

The overall reaction for electrolysis of a water–table salt solution in an electrolytic cell is

overall :

$$2H_2O(l) \rightarrow 2H_2(g) + O_2(g) \qquad (15.1.4)$$
$$2NaCl \rightarrow 2Na(s) + Cl_2(g)$$

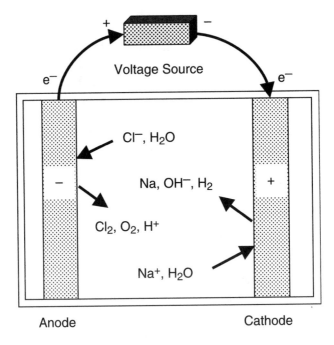

Figure 15-2. Electrolysis of water–salt solution.

Four hydrogen ions in Equation (15.1.2) combine with four hydroxyl ions in Equation (15.1.3) to form four molecules of water, which are not shown in Equation (15.1.4). The overall reaction shows that hydrogen gas is produced at the cathode of the electrolytic cell and oxygen gas is produced at the anode when enough electrical energy is supplied. The electrical energy is supplied in the form of an electrical voltage, and must be greater than the threshold energy (activation energy) of the reaction. The voltage needed to electrolyze water must be greater than the voltage that we would predict if we did not consider activation energy, consequently the actual voltage needed to electrolyze water is called *overvoltage*.

The concept of overvoltage has significant implications. For example, if we assume table salt (NaCl) is the electrolyte in the aqueous salt solution, we will produce chlorine gas (Cl_2) at the anode instead of oxygen gas (O_2) when the applied voltage is less than the required overvoltage because chlorine gas production has lower activation energy than oxygen gas production. The formation of solid sodium and gaseous chlorine are undesirable by-products: solid sodium can create a precipitate or scale that can adversely affect the efficiency of the electrolytic process, and gaseous chlorine is highly reactive and toxic. One of the problems facing society is the problem of how to produce hydrogen using an environmentally acceptable process.

Two possible sources of energy for producing hydrogen in the long-term are nuclear fusion and solar energy. Ausubel [2000] has suggested that the potential of nuclear energy will be realized when nuclear energy can be used as a source of electricity and high-temperature heat for splitting water into its constituent parts. The process of decomposing the water molecule at high temperatures is called *thermal decomposition*. Heat in the form of steam can be used to re-form hydrocarbons and produce hydrogen.

Steam re-forming exposes a hydrocarbon such as methane, natural gas, or gasoline to steam at $850°$ C and 2.5×10^6 Pa [Sørensen, 2000, pages 572–573]. The reformation requires the two reactions

$$C_nH_m + nH_2O \rightarrow nCO + \left(n + \frac{m}{2}\right)H_2$$
$$CO + H_2O \rightarrow CO_2 + H_2$$

(15.1.5)

where C_nH_m represents the hydrocarbon. Both reactions produce hydrogen. A process such as absorption or membrane separation can be used to remove the carbon dioxide by-product. The production of carbon dioxide, a greenhouse gas, is an undesirable characteristic of steam re-forming.

Some bacteria can produce hydrogen from biomass by fermentation or high-temperature gasification. The gasification process is similar to coal gasification described in Section 13.8.

FUEL CELLS

Hydrogen is considered a carrier of energy because it can be transported in the liquid or gaseous states by pipeline or in cylinders. Once produced and distributed, hydrogen can be used as a fuel for a modified internal combustion engine or as the fuel in a fuel cell. *Fuel cells* are electrochemical devices that directly convert hydrogen or hydrogen-rich fuels into electricity using a chemical rather than a combustion process.

A fuel cell consists of an electrolyte sandwiched between an anode and a cathode (Figure 15-3). The electrolyte in Figure 15-3 is a mixture of potassium hydroxide (KOH) and water. The electrolyte solution is maintained at a lower pressure than the gas cavities on either side of the porous electrodes. The pressure gradient facilitates the separation of hydrogen and oxygen molecules. The load in the figure is a circuit with amperage A and voltage V. Hydrogen is fed to the anode (negative electrode) and oxygen

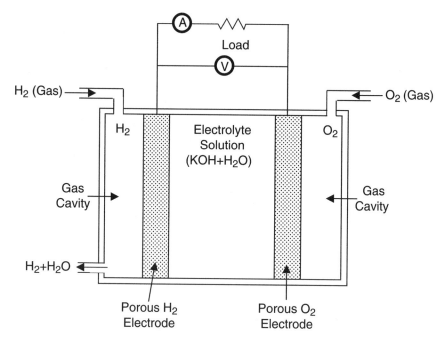

Figure 15-3. Schematic of a fuel cell [after Cassedy and Grossman [1998, page 419]; and B.J. Crowe, *Fuel Cells—A Survey*, 1973, NASA, U.S. Government Printing Office, Washington, D.C.].

is fed to the cathode (positive electrode). When activated by a catalyst, hydrogen atoms separate into protons and electrons. The charged protons and electrons take different paths to the cathode. The electrons go through the external circuit and provide an electrical current, while the protons migrate through the electrolyte to the cathode. Once at the cathode, the protons combine with electrons and oxygen to produce water and heat.

The type of electrolyte in a fuel cell distinguishes that fuel cell from other types of fuel cells. The fuel cell in Figure 15-3 is a proton exchange membrane fuel cell because it depends on the movement of protons (hydrogen nuclei) through the porous electrodes. Fuel cells produce clean energy from hydrogen by the overall reaction

$$\frac{1}{2}O_2 + H_2 \rightarrow H_2O + \text{electricity} + \text{heat} \tag{15.1.6}$$

Fuel cells do not need recharging or replacing, and can produce electricity as long as they are supplied with hydrogen and oxygen.

15.2 THE HYDROGEN ECONOMY

The historical trend toward decarbonization reflects the contention by many energy forecasters that hydrogen will be the fuel of choice in the future. These forecasters believe that power plants and motor vehicles will run on hydrogen. The economies that emerge will depend on hydrogen and are called hydrogen economies. The concept of a hydrogen economy is not new. The use of hydrogen as a significant fuel source driving a national economy was first explored in the middle of the twentieth century as a complement to the adoption of large-scale nuclear electric generating capacity. Concerns about global climate change and the desire to achieve sustainable development have renewed interest in hydrogen as a fuel.

A future that depends on hydrogen is not inevitable. Hydrogen economies will require the development of improved technologies for producing, storing, transporting, and consuming hydrogen.[2] We have already discussed some of the challenges involved in the production of hydrogen. As another example of the technological challenges that must be overcome in a transition to a hydrogen economy, let us consider the storage of hydrogen.

Hydrogen can be stored in the liquid or gaseous state, but it must be compressed to high pressures or liquefied to achieve reasonable storage volumes because of the low density of the diatomic hydrogen molecule. The energy content of hydrogen gas is less than the energy contained in methane at the same temperature and pressure. The volumetric energy density of liquid hydrogen is approximately 8700 MJ/m^3. This is about one third the volumetric energy density of gasoline. The relatively low volumetric energy density of hydrogen creates a storage problem if we want to store hydrogen compactly on vehicles.

Researchers have learned that hydrogen can be stored effectively in the form of solid metal hydrides. A metal hydride is a metal that absorbs hydrogen. The hydrogen is absorbed into the spaces, or interstices, between atoms in the metal. According to Silberberg [1996, page 246], metals such as palladium and niobium "can absorb 1000 times their volume of H_2 gas, retain it under normal conditions, and release it at high temperatures." This form of storage may be desirable for use in hydrogen-powered vehicles.

Hydrogen can be hazardous to handle. A spectacular demonstration of this fact was the destruction of the German zeppelin *Hindenburg*. The *Hindenburg* used hydrogen for buoyancy. In 1937, the *Hindenburg* burst into flames while attempting a mooring in Lakehurst, New Jersey. At first it was widely believed that the zeppelin became a ball of fire when the hydrogen ignited. An early solution to the flammability problem was to use less flammable gases such as helium in lighter-than-air ships. In the 1990s, a former NASA scientist named Addison Bain reanalyzed the data and showed that the *Hindenburg* fire was more likely started by the ignition of a flammable material that was used to coat the cloth bags that contained the hydrogen than by the ignition of hydrogen. Today it is believed that hydrogen may have contributed to the *Hindenburg* fire, but was not its cause.

Hydrogen forms an explosive mixture with air when the concentration of hydrogen in air is in the range of 4% to 75% hydrogen.[3] For comparison, natural gas is flammable in air when the concentration of natural gas in air is in the range of 5% to 15% natural gas. Furthermore, the ignition energy for hydrogen–air mixtures is approximately one-fifteenth the ignition energy for natural gas–air or gasoline–air mixtures. The flammability of hydrogen in air makes it possible to consider hydrogen a more dangerous fuel than natural gas. On the other hand, the low density of hydrogen allows hydrogen to dissipate more quickly into the atmosphere than a higher-density gas such as methane. Thus, hydrogen leaks can dissipate more rapidly than natural gas leaks. Adding an odorant to the gas can enhance the detection of gas leaks.

The environmental acceptability of hydrogen fuel cells depends on how the hydrogen is produced. If a renewable energy source such as solar energy is used to generate the electricity needed for electrolysis, vehicles powered by hydrogen fuel cells would be relatively clean because hydrogen combustion emits water vapor. Unfortunately, hydrogen combustion in air also emits traces of nitrous oxide (NO_x) compounds. Nitrogen dioxide ($x = 2$) contributes to photochemical smog and can increase the severity of respiratory illnesses.

15.3   SUMMARY OF ENERGY OPTIONS

The literature contains several publications that present a description of the energy sources that are available or are expected to be available during the twenty-first century. Energy options known today include fossil fuels, nuclear energy, solar energy, renewable fuels, and alternative sources. They are briefly described in the following paragraphs as a summary of available energy sources [Fanchi, 2000].

Fossil fuels are the dominant energy source in the modern global economy. They include coal, oil, and natural gas. Environmental concerns are motivating a change from fossil fuels to an energy supply that is clean. Clean energy refers to energy that has little or no detrimental impact on the environment. Natural gas is a source of relatively clean energy. Oil and gas fields are considered conventional sources of natural gas. Two nonconventional sources of natural gas are coalbed methane and gas hydrates.

Coalbeds are an abundant source of methane. The presence of methane gas in coal has been well known to coal miners as a safety hazard, but is now being viewed as a source of natural gas. Coalbed methane exists as a monomolecular layer on the internal surface of the coal matrix. Its composition is predominately methane, but can also include other constituents, such as ethane, carbon dioxide, nitrogen, and hydrogen. The gas, which is bound in the micropore structure of the coalbed, is able to diffuse into the natural fracture network when a pressure gradient exists between the matrix and the fracture network. The fracture network in coalbeds consists of microfractures called cleats. Gas flows through the microfractures to the production well.

Gas hydrates are chemical complexes that are formed when one type of molecule completely encloses another type of molecule in a lattice. In the case of gas hydrates, hydrogen-bonded water molecules form a cage-like structure in which mobile molecules of gas are absorbed or bound. Although gas hydrates can be found throughout the world, difficulties in cost-effective production have hampered development of the resource. Gas hydrates are generally considered troublesome for oil and gas field operations, but their potential commercial value as a clean energy resource is changing the industry perception. The potential as a gas resource is due to the relatively large volume of gas contained in the gas hydrate complex.

Nuclear energy is presently provided by nuclear fission. Nuclear fission is the process in which a large, unstable nucleus decays into two smaller fragments. It depends on a finite supply of fissionable material. Nuclear fusion is the combination, or fusing, of two small nuclei into a single larger nucleus. Many scientists expect nuclear energy to be provided by nuclear fusion sometime during the twenty-first century. Fusion reactions are the source of energy supplied by the sun. Attempts to harness and commercialize fusion energy have so far been unsuccessful because of the technical difficulties involved in igniting and controlling a fusion reaction. Nevertheless, fusion energy is expected to contribute significantly to the energy mix by the end of the twenty-first century, even though a prototype commercial-scale nuclear reactor is not expected to exist until 2015 or later

[Morrison and Tsipis, 1998]. Both fission and fusion reactions release large amounts of energy, including significant volumes of waste heat that needs to be dissipated and controlled. The decay products of the fission process can be highly radioactive for long periods of time, but the by-products of the fusion process are relatively safe.

Solar energy is available in three forms: passive, active, and electric. Passive and active solar energy are generally used for space conditioning, such as heating and cooling. Active solar energy technologies are typically mechanical devices, such as solar hot water heaters, that are used to distribute solar energy. Passive solar heating integrates building design with environmental factors that enable the capture of solar energy, a simple example being south-facing windows in a house. Solar electric devices such as photovoltaic cells convert sunlight into electricity. Groups of photovoltaic cells can provide electricity in quantities ranging from a few milliwatts to several megawatts, and thus can power devices ranging from calculators to power plants. To get an idea of the scale, a large color TV requires approximately one kilowatt of power, and a power plant for a modern city requires approximately three gigawatts [Smil, 1991].

Types of renewable fuels range from hydroelectric and wind to synfuels and biomass. The kinetic energy of wind and flowing water are indirect forms of solar energy, and are therefore considered renewable. Wind turbines harness wind energy, and hydroelectric energy is generated by the flow of water through a turbine. Both convert the mechanical energy of a rotating blade into electrical energy in a generator.

Biomass refers to wood and other plant or animal matter that can be burned directly or can be converted into fuels. Wood has historically been a source of fuel. Technologies now exist to convert plants, garbage, and animal dung into natural gas. Methanol, or wood alcohol, is a volatile fuel that has been used in racing cars for years. Another alcohol, clean-burning ethanol, which can be produced from sugarcane, can be blended with gasoline to form a blended fuel (gasohol) and used in conventional automobile engines, or used as the sole fuel source for modified engines. Synthetic fuels are fossil fuel substitutes created by chemical reactions using such basic resources as coal or biomass. Synthetic fuels are used as substitutes for conventional fossil fuels such as natural gas and oil.

There are several ways to convert biomass into synthetic fuels, or synfuels. Oils produced by plants such as rapeseed (canola), sunflowers, and soybeans can be extracted and refined into a synthetic diesel fuel that can be burned in diesel engines. Thermal pyrolysis and a series of catalytic reactions can convert the hydrocarbons in wood and municipal wastes into a synthetic gasoline. Excessive use of dung and crop residues for fuel instead of fertilizer can deprive the soil of essential nutrients that are needed for future crops.

Synthetic liquid hydrocarbon fuels can be produced from natural gas by a gas-to-liquids (GTL) conversion process. The process uses a Fischer-Tropsch (F-T) reactor. The F-T process produces a hydrocarbon mixture with a range of molecular weight components by reacting hydrogen and carbon monoxide in the presence of a catalyst.

The oceans are another solar-powered source of energy. Waves and tides can be used to drive electric generators. Temperature gradients in the ocean exist between warm surface water and cooler water below the surface. If the temperature gradient is large enough, it can be used to generate power using ocean thermal energy conversion power plants. Similarly, temperature gradients and steam generated by geothermal sources can drive electric generators as a source of energy.

Alternative sources of energy include hydrogen fuel cells and cogeneration. Hydrogen can be used as a fuel for a modified internal combustion engine or in a fuel cell. Fuel cells are electrochemical devices that directly convert hydrogen, or hydrogen-rich fuels, into electricity using a chemical process. Fuel cells do not need recharging or replacing and can produce electricity as long as they are supplied with hydrogen and oxygen. Hydrogen can be produced by the electrolysis of water, which uses electrical energy to split water into its constituent elements. Electrolysis is a net energy-consuming process.

The environmental acceptability of hydrogen fuel cells depends on how the hydrogen is produced. If a renewable energy source such as solar energy were used to generate the electricity needed for electrolysis, vehicles powered by hydrogen fuel cells would be relatively clean because hydrogen combustion emits water vapor, but it also emits NO_x compounds. Nitrogen dioxide ($x = 2$) contributes to photochemical smog and can increase the severity of respiratory illnesses. Shipping and storage of hydrogen are important unresolved issues that hinder the widespread acceptance and implementation of hydrogen fuel cell technology.

15.4 FORECAST METHODOLOGIES AND FORECASTS

Energy forecasts rely on projections of historical trends. These forecasts can be in error very quickly. For example, Schollnberger's [1999] estimate of energy consumption for the year 2000 is shown in Figure 15-4.

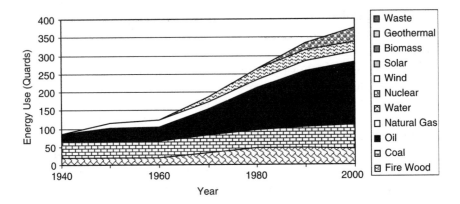

Figure 15-4. Historical energy consumption.

It was a projection from historical data that was complete through the end of 1996. According to the United States Energy Information Administration, actual oil consumption was approximately 40% of world energy consumed in 2000. This shows that Schollnberger's forecast over-estimated oil consumption even after a relatively short forecast of four years. However, Schollnberger's focus was not on short-term forecasting, but long-term trends. In addition, Schollnberger was more interested in the combined forecast of oil and gas consumption or demand because gas can be substituted for oil in many instances. If we combine oil and gas, Schollnberger forecast oil and gas consumption to be about 53% of world energy consumed in 2000, but EIA statistics showed oil and gas consumption to be about 62% of world energy consumed. From this perspective, Schollnberger underestimated oil and gas consumption in 2000.

The dominance of fossil fuels in the energy mix at the end of the twentieth century is illustrated as a percent of total energy consumed in Figure 15-5. Each percentage distribution shown in Figure 15-5 applies to the associated point in time. For example, oil accounted for approximately 22% of world energy consumed in 1940, and for approximately 45% of world energy consumed in 2000.

Forecasts of the twenty-first century energy mix show that a range of scenarios is possible. The forecast discussed here is based on Schollnberger's forecasts, which were designed to cover the entire twenty-first century and predict the contribution of a variety of energy sources to the twenty-first century energy portfolio. We do not consider Schollnberger's forecast because it is correct. We already know that it was incorrect within four years of its publication date. Schollnberger's forecast is worth studying

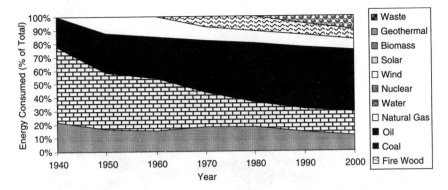

Figure 15-5. Historical energy consumption as % of total energy consumed.

because it uses more than one scenario to project energy consumption for the entire twenty-first century.

Schollnberger considered three forecast scenarios:

A. "Another Century of Oil and Gas" corresponding to continued high hydrocarbon demand
B. "The End of the Internal Combustion Engine" corresponding to a low hydrocarbon demand scenario
C. "Energy Mix" corresponding to a scenario with intermediate demand for hydrocarbons and an increasing demand for alternative energy sources

Schollnberger viewed scenario C as the most likely scenario. It is consistent with the observation that the transition from one energy source to another has historically taken several generations. Leaders of the international energy industry have expressed a similar view that the energy mix is undergoing a shift from liquid fossil fuels to other fuel sources.

There are circumstances in which scenarios A and B could be more likely than scenario C. For example, scenario B would be more likely if environmental issues led to political restrictions on the use of hydrocarbons and an increased reliance on conservation. Scenario B would also be more likely if the development of a commercially competitive fuel cell for powering vehicles reduced the demand for hydrocarbons as a transportation fuel source. Failure to develop alternative technologies would make scenario A more likely. It assumes that enough hydrocarbons will be supplied to meet demand.

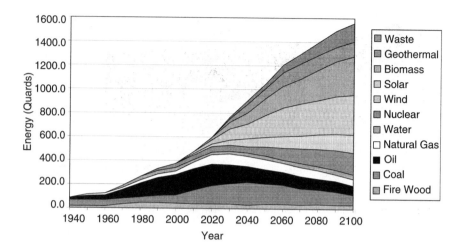

Figure 15-6. Forecast of twenty-first century energy consumption.

Scenario C shows that natural gas will gain in importance as the economy shifts from a reliance on hydrocarbon liquid to a reliance on hydrocarbon gas. Eventually, renewable energy sources such as biomass and solar energy will displace oil and gas (see Figure 15-6).

The demand by society for petroleum fuels should continue at or above current levels for a number of years, but the trend seems clear (see Figure 15-7). The global energy portfolio is undergoing a transition from an energy portfolio dominated by fossil fuels to an energy portfolio that includes a range of fuel types. Schollnberger's scenario C presents one possible energy portfolio, and the historical and projected energy consumptions trends are illustrated in Figure 15-7.

Schollnberger's forecast is based on demand. An alternative approach is to base the forecast on supply. Beginning with Hubbert [1956], several authors have noted that annual U.S. and world oil production approximately follows a bell shaped (Gaussian) curve. A Gaussian curve for calculating the average daily production of oil Q during year number T is

$$Q = Q_{max} \exp\left[-\frac{(T - T_{max})^2}{2\sigma_G}\right], \quad Q_{max} = \frac{Q_G}{\sqrt{2\pi\sigma_G}} \qquad (15.4.1)$$

where Q_G and σ_G are positive curve fit parameters for the Gaussian curve. The maximum production Q_{max} occurs at year $T = T_{max}$ in the Gaussian curve analysis. The year numbers T, T_{max} are dimensionless, such as

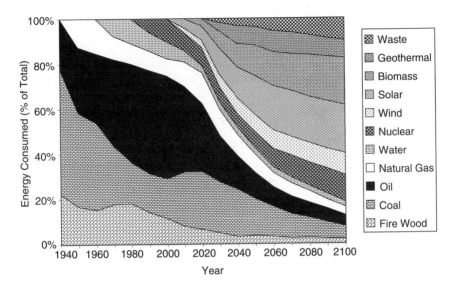

Figure 15-7. Forecast of twenty-first century energy consumption as % of energy consumed.

year 2000. Another curve that has been used to forecast average daily oil production is the logistic curve

$$Q = \frac{Q_0}{1 + \exp\left[\alpha_L(T_0 - T)\right]} \tag{15.4.2}$$

where Q_0, T_0, and α_L are curve fit parameters for the logistic curve. The parameter Q_0 equals $2Q$ at year $T = T_0$ in the logistic curve.

Forecasts based on Gaussian fits to historical data can be readily checked using publicly available data. Figure 15-8 shows a Gaussian curve fit of world oil production data from the U.S. EIA database. The fit is designed to match the most recent part of the production curve most accurately. This gives a match that is similar to results obtained by Deffeyes [2001, page 147]. The peak oil production rate in Figure 15-8 occurs in 2010 and cumulative oil production by year 2100 is a little less than 2.1 trillion barrels.

Analyses of historical data using a Gaussian curve typically predict that world oil production will peak in the first decade of the twenty-first century.[4] By integrating the area under the Gaussian curve, forecasters have claimed that cumulative world oil production will range from 1.8 to 2.1 trillion barrels. These forecasts usually underestimate the sensitivity

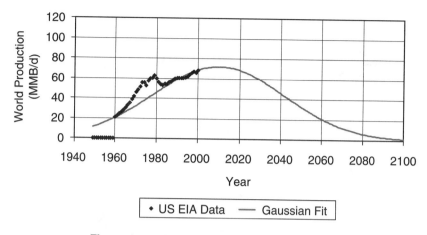

Figure 15-8. Oil forecast using Gaussian curve.

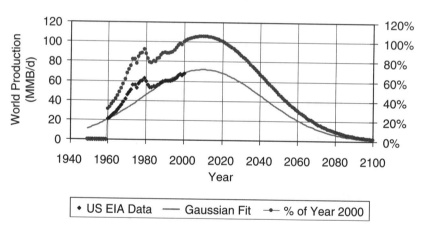

Figure 15-9. Oil forecast as % of oil produced in year 2000 using Gaussian curve.

of oil production to technical advances and price. In addition, forecasts often discount the large volume of oil that has been discovered but not yet produced because the cost of production has been too high. The sensitivity of oil recovery technology to oil price is illustrated in Table 14-4.

If we accept a Gaussian fit of historical data as a reasonable method for projecting oil production, we can estimate future oil production rate as a percentage of oil production rate in the year 2000. Figure 15-9 shows this estimate. According to this approach, world oil production rate will decline to 50% of year 2000 world oil production by the middle of the twenty-first century. For comparison, let us consider Schollnberger's scenario C.

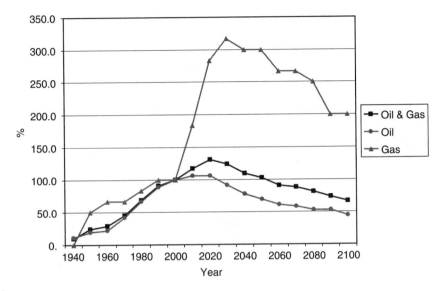

Figure 15-10. Forecast of twenty-first century oil and gas consumption as % of oil and gas consumed in year 2000.

According to scenario C, fossil fuel consumption will increase relative to its use today until about the middle of the twenty-first century, when it will begin to decline (see Figure 15-10). By the end of the twenty-first century, fossil fuel consumption will be approximately 70% of what it is today. Gas consumption will be considerably larger, but oil consumption will decline to approximately half of its use today.

A comparison of oil production as a percent of year 2000 production in Figure 15-9 with oil consumption as percent of oil consumed in year 2000 illustrates the range of uncertainty in existing forecasts. The supply-based forecast shows that oil production will approach 0% of year 2000 production by year 2100. By contrast, the demand-based forecast shown in Figure 15-10 expects oil to be consumed at about 50% of its year 2000 consumption. There is a clear contradiction between the two forecasts that can be used to test the validity of the forecasts. Another test of forecast validity is the peak of world oil production.

Forecasts of world oil production peak tend to shift as more historical data is accumulated. Laherrère [2000] pointed out that curve fits of historical data are most accurate when applied to activity that is "unaffected by political or significant economic interference, to areas having a large number of fields, and to areas of unfettered activity" (page 75).

Furthermore, curve fit forecasts work best when the inflection point (or peak) has been passed.

ALTERNATIVE FORECASTS

The forecasts presented above are just a sampling of the twenty-first century energy forecasts that are appearing in the literature. It is instructive to note two other energy scenarios: a nuclear energy scenario [Hodgson, 1999] and a renewable energy scenario [Geller, 2003]. These scenarios illustrate the range of perspectives that must be considered in deciding global energy policy.

Hodgson presented a scenario in which the world would come to rely on nuclear fission energy. He defined five objective criteria for evaluating each type of energy: capacity, cost, safety, reliability, and effect on the environment. The capacity criterion considered the ability of the energy source to meet future energy needs. The cost criterion considered all costs associated with an energy source. The safety criterion examined all safety factors involved in the practical application of an energy source. This includes hazards associated with manufacturing and operations. The reliability criterion considered the availability of an energy source. By applying the five objective criteria, Hodgson concluded that nuclear fission energy was the most viable technology for providing global energy in the future. According to Hodgson, nuclear fission energy is a proven technology that does not emit significant amounts of greenhouse gases. He argued that nuclear fission reactors have an exemplary safety record when compared in detail with other energy sources. Breeder reactors could provide the fuel needed by nuclear fission power plants, and nuclear waste could be stored in geological traps. The security of nuclear power plants in countries around the world would be assured by an international agency such as the United Nations. In this nuclear scenario, renewable energy sources would be used to supplement fission power, and fossil energy use would be minimized. Hodgson did not assume that the problems associated with nuclear fusion would be overcome. If they are, nuclear fusion could also be incorporated into the energy mix.

The nuclear fission scenario articulated by Hodgson contrasts sharply with the renewable energy scenario advocated by Geller [2003]. Geller sought to replace both nuclear energy and fossil energy with renewable energy only. An important objective of his forecast was to reduce greenhouse gas emissions to levels that are considered safe by the Kyoto protocol. Figure 15-11 summarizes Geller's energy forecast. The figure shows global

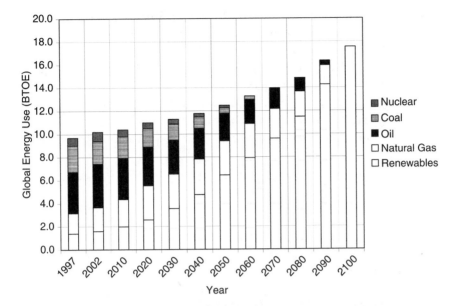

Figure 15-11. Forecast of a twenty-first century energy mix that eventually relies on renewable energy only [after Geller, 2003, page 227].

energy use as a function of time. Global energy use is expressed in billions of tons of oil equivalent (BTOE).

15.5   WHAT DOES THE FUTURE HOLD?

We have discussed several energy options that may contribute to the twenty-first century energy mix. Many influential people around the world believe that humanity should select energy options that will enable the world to move toward a global hydrogen economy. As we have seen, the hydrogen economy uses hydrogen as a carrier of energy. What will be the long-term source of energy? Will solar energy, with its low efficiencies but widespread availability, be the source? Or will nuclear fusion energy be the source, even though the technology for commercial application of nuclear fusion energy is still under development?

In a special issue of *Physics Today* devoted to energy, editor-in-chief Stephen Benka summarized the present situation: "With virtually the entire existing energy-related infrastructure designed for fossil fuels, it seems certain that the world will continue to rely heavily on hydrocarbon combustion for the foreseeable future" [Benka, 2002, page 39]. How do we transition from an economy that depends on fossil fuels to a hydrogen economy?

We have considered some representative attempts to forecast society's path toward a hydrogen economy. With the knowledge presented in previous chapters, each of us can judge these forecasts and speculate on what may happen in the years ahead. The future depends on choices and, probably, discoveries, that are yet to be made. Energy professionals will make many of those choices and discoveries. Perhaps you will be one of them.

ENDNOTES

1. Sources for electrolysis include Brown, et al. [2002], Silberberg [1996, Section 20.6], Zumdahl [1986, Sections 17.1 and 17.7], Sørensen [2000, Section 5.2], and Hoffmann [2001, Chapter 4].
2. References that discuss hydrogen, the hydrogen economy, and fuel cells include Hoffmann [2001, Chapters 1 and 12], Cassedy [2000, Chapter 9], and Ogden [2002].
3. Kraushaar and Ristinen discuss the flammability of hydrogen [1993, page 221]. Bain and Van Vorst [1999] discuss the role of hydrogen in the *Hindenburg* tragedy.
4. Additional references describing the use of Hubbert's curve include Deffeyes [2001], Campbell and Laherrère [1998], and Laherrère [2000]. Deffeyes [2001] also discusses the logistic curve.

EXERCISES

15-1. Suppose one liter of algae can produce 3 milliliters of hydrogen per hour in bright sunlight. What is the rate of hydrogen production (in L/hr) by 1 acre · foot of algae?

15-2. A. Let ΔT equal the increase in ambient temperature due to global warming. The value of ΔT is in dispute. Estimates of ΔT typically range from $0°$ C to $5°$ C. The United States was responsible for approximately 20% of ΔT in the year 2000. Suppose an international treaty is approved that requires the United States to reduce its greenhouse gas production Q_G by 30%. Assume ΔT is proportional to greenhouse gas production Q_G, and $\Delta T = 0$ at $Q_G = 0$. Estimate the percent decrease in ΔT that would result if the United States implemented the treaty.

B. Let $\Delta T = 5°$ C and estimate the reduction in global warming if the United States implemented the treaty.

15-3. A. A dam has an electric power output capacity of 2000 MWe and a capacity factor of 40%. Does the dam provide more or less power than coal-powered or nuclear power plants with power output of 1000 MWe?

B. Suppose the water in the reservoir behind the dam covers an area of 20,000 hectares and the water was collected from precipitation over a land area of 400,000 km^2. What percentage of the water collection area is covered by the reservoir?

15-4. A. Suppose the parameters for the Gaussian equation are $Q_G = 5700$, $T_{max} = 2010$, $\sigma_G = 1000$. When does the peak rate Q occur?

B. What is the maximum rate Q_{max}?

15-5. A. Fill in the following table. The parameters for the logistic equation are $Q_0 = 140$, $T_0 = 2000$, $\alpha_L = 0.04$, and the parameters for the Gaussian equation are $Q_G = 5700$, $T_{max} = 2010$, $\sigma_G = 1000$.

Historical world petroleum production analysis

	Million barrels produced per day		
Data	Data base	Logistic equation	Gaussian equation
1960	21		
1965	30		
1970	46		
1975	53		
1980	60		
1985	54		
1990	60		
1995	62		
2000	68		

B. What part of the forecasts is most accurate?

15-6. A. Plot world petroleum production rate Q versus time for the twenty-first century, that is, for the period $2000 \leq T \leq 2100$. Use the Gaussian equation parameters $Q_G = 5700$, $T_{max} = 2010$, $\sigma_G = 1000$. Express the world petroleum production rate Q in million barrels of petroleum produced per day.

B. When does the peak rate Q_{max} occur in your plot?

C. What is the value of the peak rate Q_{max} from your plot?

15-7. Suppose the world population in year 2100 is 8 billion people. Assume the amount of energy needed to provide each person an acceptable quality of life is 200,000 MJ per year. Quality of life is measured by the United Nations Human Development Index discussed in Section 1.3. How many power plants with a capacity of 1000 MW would be needed to provide this energy?

15-8. A. Fill in the following table using the data presented in Figure 15-11.

Billion tons of oil equivalent

Year	Renewable	Nat. Gas	Oil	Coal	Nuclear
1997					
2002					
2010					
2020					
2030					
2040					
2050					
2060					
2070					
2080					
2090					
2100					

B. Enter the preceding table into a spreadsheet and prepare a figure showing the decline in nuclear energy and fossil fuel energy as renewable energy increases.

Appendix A-1: Physical Constants

Physical constant	Symbol	Value
Atomic mass unit	amu	1.6605×10^{-27} kg
Avogadro's number	N_A	6.0221396×10^{23} mol^{-1}
Boltzmann's constant	k_B	1.380658×10^{-23} J·K^{-1}
Elementary charge	e	$1.60217733 \times 10^{-19}$ C
Mass of electron	m_e	9.1094×10^{-31} kg
Mass of proton	m_p	1.6726×10^{-27} kg
Mass of neutron	m_n	1.6749×10^{-27} kg
Universal gas constant	R	8.314510 J mole$^{-1\circ}$ K^{-1}
Planck's constant	h	$6.6260755 \times 10^{-34}$ J·s
Planck's constant divided by 2π	$\hbar \equiv h/2\pi$	$1.0545735 \times 10^{-34}$ J·s
Speed of light in vacuum	c	2.99792458×10^8 m/s
Stefan-Boltzmann constant	σ	5.67051×10^{-8} W·m^{-2}·K^{-4}
Permeability of vacuum	μ_0	$4\pi \times 10^{-7}$ N/A^2
Permittivity of vacuum	ε_0	$1/(\mu_0 c^2)$

Source: Cohen, 1996; Particle Data Group, 2000.

Appendix A-2: Astronomical Constants

Physical constant	Symbol	Value
Mass of sun	m_s	1.9891×10^{30} kg
Mass of earth	m_e	5.9742×10^{24} kg
Mean distance, earth to sun (Astronomical unit)	1 A.U.	1.496×10^{11} m
Newton's gravitational constant	G	6.67259×10^{-11} m^3 kg^{-1} s^{-2}
Light-year	ly	9.461×10^{15} m
Parsec	pc	3.086×10^{16} m $= 3.261$ ly
Radius of earth (equatorial)	a_e	6.378×10^6 m
Radius of sun		6.96×10^8 m
Solar constant		1370 W/m^2
Solar luminosity		3.85×10^{26} W

Source: Cohen, 1996; Particle Data Group, 2000.

Appendix B-1: SI Units

Physical quantity	Unit	Abbreviation	Comment
Length	meter	m	
Mass	kilogram	kg	
Time	second	s	
Force	Newton	N	$1\ N = 1\ kg{\cdot}m/s^2$
Pressure	Pascal	Pa	$1\ Pa = 1\ N/m^2$
Energy	Joule	J	$1\ J = 1\ kg{\cdot}m^2/s^2$
Power	Watt	W	$1\ W = 1\ J/s$
Temperature	Kelvin	K	
Amount of substance	mole	mol	
Frequency	Hertz	Hz	$1\ Hz = 1\ cycle/s$
Electric charge	Coulomb	C	
Electric current	Ampere	A	$1\ A = 1\ C/s$
Electric potential	Volt	V	$1\ V = 1\ J/C = 1\ W/A$
Resistance	Ohm	Ω	$1\ \Omega = 1\ V/A$
Capacitance	Farad	F	$1\ F = 1\ C/V$
Inductance	Henry	H	$1\ H = 1\ V{\cdot}s/A$
Magnetic induction $\vec{B}$	Tesla	T	$1\ T = 1\ N/(A{\cdot}m)$
Magnetic flux	Weber	Wb	$1\ Wb = 1\ T{\cdot}m^2 = 1\ V{\cdot}s$
Luminous intensity	candela	cd	

Appendix B-2: Powers of 10

Prefix	Symbol	Value	Prefix	Symbol	Value
atto	a	10^{-18}	kilo	k	10^3
femto	f	10^{-15}	mega	M	10^6
pico	p	10^{-12}	giga	G	10^9
nano	n	10^{-9}	tera	T	10^{12}
micro	μ	10^{-6}	peta	P	10^{15}
milli	m	10^{-3}	exa	E	10^{18}

Appendix C: Unit Conversion Factors

TIME
1 hour = 1 hr = 3600 s
1 day = 8.64×10^4 s
1 year = 1 yr = 3.1536×10^7 s

LENGTH
1 foot = 1 ft = 0.3048 m
1 kilometer = 1 km = 1000 m
1 mile = 1 mi = 1.609 km

VELOCITY
1 foot per second = 0.3048 m/s
1 kilometer per hour = 1 kph = 1000 m/hr = 0.278 m/s
1 mile per hour = 1 mph = 1.609 km/hr = 1609 m/hr = 0.447 m/s

AREA
1 square foot = 1 ft^2 = 0.0929 m^2
1 square mile = 1 mi^2 = 2.589 km^2 = 2.589×10^6 m^2
1 square mile = 1 mi^2 = 640 acres
1 acre = 1 ac = 4047 m^2
1 hectare = 1 ha = 1.0×10^4 m^2
1 millidarcy = 1 md = 0.986923×10^{-15} m^2
1 Darcy = 1000 md = 0.986923×10^{-12} m^2
1 barn = 1.0×10^{-24} cm^2 = 1.0×10^{-28} m^2

VOLUME
1 liter = 1 L = 0.001 m^3
1 cubic foot = 1 ft^3 = 2.83×10^{-2} m^3
1 standard cubic foot = 1 SCF = 1 ft^3 at standard conditions

1 acre-foot = 1 ac-ft = 1233.5 m^3
1 barrel = 1 bbl = 0.1589 m^3
1 gallon (U.S. liquid) = 1 gal = 3.785 × 10^{-3} m^3
1 barrel = 42 gallons = 0.1589 m^3

MASS

1 gram = 1 g = 0.001 kg
1 pound (avoirdupois) = 1 lb (avdp) = 1 lbm = 0.453592 kg
1 tonne = 1000 kg

MASS DENSITY

1 g/cm^3 = 1000 kg/m^3

FORCE

1 pound-force = 1 lbf = 4.4482 N

PRESSURE

1 Pascal = 1 Pa = 1 N/m^2 = 1 kg/m·s^2
1 Megapascal = 1 MPa = 10^6 Pa
1 Gigapascal = 1 GPa = 10^9 Pa
1 pound-force per square inch = 1 psi = 6894.8 Pa
1 atmosphere = 1 atm = 1.01325 × 10^5 Pa
1 atmosphere = 1 atm = 14.7 psi

ENERGY

1 megajoule = 1 MJ = 1.0 × 10^6 J
1 gigajoule = 1 GJ = 1.0 × 10^9 J
1 exajoule = 1 EJ = 1.0 × 10^{18} J
1 eV = 1.6022 × 10^{-19} J
1 MeV = 10^6 eV = 1.6022 × 10^{-13} J
1 erg = 10^{-7} J
1 BTU = 1055 J
1 calorie (thermochemical) = 1 cal = 4.184 J
1 kilocalorie = 1 kcal = 1000 calories = 4.184 × 10^3 J
1 Calorie = 1000 calories = 4.184 × 10^3 J
1 kilowatt-hour = 1 kWh = 1 kW · 1 hr = 3.6 × 10^6 J
1 quad = 1 quadrillion BTU = 1.0 × 10^{15} BTU = 1.055 × 10^{18} J
1 quad = 2.93 × 10^{11} kWh = 1.055 × 10^{12} MJ
1 quad = 1.055 exajoule = 1.055 EJ
1 barrel of oil equivalent = 1 BOE = 5.8 × 10^6 BTU = 6.12 × 10^9 J
1 quad = 1.72 × 10^8 BOE = 172 × 10^6 BOE

ENERGY DENSITY
1 BTU/lbm = 2326 J/kg

1 BTU/SCF = 3.73×10^4 J/m^3

POWER
1 Watt = 1 W = 1 J/s

1 Megawatt = 10^6 W = 10^6 J/s

1 kilowatt-hour per year = 1 kWh/yr = 0.114 W = 0.114 J/s

1 horsepower = 1 hp = 745.7 W

VISCOSITY
1 centipoise = 1 cp = 0.001 Pa ·s

1 mPa·s = 0.001 Pa ·s =1 cp = 10^{-3} Pa·s

1 poise = 100 cp = 0.1 Pa ·s

RADIOACTIVITY
1 Curie = 1 Ci = 3.7×10^{10} decays/s

1 Roentgen = 1 R = 2.58×10^{-4} C/kg

1 Radiation Absorbed Dose = 1 rad = 100 erg/g = 0.01 J/kg

1 Gray = 1 Gy = 1 J/kg

100 rems = 1 sievert = 1 Sv

TEMPERATURE
Kelvin to Centigrade: $^\circ$C = $^\circ$K – 273.15

Centigrade to Fahrenheit: $^\circ$F = (9/5) $^\circ$C + 32

Source: Cohen, 1996; Particle Data Group, 2000.

Appendix D: Energy Density

Material	Energy density MJ kg^{-1}	Energy density MJ m^{-3}
Crude oil[a]	42	37,000
Coal[a]	32	42,000
Dry wood[a]	12.5	10,000
Hydrogen, gas[a]	120	10
Hydrogen, liquid[a]	120	8,700
Methanol[a]	21	17,000
Ethanol[a]	28	22,000
Methane, gas[b]	53	38
Gasoline, liquid[c]	46	35,000

[a] Sørensen, 2000, page 552.
[b] Ramage and Scurlock [1996, Box 4.8, page 152].
[c] Ogden [2002, Box 2, page 71].

References

Abell, G.O., Morrison, D., and Wolff, S.C., 1991, *Exploration of the Universe*, Saunders, Philadelphia.

Abramowitz, M.J., and Stegun, I.A., 1972, *Handbook of Mathematical Functions*, Dover, New York.

Ahmed, T., 2000, *Reservoir Engineering Handbook*, Gulf Publishing, Houston.

Ahrens, T.J., 1994, "The Origin of the Earth," *Physics Today* (August), pages 38–45.

Alvarez, L.W., 1987, "Mass extinctions caused by large bolide impacts," *Physics Today* (July), pages 24–33.

Arms, K., and Camp, P.S., 1982, *Biology*, 2nd Edition, Saunders College Publishing, Philadelphia, Pennsylvania.

Attenborough, D., 1979, *Life on Earth*, Little, Brown and Company, Boston, Massachusetts.

Aubrecht, Gordon J., 1995, *Energy*, 2nd Edition, Prentice-Hall, Inc., Upper Saddle River, New Jersey.

Ausubel, J.H., 2000, "Where is Energy Going?", *Industrial Physicist* (February), pages 16–19.

Bain, A., and Van Vorst, W.D., 1999, "The *Hindenburg* tragedy revisited: the fatal flaw found," *International Journal of Hydrogen Energy* Volume 24, pages 399–403.

Ballentine, L.E., 1970, "The Statistical Interpretation of Quantum Mechanics," *Reviews of Modern Physics*, Volume 42, page 358 ff.

Barrow, J.D., and Tipler, F.J., 1986, *The Anthropic Cosmological Principle*, Oxford University Press, Oxford, United Kingdom.

Bartlett, D.A., 1996, "The Fundamentals of Heat Exchangers," *Industrial Physicist*, pages 18–21.

Baumann, R.P., 1992, *Modern Thermodynamics with Statistical Mechanics*, Macmillan, New York.

Bear, J., 1972, *Dynamics of Fluids in Porous Media*, Elsevier, New York.

Beggs, H.D., 1991, *Production Optimization Using Nodal Analysis*, OGCI Publications, Tulsa, Oklahoma.

Bell, J.S., 1987, *Speakable and Unspeakable in Quantum Mechanics*, Cambridge University Press, Cambridge, United Kingdom.

Benka, S.G., 2002, "The Energy Challenge," *Physics Today* (April), pages 38–39.

Bennett, C.L., Hinshaw, G.F., and Page, L., 2001, "A Cosmic Cartographer," *Scientific American* (January), pages 44–45.

Bergstrom, L., and Goobar, A., 1999, *Cosmology and Particle Physics*, Wiley, New York.

Bernal, J.D., 1997, *A History of Classical Physics*, Barnes and Noble, New York.

Bernstein, J., Fishbane, P.M., and Gasiorowicz, S., 2000, *Modern Physics*, Prentice Hall, Upper Saddle River, New Jersey.

Binzel, R.P., Barucci, M.A., and Fulchignoni, M., 1991, "The Origins of the Asteroids," *Scientific American* (October), pages 88–94.

Blandford, R., and Gehrels, N., 1999, "Revisiting the Black Hole," *Physics Today* (June), pages 40–46.

Bohm, D., and Hiley, B.J., 1993, *The Undivided Universe*, Routledge, London, United Kingdom.

Borbely, A., and Kreider, J.F., 2001, *Distributed Generation: The Power Paradigm for the New Millenium*, CRC Press, New York.

Börner, G., 1993, *The Early Universe*, Springer-Verlag, Berlin, Germany.

Boyle, G. (editor), 1996, *Renewable Energy: Power for a Sustainable Future*, Oxford University Press, Oxford, United Kingdom.

Brackman, A.C., 1980, *A Delicate Arrangement*, Times Books, New York.

Brennan, T.J., Palmer, K.L., Kopp, R.J., Krupnick, A.J., Stagliano, V., and Burtraw, D., 1996, *A Shock to the System: Restructuring America's Electricity Industry*, Resources for the Future, Washington, D.C.

Brill, J.P., and Mukherjee, H., 1999, *Multiphase Flow in Wells*, Society of Petroleum Engineers, Richardson, Texas.

Brown, G., 1996, "Geothermal Energy," *Renewable Energy: Power for a Sustainable Future*, edited by G. Boyle, Oxford University Press, Oxford, United Kingdom.

Brown, T.L., LeMay, H.E., Jr., Bursten, B.E., and Burdge, J.R., 2002, *Chemistry—The Central Science*, 9th Edition, Prentice Hall, Upper Saddle River, New Jersey.

Bucher, M.A., and Spergel, D.N., 1999, "Inflation in a Low-Density Universe," *Scientific American* (January), pages 63–69.

Burbridge, G., Hoyle, F., and Narlikar, J.V., 1999, "A Different Approach to Cosmology," *Physics Today*, pages 38–44. A. Albrecht appends a reply that represents the mainstream view supporting the Big Bang, ibid, pages 44–46.

Burke, J., 1985, *The Day the Universe Changed*, Little, Brown and Company, Boston.

Campbell, B.G., 1985, *Humankind Emerging*, 4th Edition, Little, Brown and Company, Boston, Massachusetts.

Campbell, C.J., and Laherrere, J.H., 1998, "The End of Cheap Oil," *Scientific American* (March), pages 78–83.

Carbon, M.W., 1997, *Nuclear Power: Villain or Victim?*, Pebble Beach Publishers, Madison, Wisconsin.

Cassedy, E.S., 2000, *Prospects for Sustainable Energy*, Cambridge University Press, Cambridge, United Kingdom.

Cassedy, E.S., and Grossman, P.Z., 1998, *Introduction to Energy*, 2nd Edition, Cambridge University Press, Cambridge, United Kingdom.

Çengel, Y.A., and Boles, M.A., 2002, *Thermodynamics*, 4th Edition, McGraw Hill, Boston, Massachusetts.

Challoner, J., 1993, *Energy*, Dorling Kindersley Publishing, New York.

Clark, B., and Kleinberg, R., 2002, "Physics in Oil Exploration," *Physics Today* (April), pages 48–53.

Cohen, E.R., 1996, *The Physics Quick Reference Guide*, American Institute of Physics, Woodbury, New York.

Collins, R.E., 1961, *Flow of Fluids through Porous Materials*, Petroleum Publishing, Tulsa, Oklahoma.

Cook, E., 1971, "The Flow of Energy in an Industrial Society," *Scientific American* (September), pages 135–144.

Copeland, T., Koller, T., and Murrin, J., 2000, *Valuation: Measuring and Managing the Value of Companies*, 3rd edition, Wiley, New York.

Couvaris, G., 1999, "Gas to liquids: A paradigm shift for the oil industry?," *Oil and Gas Journal* (13 December), pages 124–126.

Cowen, R., 2000, *History of Life*, 3rd Edition, Blackwell Science, Malden, Massachusetts.

D'Abro, A., 1951, *The Rise of the New Physics*, Dover, New York (2 volumes).

Darwin, C., 1952, *The Origin of Species* and *The Descent of Man*, Great Books Volume 49, Encyclopedia Britannica, Chicago, Illinois.

Darwin, C., 1959, *The Voyage of the Beagle*, Harper and Brothers, New York.

Dawson, J., 2002, "Fusion Energy Panel Urges US to Rejoin ITER," *Physics Today* (November), pages 28–29.

de Broglie, L., 1966, *Physics and Metaphysics*, Grosset's Universal Library, Grosset and Dunlap, New York.

Decker, R., and Decker, B., 1998, *Volcanoes*, 3rd Edition, W.H. Freeman, New York.

Deffeyes, K.S., 2001, *Hubbert's Peak—The Impending World Oil Shortage*, Princeton University Press, Princeton, New Jersey.

Devaney, R.L., 1992, *A First Course in Chaotic Dynamical Systems*, Addison-Wesley, Reading, Massachusetts.

Dixit, A.K., and Pindyck, R.S., 2000, *Investment Under Uncertainty*, Princeton University Press, Princeton, New Jersey.

DoE Biomass, 2002, "Electricity from Biomass," United States Department of Energy website (Accessed 24 October 2002) http://www.eren.doe.gov/biopower/basics/ba_efb.html.

DoE Geothermal, 2002, "Geothermal Energy Basics," United States Department of Energy website (Accessed 23 October 2002) http://www.eren.doe.gov/geothermal/geobasics.html.

DoE Hydropower, 2002, "Hydropower," United States Department of Energy website (Accessed 24 October 2002) http://www.eren.doe.gov/RE/hydropower.html.

DoE Ocean, 2002, "Ocean," United States Department of Energy website (Accessed 24 October 2002) http://www.eren.doe.gov/RE/Ocean.html.

Economides, M.J., Hill, A.D., and Ehlig-Economides, C., 1994, *Petroleum Production Systems*, Prentice Hall, Upper Saddle River, New Jersey.

EIA Table 6.2, 2002, "World Total Net Electricity Consumption, 1980–2000," United States Energy Information Administration website (Accessed 10 June 2002) http://www.eia.doe.gov/emeu/international/ electric.html#IntlConsumption.

EIA Table 11.1, 2002, "World Primary Energy Production by Source," 1970–2000, United States Energy Information Administration website (Accessed 10 November 2002) http://www.eia.doe.gov/emeu/international/ electric.html#IntlProduction.

EIA Table E.1, 2002, "World Primary Energy Consumption (Btu), 1980–2000," United States Energy Information Administration website (Accessed 10 June 2002) http://www.eia.doe.gov/emeu/international/ total.html#IntlConsumption.

Elliot, David, 1997, *Energy, Society, and Environment*, Routledge, New York.

Fanchi, J.R., 1986, "Local Effects of Nuclear Weapons," *Byte Magazine* (December), pages 143–155.

Fanchi, J.R., 1988, "Cosmological Implications of the Gibbs Ensemble in Parametrized Relativistic Classical Mechanics," *Physical Review* A37, pages 3956–3962.

Fanchi, J.R., 1990, "Tachyon Kinematics in Parametrized Relativistic Quantum Theories," *Foundations of Physics*, Volume 20, pages 189–224.

Fanchi, J.R., 1993, *Parametrized Relativistic Quantum Theory*, Kluwer, Dordrecht, Netherlands.

Fanchi, J.R., 1998, "The Mass Operator and Neutrino Oscillations," *Foundations of Physics*, Volume 28, pages 1521–1528.

Fanchi, J.R., 2000a, "Oil and Gas in the Energy Mix of the 21st Century," *Journal of Petroleum Technology* (December), pages 40–46.

Fanchi, J.R., 2000b, *Math Refresher for Scientists and Engineers*, 2nd Edition, Wiley, New York.

Fanchi, J.R., 2001, *Principles of Applied Reservoir Simulation*, 2nd Edition, Butterworth–Heinemann, Boston, Massachusetts.

Fanchi, J.R., 2002, *Shared Earth Modeling*, Butterworth–Heinemann, Boston, Massachusetts.

Fay, J.A., and Golomb, D.S., 2002, *Energy and the Environment*, Oxford University Press, New York.

Ferguson, H.C., Williams, R.E., and Cowie, L.L., 1997, "Probing the Faintest Galaxies," *Physics Today* (April), pages 24–30.

Feynman, R.P., 1942, *The Principle of Least Action in Quantum Mechanics*, Ph.D. Dissertation, Princeton University, New Jersey.

Feynman, R.P., Leighton, R.B., and Sands, M., 1963, *The Feynman Lectures on Physics*, Addison-Wesley, Reading, Massachusetts.

Folsome, C.E., 1979, *Life: Origin and Evolution, Readings from Scientific American*, W.H. Freeman, San Francisco, California.

Fowles, G.R., 1970, *Analytical Mechanics*, 2nd Edition, Holt, Rinehart and Winston, New York.

Freedman, W.L., 1992, "The Expansion Rate and Size of the Universe," *Scientific American* (November), pages 54–60.

Friedlander, M.W., 1995, *At the Fringes of Science*, Westview Press, Boulder, Colorado.

Fukugita, M., and Hogan, C.J., 2000, "Global Cosmological Parameters," *European Physical Journal C*, Volume 15, pages 136–142.

Gamble, C., 1993, *Timewalkers*, Harvard University Press, Harvard, Massachusetts.

Garrod, C., 1984, *Twentieth Century Physics*, Faculty Publishing, Davis, California.

Garwin, R.L., and Charpak, G., 2001, *Megawatts and Megatons: A Turning Point in the Nuclear Age?*, Alfred A. Knopf, New York.

Geller, H., 2003, *Energy Revolution*, Island Press, Washington.

Gjertsen, D., 1984, *The Classics of Science*, Lilian Barber Press, New York.

Glasstone, S., and Dolan, P.J. (editors), 1977, *The Effects of Nuclear Weapons*, Superintendent of Documents, United States Government Printing Office, Washington, D.C.

Gold, Thomas, 1999, *The Deep Hot Biosphere*, Springer-Verlag New York, Inc., New York.

Goldsmith, D., 2000, *The Runaway Universe*, Perseus Books, Cambridge, Massachusetts.

Goldstein, H., 1980, *Classical Mechanics*, 2nd Edition, Addison-Wesley, Reading, Massachusetts.

Goswami, D.Y., Kreith, F., and Kreider, J.F., 2000, *Principles of Solar Engineering*, George H. Buchanan Co., Philadelphia, Pennsylvania.

Gould, S.J., 1989, *Wonderful Life*, Norton, New York.

Gould, S.J. (editor), 1993, *The Book of Life*, W.W. Norton, New York.

Gould, S.J., 1996, *Full House*, Harmony Books, New York.

Gould, S.J., 2002, *The Structure of Evolutionary Theory*, Harvard University Press, Cambridge, Massachusetts.

Greene, B., 2000, *The Elegant Universe*, Vintage Books, New York.

Greenberg, O.W., 1985, "A New Level of Structure," *Physics Today* (September), pages 22–30.

Greenstein, G., and Zajonc, A.G., 1997, *The Quantum Challenge*, Jones and Bartlett, Sudbury, Massachusetts.

Green, D.W., and Willhite, G.P., 1998, *Enhanced Oil Recovery*, Society of Petroleum Engineers, Richardson, Texas.

Greiner, W., Neise, L., and Stöcker, H., 1995, *Thermodynamics and Statistical Mechanics*, Springer, New York.

Gribbin, J., 1991, *Blinded by the Light*, Harmony Books, New York.

Gurnis, M., 2001, "Sculpting the Earth from Inside Out," *Scientific American* (March), pages 40–47.

Hafemeister, D.W., 1983, "Science and Society test VIII: The arms race revisited," *American Journal of Physics*, Volume 51 (March), pages 215–225.

Halliday, D., and Resnick, R., 1981, *Fundamentals of Physics*, 2nd Edition, Wiley, New York.

Hamto, M., 2000, "Fuel-cell Technology to Provide Clean, Sustainable Energy," *Journal of Petroleum Technology* (April), pages 26–30.

Hayden, H.C., 2001, *The Solar Fraud: Why Solar Energy Won't Run the World*, Vales Lake Publishing, LLC, Pueblo West, Colorado.

Hellman, H., 1998, *Great Feuds in Science*, Wiley, New York.

Herbert, S., 1986, "Darwin as a Geologist," *Scientific American* (May), pages 116–123.

Hester, J., Burstein, D., Blumenthal, G., Greeley, R., Smith, B., Voss, H., and Wegner, G., 2002, *21st Century Astronomy*, W.W. Norton, New York.

Hirsch, M.W., and Smale, S., 1974, *Differential Equations, Dynamical Systems, and Linear Algebra*, Academic Press, New York.

Hodgson, P.E., 1999, *Nuclear Power, Energy and the Environment*, Imperial College Press, London, United Kingdom.

Hoffmann, P., 2001, *Tomorrow's Energy: Hydrogen, Fuel Cells, and the Prospects for a Cleaner Planet*, MIT Press, Cambridge, Massachusetts.

Hogan, C.J., 1998, *The Little Book of the Big Bang*, Springer-Verlag, New York.

Horowitz, N., 1986, *To Utopia and Back*, W.H. Freeman, New York.

Horvitz, L.A., 2002, *Eureka!*, Wiley, New York.

Hoyle, F., 1963, *Frontiers of Astronomy*, New American Library, New York.

Hubbert, M.K., 1956, "Nuclear Energy and the Fossil Fuels," American Petroleum Institute Drilling and Production Practice, Proceedings of the Spring Meeting, San Antonio, pages 7–25.

Huntington, S.P., 1996, *The Clash of Civilizations*, Simon and Schuster, London, United Kingdom.

Jackson, J.D., 1999, *Classical Electrodynamics*, Third Edition, Wiley, New York.

Jammer, M., 1966, *The Conceptual Development of Quantum Mechanics*, McGraw-Hill, New York.

Jeanloz, R., and Romanowicz, B., 1997, "Geophysical Dynamics at the Center of the Earth," *Physics Today* (August), pages 22–27.

Jensen, R.V., 1987, "Classical Chaos," *American Scientist*, Volume 75, pages 168–181.

Jet Propulsion Laboratory Fact Sheet, 1984, IRAS: Infrared Astronomical Satellite, Office of Public Information (February), NASA Jet Propulsion Laboratory, Pasadena, California.

Johanson, D.C., and O'Farrell, K., 1990, *Journey from the Dawn: Life with the World's First Family*, Villard Books, New York.

Kane, G., 2000, *Supersymmetry*, Perseus Books, Cambridge, Massachusetts.

Kane, H.K., 1996, *Pele: Goddess of Hawaii's Volcanoes*, Expanded Edition, Kawainui Press, Captain Cook, Hawaii.

Kasting, J.F., Toon, O.B., and Pollack, J.B., 1988, "How Climate Evolved on the Terrestrial Planets," *Scientific American* (February), pages 90–97.

Kimball, J.W., 1968, *Biology*, 2nd Edition, Addison-Wesley, Reading, Massachusetts.

Kolb, E.W., and Turner, M.S., 1990, *The Early Universe*, Addison-Wesley, Reading, Massachusetts.

Kolb, E.W., and Turner, M.S., 2000, "The Pocket Cosmology," *European Physical Journal C*, Volume 15, pages 125–132.

Kraushaar, J.J., and Ristinen, R.A., 1993, *Energy and Problems of a Technical Society*, 2nd Edition, Wiley, New York.

Krauss, L.M., 1999, "Cosmological Antigravity," *Scientific American* (January), pages 53–59.

Kreyszig, E., 1999, *Advanced Engineering Mathematics*, 8th Edition, Wiley, New York.

Kuhn, T.S., 1970, *The Structure of Scientific Revolutions*, 2nd Edition, University of Chicago Press, Chicago.

Ladbury, R., 1996, "Martians Invaded Earth 13,000 Years Ago—Maybe," *Physics Today*, page 18.

Lahav, N., 1999, *Biogenesis*, Oxford University Press, Oxford, United Kingdom.

Laherrère, J.H., 2000, "Learn strengths, weaknesses to understand Hubbert curves," *Oil and Gas Journal*, pages 63–76 (17 April); see also Laherrère's earlier article "World oil supply—what goes up must come down, but when will it peak?" *Oil and Gas Journal*, pages 57–64 (1 February 1999) and letters in *Oil and Gas Journal* (1 March 1999).

Lakatos, I., 1970, *Criticism and the Growth of Knowledge*, edited by Imre Lakatos and A. Musgrave, Cambridge University Press, Cambridge, United Kingdom.

Lake, L.W., 1989, *Enhanced Oil Recovery*, Prentice Hall, Englewood Cliffs, New Jersey.

Lawrie, I.D., 1990, *A Unified Grand Tour of Theoretical Physics*, Adam Hilger, Bristol, United Kingdom.

Leakey, R., 1994, *The Origin of Humankind*, Basic Books, New York.

Levin, H.L., 1991, *The Earth Through Time*, 4th Edition, Harcourt-Brace-Jovanovich, New York.

Levy, D.H., 1998, *The Ultimate Universe*, Pocket Books, New York.

Lichtenberg, A.J., and Lieberman, M.A., 1983, *Regular and Stochastic Motion*, Springer-Verlag, New York.

Liddle, A., 1999, *An Introduction to Modern Cosmology*, Wiley, New York.

Lide, D.R., 2002, *CRC Handbook of Chemistry and Physics*, 83rd Edition, CRC Press, Boca Raton, Florida.

Lilley, J., 2001, *Nuclear Physics*, Wiley, New York.

Linde, A., 1987, "Particle Physics and Inflationary Cosmology," *Physics Today* (September), pages 61–68.

Lindley, D., 1987, "Cosmology from Nothing," *Nature* (December), pages 603–604.

Lowrie, W., 1997, *Fundamentals of Geophysics*, Cambridge University Press, Cambridge, United Kingdom.

Ludvigsen, M., 1999, *General Relativity*, Cambridge University Press, Cambridge, United Kingdom.

MacDonald, R., 1999, "Finding a Job After 45 Years of Age," *Journal of Petroleum Technology* (December), page 14.

Mandelbrot, B.B., 1967, "How long is the Coast of Britain? Statistical Self-Similarity and Fractional Dimension," *Science*, Volume 155, pages 636–638.

Mandelbrot, B.B., 1983, *The Fractal Geometry of Nature*, Freeman, New York.

Mather, J.C., and Boslough, H., 1996, *The Very First Light*, Basic Books, New York.

Mathieu, J., and Scott, J., 2000, *An Introduction to Turbulent Flow*, Cambridge University Press, Cambridge, United Kingdom.

Maxwell, J.C., 1985, "What is the Lithosphere?" *Physics Today* (September), pages 32–40.

May, R.M., 1976, "Simple Mathematical Models with Very Complicated Dynamics," *Nature*, Volume 261, page 459.

McCauley, J.L., 1997, *Classical Mechanics*, Cambridge University Press, Cambridge, United Kingdom.

McCay, D.R. (correspondent), 2002, "Global Scenarios 1998–2020," Summary Brochure, Shell International, London, United Kingdom.

Merzbacher, E., 2002, "The Early History of Quantum Tunneling," *Physics Today* (August), pages 44–49.

Mihelcic, J.R., 1999, *Fundamentals of Environmental Engineering*, Wiley, New York.

Misner, C.W., Thorne, K.S., and Wheeler, J.A., 1973, *Gravitation*, W.H. Freeman, San Francisco, California.

Montgomery, C.W., 1990, *Physical Geology*, 2nd Edition, Wm. C. Brown, Dubuque, Iowa.

Moore, R., 1971, *Evolution*, Time-Life Books, New York.

Morrison, P., and Tsipis, K., 1998, *Reason Enough to Hope*, MIT Press, Cambridge, Massachusetts, especially Chapter 9.

Muller, R., 1988, *Nemesis–The Death Star*, Weidenfeld & Nicolson, New York.

Murray, R.L., 2001, *Nuclear Energy: An Introduction to the Concepts, Systems, and Applications of Nuclear Processes*, 5th Edition, Butterworth–Heinemann, Boston, Massachusetts.

Nef, J.U., 1977, "An Early Energy Crisis and its Consequences," *Scientific American* (November), pages 140–151.

Newendorp, P.D., and Schuyler, J.R., 2000, *Decision Analysis for Petroleum Exploration*, 2nd Edition, Planning Press, Aurora, Colorado.

Newton, R.G., 1980, "Probability Interpretation of Quantum Mechanics," *American Journal of Physics*, pages 1029–1034.

Ogden, J.M., 2002, "Hydrogen: The Fuel of the Future?" *Physics Today* (April), pages 69–75.

Omnes, R., 1994, *The Interpretation of Quantum Mechanics*, Princeton University Press, Princeton, New Jersey.

Orgel, L.E., 1973, *The Origins of Life*, Wiley, New York.

Orr, K., and Cook, M., 2000, *Discover Hawaii's Birth by Fire Volcanoes*, Island Heritage, Aiea, Hawaii.

Pais, A., 1982, *Subtle is the Lord...*, Oxford University Press, Oxford, United Kingdom.

Park, R., 2000, *Voodoo Science*, Oxford University Press, Oxford, United Kingdom.

Particle Data Group, 2000, "Review of Particle Properties," *European Physical Journal C*, Volume 15.

Pavŝiĉ, M., 2001, *The Landscape of Theoretical Physics: A Global View*, Kluwer, Dordrecht, Netherlands.

Peebles, P.J.E., 1992, *Quantum Mechanics*, Princeton University Press, Princeton, New Jersey.

Peebles, P.J.E., 1993, *Principles of Physical Cosmology*, Princeton University Press, Princeton, New Jersey.

Peebles, P.J.E., 2001, "Making Sense of Modern Cosmology," *Scientific American* (January), pages 54–55.

Perlmutter, S., 2003, "Supernovae, Dark Energy, and the Accelerating Universe," *Physics Today* (April), pages 53–60.

Ponnamperuma, C., 1974, *Origins*, American Association for the Advancement of Science, Washington, D.C.

Press, F., and Siever, R., 1982, *Earth*, 3rd Edition, W.H. Freeman, San Francisco, California.

Press, F., and Siever, R., 2001, *Understanding Earth*, 3rd Edition, W.H. Freeman, San Francisco, California.

Purves, W.K., Sadava, D., Orians, G.H., and Heller, H.C., 2001, *Life: The Science of Biology*, W.H. Freeman, Sunderland, Massachusetts.

Ramage, J., and Scurlock, J., 1996, "Biomass," *Renewable Energy: Power for a Sustainable Future*, edited by G. Boyle, Oxford University Press, Oxford, United Kingdom.

Rees, J.V., 1994, *Hostages of Each Other: The Transformation of Nuclear Safety Since Three Mile Island*, University of Chicago, Chicago, Illinois.

Ridley, M., 1996, *Evolution*, Blackwell Science, Cambridge, Massachusetts.

Ringwood, A.E., 1986, "Terrestrial Origin of the Moon," *Nature* (July), pages 323–328.

Ristinen, R.A., and Kraushaar, J.J., 1999, *Energy and the Environment*, Wiley, New York.

Rockefeller, G., and Patrick, R., 1999, "Keeping Good Company in the Energy Industry: Extrapolating Trends in Energy Convergence," *Oil and Gas Journal* (13 December), pages 149–152.

Root-Bernstein, R.S., 1989, *Discovering*, Harvard University Press, Cambridge, Massachusetts.

Ross, H., 1993, *The Creator and the Cosmos*, NavPress, Colorado Springs, Colorado.

Russell, D.A., 1982, "The Mass Extinctions of the Late Mesozoic," *Scientific American* (January), pages 48–55.

Sagan, C., 1980, *Cosmos*, Random House, New York.

Sahimi, M., 1995, *Flow and Transport in Porous Media and Fractured Rock*, VCH, New York.

Sambursky, S., 1975, *Physical Thought from the Presocratics to the Quantum Physicists*, Pica Press, New York.

Sargent, A.I., and Beckwith, S.V.W., 1993, "The Search for Forming Planetary Systems," *Physics Today* (April), pages 22–29.

Sartori, L., 1983, "Effects of Nuclear Weapons," *Physics Today* (March), pages 32–41.

Scheinbein, L.A., and Dagle, J.E., 2001, "Electric Power Distribution Systems," in Borbely, A. and Kreider, J.F., 2001, *Distributed Generation: The Power Paradigm for the New Millenium*, CRC Press, New York.

Schollnberger, W.E., 1999, "Projection of the World's Hydrocarbon Resources and Reserve Depletion in the 21st Century," *The Leading Edge*, 622–625 (May 1999); also in *Houston Geological Society Bulletin* (November), pages 31–37.

Schopf, J.W., 1999, *The Cradle of Life*, Princeton University Press, Princeton, New Jersey.

Schwarzschild, B., 2003, "WMAP Spacecraft Maps the Entire Cosmic Microwave Sky with Unprecedented Precision," *Physics Today* (April), pages 21–24.

Selley, R.C., 1998, *Elements of Petroleum Geology*, Academic Press, San Diego, California.

Serway, R.A., and Faughn, J.S., 1985, *College Physics*, Saunders, Philadelphia, Pennsylvania.

Serway, R.A., Moses, C.J., and Moyer, C.A., 1997, *Modern Physics*, 2nd Edition, Harcourt Brace, New York.

Shapiro, R., 1987, *Origins: A Skeptic's Guide to the Creation of Life on Earth*, Bantam, New York.

Shepherd, W., and D.W. Shepherd, 1998, *Energy Studies*, Imperial College Press, London, United Kingdom.

Shermer, M., 2002, "Why ET Hasn't Called," *Scientific American* (August), page 33.

Siever, R., 1983, "The Dynamic Earth," *Scientific American* (September), pages 46–55.

Silberberg, M., 1996, *Chemistry*, Mosby, St. Louis, Missouri.

Silk, J., 1987, "The Formation of Galaxies," *Physics Today* (April), pages 28–35.

Silk, J., 2001, *The Big Bang*, 3rd Edition, W.H. Freeman, New York.

Singer, S.F., 2001, *Symmetry in Mechanics*, Birkhäuser, Boston, Massachusetts.

Skinner, B.J., 1986, "Can You Really Believe the Evidence? Two Stories from Geology," *American Scientist* (July-August), pages 401–409.

Smil, V., 1991, *General Energetics*, Wiley, New York.

Smith, J.M., and Szathmáry, E., 1999, *The Origins of Life*, Oxford University Press, Oxford, United Kingdom.

Smoot, G.F., and Scott, D., 2000, "Cosmic Background Radiation," *European Physical Journal C*, Volume 15, pages 145–149.

Sørensen, Bent, 2000, *Renewable Energy: Its Physics, Engineering, Environmental Impacts, Economics & Planning*, 2nd Edition, Academic Press, London, United Kingdom.

Srednicki, M., 2000, "Dark Matter," *European Physical Journal C*, Volume 15, pages 143–144; article written in 1999.

Stahler, S.W., 1991, "The Early Life of Stars," *Scientific American* (July), pages 48–52.

Stokes, W.L., 1960, *An Introduction to Earth History*, Prentice Hall, Englewood Cliffs, New Jersey.

Strogatz, S.H., 1994, *Nonlinear Dynamics and Chaos*, Perseus Books, Reading, Massachusetts.

Taylor, D., 1996, "Wind Energy," *Renewable Energy: Power for a Sustainable Future*, edited by G. Boyle, Oxford University Press, Oxford, United Kingdom.

Taylor, G.J., 1994, "The Scientific Legacy of Apollo," *Scientific American* (July), pages 40–47.

Taylor, S.R., 2001, *Solar System Evolution—A New Perspective*, 2nd Edition, Cambridge University Press, Cambridge, United Kingdom.

Tegmark, M., and Wheeler, J.A., 2001, "100 Years of Quantum Mysteries," *Physics Today* (February), pages 68–73.

Thompson, R.S., and Wright, J.D., 1985, *Oil Property Evaluation*, Thompson-Wright Associates, Golden, Colorado.

Thuan, T.X., 1995, *The Secret Melody*, Oxford University Press, Oxford, United Kingdom.

Thumann, A., and Mehta, D.P., 1997, *Handbook of Energy Engineering*, 4th Edition, Fairmont Press, Lilburn, Georgia.

Tipler, F.J., 1981, "Extraterrestrial Intelligent Beings Do Not Exist," *Physics Today* (April), page 9 ff.

Traweek, S., 1988, *Beamtimes and Lifetimes*, Harvard University Press, Cambridge, Massachusetts.

Trefil, J.S., 1985, *Space Time Infinity*, Smithsonian Books, Washington, D.C.

Tudge, C., 2000, *The Variety of Life*, Oxford University Press, Oxford, United Kingdom.

Turner, M.S., 2003, "Dark Energy: Just What Theorists Ordered," *Physics Today* (April), pages 10–11.

UNDP, 2001, *Human Development Report 2001: Making New Technologies Work for Human Development*, United Nations Development Program, Oxford University Press, New York.

van den Bergh, S., and Hesser, J.E., 1993, "How the Milky Way Formed," *Scientific American* (January), pages 72–78.

van Dyke, K., 1997, *Fundamentals of Petroleum*, 4th Edition, Petroleum Extension Service, University of Texas, Austin, Texas.

van Sciver, S.W., and Marken, K.R., 2002, "Superconducting Magnets above 20 Tesla," *Physics Today* (August), pages 37–43.

Verleger, P.K. Jr., 2000, "Third Oil Shock: Real or Imaginary?" *Oil & Gas Journal* (12 June), pages 76–88.

Walesh, S.G., 1995, *Engineering Your Future*, Prentice Hall PTR, Englewood Cliffs, New Jersey.

Wallace, R.A., 1990, *Biology: The World of Life*, HarperCollins, New York.

Ward, P.D., and Brownlee, D., 2000, *Rare Earth*, Springer-Verlag, New York.

WCED (World Commission on Environment and Development), Brundtland, G., Chairwoman, 1987, *Our Common Future*, Oxford University Press, Oxford, United Kingdom.

Weart, S.R., 1997, "The Discovery of the Risk of Global Warming," *Physics Today* (January), pages 34–59.

Weinberg, S., 1977, *The First Three Minutes*, Bantam Books, New York.

Weinberg, S., 1993, *The First Three Minutes*, Updated Edition, Basic Books, New York.

Weinberg, S., 1995, *The Quantum Theory of Fields*, Cambridge University Press, New York, Volume I: Foundations.

Weinberg, S., 1999, *The Quantum Theory of Fields*, Cambridge University Press, New York, Volume III: Supersymmetry.

Whittaker, M., 1999, "Emerging 'triple bottom line' model for industry weighs environmental, economic, and social considerations," *Oil and Gas Journal*, pages 23–28 (20 December).

Whittaker, R.H., 1969, "New Concepts of Kingdoms of Organisms," *Science* (January), pages 150–159.

Wick, D., 1995, *The Infamous Boundary*, Birkhauser, Boston, Massachusetts.

Wigley, T.M.L., Richels, R., and Edmonds, J.A., 1996, "Economic and environmental choices in the stabilization of atmospheric CO_2 concentrations," *Nature* (18 January), pages 240–243.

Williams, W.S.C., 1991, *Nuclear and Particle Physics*, Oxford University Press, Oxford, United Kingdom.

Winchester, S., 2001, *The Map that Changed the World*, HarperCollins, New York.

Wiser, Wendell H., 2000, *Energy Resources: Occurrence, Production, Conversion, Use*, Springer-Verlag, New York.

Wolff, P., 1965, *Breakthroughs in Physics*, New American Library, New York.

Wolff, P., 1967, *Breakthroughs in Chemistry*, New American Library, New York.

Wolpert, L.,1992, *The Unnatural Nature of Science*, Faber and Faber, London, United Kingdom.

Yergin, D., 1992, *The Prize*, Simon and Schuster, New York.

York, D., 1993, "The Earliest History of the Earth," *Scientific American* (January), pages 90–96.

Young, H.D., and Freedman, R.A., 2000, *Sears and Zemansky's University Physics*, 10th Edition, Addison Wesley, San Francisco, California.

Zumdahl, S.S., 1986, *Chemistry*, D.C. Heath, Lexington, Massachusetts.

Index